Palgrave Studies in Educational Futures

Series Editor
jan jagodzinski
Department of Secondary Education
University of Alberta
Edmonton, AB, Canada

The series Educational Futures would be a call on all aspects of education, not only specific subject specialist, but policy makers, religious education leaders, curriculum theorists, and those involved in shaping the educational imagination through its foundations and both psychoanalytical and psychological investments with youth to address this extraordinary precarity and anxiety that is continually rising as things do not get better but worsen. A global de-territorialization is taking place, and new voices and visions need to be seen and heard. The series would address the following questions and concerns. The three key signifiers of the book series title address this state of risk and emergency:

1. **The Anthropocene**: The 'human world,' the world-for-us is drifting toward a global situation where human extinction is not out of the question due to economic industrialization and over-development, as well as the exponential growth of global population. How do we address this ecologically and educationally to still make a difference?

2. **Ecology**: What might be ways of re-thinking our relationships with the non-human forms of existence and in-human forms of artificial intelligence that have emerged? Are there possibilities to rework the ecological imagination educationally from its over-romanticized view of Nature, as many have argued? Nature and culture are no longer tenable separate signifiers. Can teachers and professors address the ideas that surround differentiated subjectivity where agency is no long attributed to the 'human' alone?

3. **Aesthetic Imaginaries**: What are the creative responses that can fabulate aesthetic imaginaries, which are viable in specific contexts where the emergent ideas, and which are able to gather heterogeneous elements together to present projects that address the two former descriptors: the Anthropocene and the every changing modulating ecologies. Can educators draw on these aesthetic imaginaries to offer exploratory hope for what is a changing globe that is in constant crisis?

The series Educational Futures: Anthropocene, Ecology, and Aesthetic Imaginaries attempts to secure manuscripts that are aware of the precarity that reverberates throughout all life, and attempts to explore and experiment to develop an educational imagination which, at the very least, makes conscious what is a dire situation.

More information about this series at
http://www.palgrave.com/gp/series/15418

jan jagodzinski
Editor

Interrogating
the Anthropocene

Ecology, Aesthetics, Pedagogy, and the Future
in Question

Editor
jan jagodzinski
Department of Secondary Education
University of Alberta
Edmonton, AB, Canada

Palgrave Studies in Educational Futures
ISBN 978-3-030-08777-7   ISBN 978-3-319-78747-3   (eBook)
https://doi.org/10.1007/978-3-319-78747-3

This Palgrave Macmillan imprint is published by the registered company Springer International
Publishing AG part of Springer Nature
The registered company address is: Gewerbestrasse 11, 6330 Cham, Switzerland

This book is dedicated to Jessie and Jason
friends

ACKNOWLEDGEMENTS

This book emerged from a series of fifteen lectures entitled Anthropocene, Ecology, Pedagogy: The Future in Question, which concluded with a 'Sounding the Anthropocene' symposium with special keynote guests Hanjo Berressem (University of Koln, Department of American Studies) and Bernd Herzogenrath (Goethe University, Department of English and American Studies) in 2015–2016. The lecture series was presented at the University of Alberta, with support from the Faculty of Education, curated by myself, with the energetic and impassioned assistance of Jessie Beier, as well as a number of dedicated graduate students: Ron Wigglesworth who presented an art exhibition of photographs in relation to the Anthropocene, and the consistent help and support of Adriana Boffa and Cathryn van Kessel. Many thanks to Diane Conrad who enabled the lecture series to take place in her theatre space where many presentations concerning Arts Based Research take place. Thanks also to the video recording work by 'Eb' (Ebenezer Militsala) who was diligent in editing and positing the lectures. The lecture series can be viewed on YouTube:

https://www.youtube.com/channel/UCiYpb7adp3lvjdi-F8RktrQ.

I also want to thank other members of the support staff who made sure rooms and flights were booked (Scott Mayo) and payments received and reimbursed (Debra Mallett), so important for events such as this. However, again, without Jessie's help the series would not have been so successful, and running so smoothly. Her series of posters that she did for the event appear both on the front cover and in this acknowledgment.

There are many contributors to thank, beginning with Bradley Necyk. Bradley was brave enough to start the series and, along with Daniel Harvey contributed to this collection. I very much appreciate Andony Melanthropolus' efforts to make the trip from Calgary on what is a busy schedule, to both present, and together with Alexander Stoner contribute to this volume. Many thanks to Matthew Tiessen whose work I very much admire. Matthew's clear-headed grasp of the circulation of money and his profound grasp of Deleuze is greatly appreciated. I am also so appreciative of Nick Dyer-Witheford contribution to this collection. Nick, for me, has never lost sight of the worker's plight in capitalism. He has written seminal books in this area; his chapter simply confirms his strength in social Marxist theory, history, and in recognizing the importance of "species becoming." My thanks to Janae Sholtz, a remarkable philosopher whose grasp of Heidegger and Deleuze is well-know. Few can offer a comprehensive projection as to "what needs to be done" given the precarity of the world order. Mickey Vallee is another one of those rare individuals who plays with "sound" imaginaries. One cannot easily pinpoint Mickey's disciplinary activities as he is a sociologist, musician, sound theoretician, pedagogue, all rolled into one. In brief, I admire Mickey for his interdisciplinary theorizing. Thanks to Michael Trucello whom I had the pleasure of meeting at Brock University in the context of a Deleuze conference. Despite an extreme schedule of deadlines, Michael managed to develop a superb chapter on environmentalist documentary film, which he presented for the lecture series. Thank you, Michael, for sticking with it and seeing the chapter through. I met Nathan Snaza for the first time at the "Bergamo conference," a year prior to the lecture series. Nathan is a consummate teacher and theoretician who is acutely aware of the Anthropocene issues and with their consequences for education. I thank him for providing a unique perspective on this position. His was one of the earliest chapters finished, and I appreciate his patience as this collection finally came to fruition.

The collection also includes authors who did not participate in the lecture series. I personally approached a number of these authors and was pleased that they accepted my invitation. I have never met Ted Stolze in person, but he has an amazing reputation as a teacher and philosopher who constantly challenges his students and his readership. I am so happy he accepted the invitation to address the Anthropocene in such a way few can. Unfortunately, I have not met David Fancy as we seem to have just missed crossing path at Brock University, and there was no way to meet at a Deleuze conference in Toronto. David's reputation as a Deleuze-Guattarian scholar is well known as he has written widely in this area, drawing on his theatre-drama background. I so appreciate his contribution on a particularly interesting topic on human-non-human relations.

A special thanks to my good artist friend, Mia Feuer for allowing me to generously utilize her images throughout this collection. Mia's eco-logical consciousness holds no bounds, as she thinks deeply and ethically when it comes to developing her projects, choosing her materials, and staging her installations. I so appreciate her coming to Edmonton to talk about her ecological undertakings. An expectant mother at the time, it was a joy to host her, Constantine and Galileo who came into the world later. An image of her sculpture appears on the front cover.

I close with many thanks to colleagues who have contributed to this collection. First, to Patti Pente, my colleague in art education, who has written such an interesting chapter to rethink landscape within the parameters of the Anthropocene. An especially warm thank you to two friends and colleagues whom this book is dedicated to: Jessie Beier and Jason Wallin. Both are artists and experimenters, both push the limits of the imagination to make you think, Jessie with an intriguing chapter that addresses the near future, and Jason exploring Pokemon in a way few could envision.

Last but not least, I want to thank our former Dean of the Faculty of Education, Fern Snart. Without her the lecture series event would have never happened. This collection is the culminating point of that initiative.

Three Posters by Jessie Beier

Poster 1   Future in question

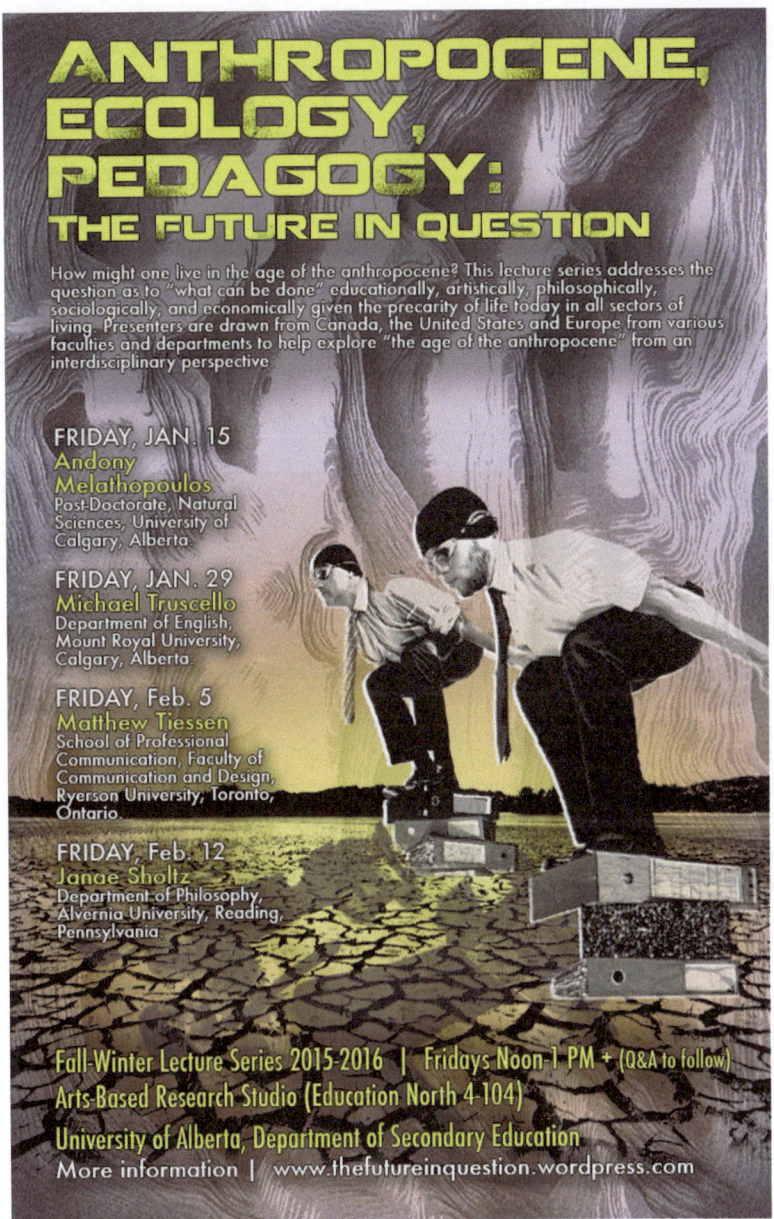

Poster 2 Future in question

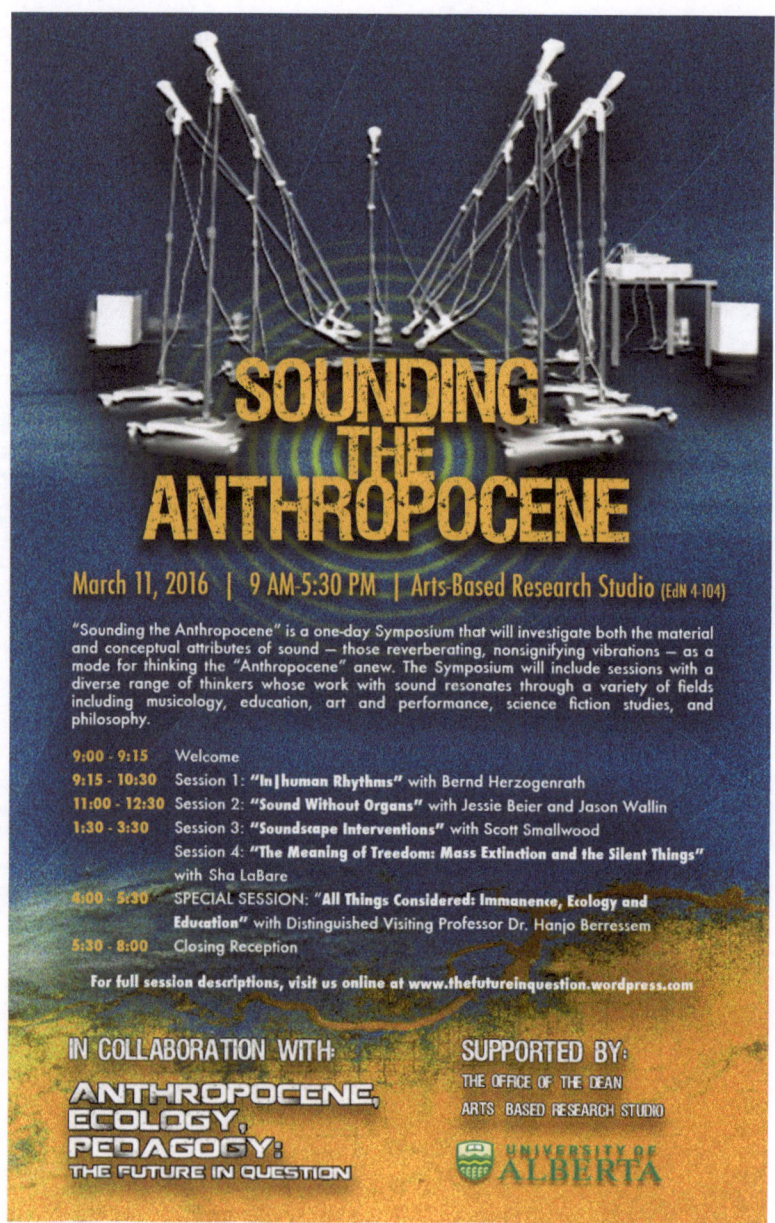

Poster 3   Sounding the anthropocene

CONTENTS

Part IV   Pedagogical Responses

NOTES ON CONTRIBUTORS

Jessie Beier is a teacher, artist, and Ph.D. student at the University of Alberta in Edmonton, Alberta. Beier's interests in visual and sonic ecologies have led to a research-creation practice that works to think pedagogy, in its many forms, as a power for overturning cliché and dismantling common-sense habits of interpretation and understanding. Beier's current research investigates how education systems produce resources for thinking the future in light of what has been termed the "Anthropocene," with the aim of fabulating alternative speculations on how pedagogical life *might* be thought otherwise.

Nick Dyer-Witheford an Associate Professor in the Faculty of Information and Media Studies at the University of Western Ontario, is the author of *Cyber-Marx: Cycles and Circuits of Struggle in High Technology Capitalism* (Chicago: University of Illinois, 1999) and *Cyber-Proletariat: Global Labour in the Digital Vortex* (London: Pluto Press, 2015), and also writes on the video and computer game industry, the uses of the Internet by social movements and theories of technology.

David Fancy is an associate professor in the Department of Dramatic Arts at Brock, University. His research interests and current publishing deals with questions of ontology, immanence and performance, with a specific interest in immanence and performativity, immanence and performance training, and immanence and technology. Fancy has an extensive creative practice as a playwright and director or theatre, opera, and circus.

Mia Feuer's practice since 2007 has been fueled by an ongoing chain reaction of on-site field research that lead to projects that in turn lead to further on site investigations. She is interested in the post-natural landscape, visible sites where human interaction—be it personal, social, political or financial—has altered or is in the process of rapidly changing the land, and thus our relationship to it. Her sculptural practice is driven by a causal sequence of explorative inquiries, prolific discoveries, and fascinating collaborations. Her work makes connections between our intense material dependency and the accelerated environmental impact this creates. What started with a naive love affair of manipulating Styrofoam to build lightweight sculptures turned into almost a decade of exploring the environmental and sociopolitical implications of extracting fossil fuels from the Earth and engaging with damaged, marginalized, or threatened places.

Born in Winnipeg, Manitoba, Canada, Mia received her BFA from the University of Manitoba in 2004 and her MFA in 2009 from the Department of Sculpture + Extended Media at Virginia Commonwealth University. Mia has received numerous travel, research, production and creation grants from the Manitoba Arts Council; the Winnipeg Arts Council; The Canada Council for the Arts and The Lila Acheson Readers Digest Foundation. In 2007, with the support of The Winnipeg Arts Council, she traveled to Palestine to facilitate sculptural research and puppet workshops in the West Bank with Palestinian children. Mia received several fellowships including: Vermont Studio Center; Seven Below Arts Initiative, The Mid Atlantic Foundation for the Arts, The Millay Colony, The Macdowell Colony; The Bemis Center for Contemporary Art and Sculpture Space. Mia Feuer was a 2011 District of Columbia Center for the Arts and Humanities fellow and won the 2011 Trawick Prize. In 2012, she received the prestigious Joseph S. Stauffer Prize from the Canada Council for the Arts. She conducted sculptural research in Egypt during the 2011 Revolution and in 2012 and 2013, Mia gained unprecedented access to the Alberta Tar Sands of Fort McMurray to conduct Sculptural research. In 2013, Mia participated in an Arts and Science expedition to the Arctic Circle and in 2014, she was a visiting artist at The Banff Centre during the thematic residency titled: Making. Solo Exhibitions include FLUXspace, Philadelphia, PA, Transformer Gallery, Washington DC, Arlington Arts Center, Arlington, VA, The Firehouse Gallery, Burlington, VT, The

Atlanta Contemporary Arts Center, Atlanta, GA, CONNERSMITH Gallery, Washington, DC, Goodyear Gallery in Carslile, PA, RAW: Gallery of Architecture and Design, Winnipeg, Manitoba, The Corcoran Museum in Washington, DC and University of Mary Washington Gallery in Fredericksburg VA. Mia created and curated The free Flooded Lecture Series which took place on the Anacostia River in fall 2014 in Washington DC. Recently, Mia has been conducting Sculptural research in the devastated bayous of the Gulf Coast and is currently developing projects with federally unrecognized Indigenous Pointe au Chien and Isle de Jean Charles communities. Most recently, Mia exhibited a massive solo exhibition at The Esker Foundation in Calgary, Alberta, and Locust Projects in Miami FL. In the summer of 2015, Mia was an artist in residence at the Va Space Residency in Isfahan, Iran and in 2016 exhibited work in Champagne Life- the first female only exhibition at The Saatchi Gallery in London, UK. Mia received a major production and research grant from The Canada Council for the Arts in 2016 and is currently a semi-finalist for the prestigious 2016 Canadian Sobey Award.

Daniel Harvey is a Ph.D. candidate in English and Film Studies at the University of Alberta. His research areas include contemporary cultural studies, finance, and the politics of the Anthropocene. His dissertation, "On entrepreneurship: Immaterial labour and subjectivity in the twenty-first century," examines the ways that entrepreneurial endeavor has become a cultural common sense, and the creation of entrepreneurial subjects an increasingly common trope of education, both formal and informal.

jan jagodzinski is Professor of Visual Art and Media Education, University of Alberta in Edmonton, Alberta, Canada. He is the author of seventeen books and numerous journal contributions and chapters. He is cofounder and former editor of *The Journal of Social Theory in ArtEducation (JSTE)*. Current series Editor of *Educational Futures* (Palgrave Press), and former co-series book editor with Mark Bracher, *Pedagogy, Psychoanalysis, Transformation* (Palgrave Press), Sample book credits include: *Youth Fantasies: The Perverse Landscape of the Media* (Palgrave, 2004); *Musical Fantasies: A Lacanian Approach* (Palgrave, 2005); *Television and Youth: Televised Paranoia* (Palgrave, 2008); *The Deconstruction of the Oral Eye: Art and Its Education in an Era of Designer Capitalism* (Palgrave, 2010), *Misreading Postmodern Antigone:*

Marco Bellocchio's Devil in the Flesh (Diavolo in Corpo) (Intellect Books, 2011), Editor of *Psychoanalyzing Cinema: A Productive Encounter Between Lacan, Deleuze, and Žižek* (Palgrave, 2012); *Arts Based Research: A Critique and Proposal* (with Jason Wallin) (Sense, 2013). Editor: *The Precarious Future of Education* (Palgrave, 2017). Editor: *What is Art Education? After Deleuze and Guattari* (Palgrave, 2017).

Andony Melathopoulos is an Assistant Professor of Pollinator Health Extension in the Department of Horticulture at OSU. He holds an Interdisciplinary Ph.D. from Dalhousie University. His work combines critical theory with ecological research into wild bee conservation and management. His contributions have been recognized by the Entomological Society of Canada's Postgraduate Award (2015) and the Natural Sciences and Engineering Research Council Alexander Graham Bell Graduate Scholarship (2011–2014). He was also one of two expert reviewers on the economic sections of the recently published Intergovernmental Panel on Biodiversity and Ecosystem Services' *Thematic Assessment of Pollinators, Pollination and Food Production.* He is also coauthor with Alex Stoner of *Freedom in the Anthropocene, Twentieth Century Helplessness in the Face of Climate Change.*

Bradley Necyk is a Canadian visual artist working through the mediums of drawing, photography, video, film, sculpture, and performance. He recently finished as the Artist in Residence with AHS Transplant Services for the length of 2015–16, is an artist/researcher in a project on Head and Neck Cancer, and is completing an arts-based Ph.D. in Psychiatry at the University of Alberta. Currently, he is a visiting artist/researcher at the Centre for Addiction and Mental Health in Toronto and has a studio residency at Workman Arts, Toronto. His work focuses on patient experience, auto-ethnography, psychiatry, pharmaceutics, and biopolitics. He has been shown internationally, was an artist in the 2015 Alberta Biennial, participates in artists' residencies, delivers academic papers internationally, is a committee member on several professional bodies, and is a Scholar in the Integrative Health Institute at the University of Alberta.

Patti Pente is an artist and associate professor in the Faculty of Education at the University of Alberta, Edmonton, Canada. She obtained her Doctor of Philosophy degree in Curriculum Studies at the University of British Columbia, Canada. She researches into the pedagogical shifts that occur when examining philosophical texts through art process. Her interests include contemporary notions of landscape art and the subject, with particular emphasis on educational potential within relationality and materiality. She continues to investigate the aesthetic nature of physical and virtual relationships to space and place through various technologies and within different populations. As a practicing artist, she has exhibited both locally and internationally.

Leslie Sharpe's artwork addressing animal–human relationships, environment, colonialism, and digital histories has been exhibited in Canada, Colombia, Finland, Germany, Spain, France, United Kingdom, and USA, and her writing included in *Far Field: Digital Culture, Climate Change and the Poles, Textile Messages, Transmission Arts*, and Leonardo. She has been an artist-in-residence at P.S. 1, New York, the Banff Centre, Alberta, and lvvavik National Park, Yukon. Sharpe is Chair of the Department of Art and Design at MacEwan University, and previously taught at Pratt Institute, New York, Indiana University, Bloomington, and University of California, San Diego, where she was a Faculty Fellow.

Janae Sholtz is Associate Professor of Philosophy at Alvernia University, Coordinator of Women's ad Gender Studies, and Alvernia Neag Professor. She received her Ph.D. from University of Memphis and MA from New School for Social Research. She is the author *The Invention of a People, Heidegger and Deleuze on Art and the Political*, Edinburgh Press (2015), in which she contemplates the potential for new political futures from reconceptualizing ontology through the imaginative, creative paradigms opened through the aesthetic considerations of Heidegger and Deleuze. Her research focus is Continental Philosophy, avant-garde art and contemporary aesthetics, social and political philosophy, and feminist theory, with particular focus on art as a form of social/political resistance. She has written articles for *Deleuze Studies, Journal of Continental and Comparative Philosophy, philoSOPHIA*, and is currently coediting a volume entitled *Deleuze and the Schizoanalysis of Feminism: Alliances and Allies for Bloomsbury*.

Nathan Snaza teaches modern literature, educational foundations, and cultural theory at the University of Richmond. He is the coeditor of *Pedagogical Matters: New Materialisms and Curriculum Studies* (Peter

Lang, 2016) and *Posthumanism and Educational Research* (Routledge, 2014) and his essays have appeared in journals such as *Journal of Curriculum and Pedagogy,Journal of Curriculum Theorizing,Educational Philosophy and Theory,Educational Researcher,Parallax,Angelaki, Symploke,Cultural Critique,LIT: Literature Interpretation Theory,* and *Journal for Critical Animal Studies.* He is presently coediting, with Aparna Mishra Tarc, a special issue of *Curriculum Inquiry* entitled "Sylvia Wynter, the Human, and Curriculum Studies" and is finishing a book entitled *AnimaLiterature.*

Ted Stolze teaches philosophy at Cerritos College. He has coedited (with Warren Montag) *The New Spinoza* and translated *In a Materialist Way,* an anthology of writings by Pierre Macherey. He has engaged in research primarily in the areas of early modern and contemporary continental philosophy, and he has published articles on such figures as Hobbes, Spinoza, Deleuze, Habermas, Althusser, and Negri. His book *Becoming Marxist: Studies in Philosophy, Struggle, and Endurance* is forthcoming from Brill.

Alexander M. Stoner is Assistant Professor in the Department of Sociology at Salisbury University. His primary areas of expertise are environmental sociology, social theory, and political economy. His work has appeared in journals such as *Ecological Economics,Critical Sociology,* and the *Journal of World-Systems Research.* His latest book (coauthored with Andony Melathopoulos), *Freedom in the Anthropocene: Twentieth-Century Helplessness in the Face of Climate Change* (Palgrave, 2015) outlines the contribution of critical theory to understanding our current ecological predicament.

Matthew Tiessen is an Assistant Professor in the School of Professional Communication at Ryerson University (Toronto) and a Research Associate at The Infoscape Research Lab. Dr. Tiessen holds a SSHRC Insight Grant in the area of "Digital Economy" to support research on the social implications of algorithmically-driven digital technologies.

Michael Truscello is an Associate Professor in English and General Education at Mount Royal University in Calgary, Alberta, Canada. He is author of *Infrastructural Brutalism* (MIT Press, forthcoming) and coeditor of *Why Don't the Poor Rise Up? Organizing the Twenty-First Century*

Resistance (AK Press, 2017). His recent publications include chapters on petrocultures and radical politics in *Petrocultures: Oil, Politics, Culture* (MQUP, 2017) and *Fueling Culture: 101 Words for Energy and Environment* (Fordham UP, 2017). He coedited an issue of Anarchist Studies on the topic of "anarchism and technology," and in 2011 directed the documentary film Capitalism Is the Crisis: Radical Politics in the Age of Austerity.

Mickey Vallee is a Canada Research Chair (Tier II) and Associate Professor in the Centre for Interdisciplinary Studies at Athabasca University. His research is generally about the complex relationships between sound, embodiment, and emerging technologies as they intersect with contemporary social problems. He has published in *Body & Society, The Sociological Review, Theory, Culture & Society, Parallax, Deleuze Studies*, and numerous other places.

Jason J. Wallin is an amateur Pokemon trainer working out of the Ivory gym in Edmonton, Canada. While Jason has never battled a Pokemon, he is known for his work on the ecology of Gogoat and the morphogenic characteristics of Sunkern.

LIST OF FIGURES

CHAPTER 1

Introduction: Interrogating the Anthropocene

jan jagodzinski

One hopes that this is another 'untimely' book that adds to the many voices of artists, poets, academics, politicians, and leaders around the world who have embraced the necessity of addressing the precarity of the Earth and the crisis of our species in what is has been arguably termed the Anthropocene; its euphemism, 'climate change' is certainly the more common term, but no better understood. There is no part of the Earth that has not been touched by anthropogenic activity. Strontium-90 did not exist before 5:29 a.m. on July 16, 1945; nor did the manufacture of long-lived quantities of halogenated gases; plastic has penetrated the deepest of ocean trenches (Galloway et al. 2017); plastic-eating bacteria have now been discovered (Yoshida et al. 2016), and even a new rock, the plastiglomerate, a stone containing a mixture of sedimentary grains of melted plastic, beach sediment, basaltic lava, and organic debris has been proposed as marking our species presence in the geological record and, therefore become recognizable in the stratigraphic record at some future date as a global boundary marker of a formal geologic unit of time (Corcorn et al. 2014). The depths of

j. jagodzinski (✉)
Department of Secondary Education, University of Alberta,
Edmonton, AB, Canada

© The Author(s) 2018   1
j. jagodzinski (ed.), *Interrogating the Anthropocene*, Palgrave Studies in
Educational Futures, https://doi.org/10.1007/978-3-319-78747-3_1

the ocean floor do not escape human intervention of one kind or other, a concern deeply explored by Stacy Alaimo (2017) who alerts us to the fragility of underwater creatures, whilst Heather Davis (2015) articulates how plastic micro-polymer particles are killing river, sea, and ocean life. The Anthropocene now becomes the Plasticene. As Camilio Mora et al. (2011) point out, 86% of the species on Earth have been catalogued, whilst it is estimated that 91% of those in the oceans still await description.

The central problem, as many scholars have astutely pointed out (e.g., Chernilio 2017a, b) is the generic term *anthropos* (Greek for man, human being) that is embedded within its nomenclature (see Chakrabarty 2015). Whilst it is the anthropogenic impact of our species on the Earth—on its resources and on its biosphere—that marks the transition into this new era that leaves the Holocene behind, the Anthropocene's description and resemblance to the hegemonic model of the 'human' both exemplifies and, at the same time, problematizes the collapse and enfoldment of humanity within the term Man. At the same time, it displaces the overemphasis on its narrative strictly in terms dominated by the natural sciences, primarily geology and climatology, which offer conclusive evidence of our species impact on the planet, and yet remain weak regarding the shaping of its sociopolitical narrative as contested by various ideological interests. The Anthropocene directly equates the agent of incumbent responsibility for this global crisis to the 'white Man' of European Enlightenment, and to the emergence of scientism and the largely instrumentalist legacy of progressive modernity that is as much entangled with hierarchy and enslavement, which pervaded the colonialist mentality of conquest in the name of Man, bringing with it the spread of infectious diseases of one sort or another (e.g., smallpox, measles, influenza, flu), and the death of approximately 50 million people. Such colonization eventually led to the entitlement of appropriating the material world in the name of progressive global Capitalism; Earth became simply matter, our 'standing reserve' as currently still supported through a neoliberalist philosophical and political agenda that ensures the continuation of profit at the expense of a sustainable planet, forwarding an ideal of a sovereign subject with certain 'rights and freedoms,' a subject that is already predefined by the symbolic order that is in place.

The 'Great Acceleration' is taken by many social scientists as one of the possible arguments for the rise of the Anthropocene era as the statistical documentation of the International Geosphere-Biosphere Programme (IGBP 2015) shows. Roughly, beginning in 1950, shortly after WW2, it is marked by a major expansion in human population, changes in natural processes, and the development of novel materials from minerals to plastics that led to persistent organic pollutants as well as inorganic compounds. Rachel Carson's *Silent Spring* remains a standing testament, documenting the environmental detriment. This increased industrialization is attributed and confined to the world dominated by the post-industrial OECD countries in relation to consumption and economic production, whilst most of the population growth is attributed to the non-OECD world (Steffen et al. 2015). The injustice and inequality of climate change responsibility has been well established. It is not 'humanity's fault,' but a question of the prevalence of global injustices that continue to maintain the current status quo. In the early twenty-first century, the poorest 45% of the human population accounted for 7% of emission, contrasted to the richest 7% who produced 50% of emission; studies conducted between 1990 and 1998 by the World Bank found that 94% of the world's disaster deaths occurred in developing countries, a major North–South divide (Parks and Timmons 2007). Disasters like Hurricane Katrina that hit the Gulf Coast of the United States in August of 2005 show the differences of economic support and possibilities of renewal between black and white neighbourhoods (Tauna 2008). The economic and social devastation caused by Hurricane Sandy to Manhattan is pale in comparison to what happened in Haiti. With the continual rising of the coastal sea levels due to the melting Polar Ice caps and thermal expansion of the oceans, the devastation will not have the same impact on the Netherlands, as it will along India's Bay of Bengal or on the Nile Delta coastline (Malm 2013). The bottom line is that it is the wealthiest 'few' relative to the 7.5 billion-population who are responsible for the impending anthropogenic made disaster; they will also have the resources to survive the longest (Satterthwaite 2009; Hornborg 2017).

There are other claims that squarely challenge this view of the Great Acceleration. Lewis and Maslin (2015) in a highly researched article in *Nature* present a convincing alternative *if* the criteria set out by the International Commission on Stratigraphy (ICS) are met. Such criteria, known as Global Standard Stratigraphic Age (GSSA) enable an assessment of various global markers known as 'global spikes' or Global Stratotype Section and Points (GSSPs). Given this standard, a different picture emerges with other consequences to ponder. Lewis and Maslin, after reviewing the range of other proposed dates (agricultural farming, rice production, industrial revolution, and so on) identify two contending dates: 1610 and 1964 that have identifiable golden spikes (GSSP markers). (For an artistic response on golden spikes, see Hannah and Krajewski 2015.) The first, 1610, is what they call their 'Orbis Hypothesis.' This is when there is a meeting of Old and New World human populations through discovery and colonialization, registering a dip in CO_2 levels due to the growing of new crops, the homogenization of the Earth's biota in terms of the transoceanic spreading of species, and the breakout of extreme diseases, famine, war, and enslavement: an estimated 50 million people died in the Americas (Mann 2006, 2011). This 'Orbis spike' meant that the two hemispheres were connected, furthering global trade and the beginning of the modern 'world-system' (Wallerstein 1976). The second date—1964, refers to the peak registration of Carbon-14 from the nuclear explosions dating from 1945 until late 1950s when there was a decline thanks to the Partial Test Ban Treaty of 1963. Selecting 1964 is more consonant with the Great Acceleration where there is unambiguous anthropogenic activity. Lewis and Maslin argue that the radionuclide spike is a good GSSP boundary marker, however, the disadvantage is that this was not an Earth-changing event, which holds as an argument only within the criteria as formulated by the ICS. The nuclear bomb changed the fundamental ontology of our species as self-extinction was now on the table. For Lewis and Maslin, the 'Great Acceleration' is diachronous and open to challenge as being too much of an arbitrary marker: the GSSA suggested date could be 1950, 1954, or 1955. Hence, they settle the argument for the Orbis spike that includes colonializaton, species exchanges, global trade, and coal as the transformative changes that had brought about the Anthropocene. Choosing 1964 instead, marks the advancement of technological weaponry—from hand axes to spears to nuclear weapons. At the same time, this history of aggression and violence underscores the fundamental question as to how this 'stain' on our species psyche can be managed, if never eradicated. The Partial Test Ban Treaty (PTBT) provided once such marker, but has never assured a solution as nuclear warheads spread (e.g., Israel, India, Pakistan), averted perhaps in Iran and waiting to cross the threshold by North Korea as ICBM's are being readied.

Be as it may, it shows that 'some' global cooperation is possible because of this danger, yet the Doomsday Clock has recently been moved to two and a half minutes to midnight. The physicist Michio Kaku is interesting to note here when he (wildly) speculates that our species is a Type 0 civilization; it is transitionally balanced towards advancement or suicide. We either advance to a Type 1 Planetary civilization where 'Nature' is controlled and managed through technological scientific means, or planetary suicide is on our agenda in one way or another, due to our inability to cooperate as a species and potentially sink into various forms of barbarism as Isabelle Stengers (2015) outlines. He imaginatively projects that perhaps there are many Type 0 civilizations in the universe who never made the transition, yet alone reaches the sci-fi capabilities of a Stellar Type II civilization (e.g., Star Trek's Federation of Planets) or even a Type III Galactic civilization (e.g., the Borg).

Lewis and Maslin's 1610 argument highlights the emergence of the social concerns, the unequal power relationships amongst different groups, the economic disparity, the further impact of global trade, and the continual and current reliance on fossil fuels as the crisis that is faced today. Social scientists, such as Alf Hornborg (2015), especially those coming from a Marxist grounding, are quite clear in their claims that the Anthropocene narrative must recognize the unequal global exchange of labour energy and biophysical resources that shape mainstream capitalist economic policies, and precisely why the obsession with 'technological progress,' especially those technologies that are dependent on fossil fuels; they coincide with obscene purchasing power, cover-ups surrounding access to oil, military involvement, and the avoidance of the fetishistic account of its consumption that obscures the material and social dimensions of power. For these social scientists, the Technocene, Capitalocene, or Econocene are far better and the more accurate synonyms for the dystopic teleological grand narrative that has now established itself, a rather bitter irony when François Lyotard's persuasive argument that the more utopian grand or master narratives (of the Enlightenment and Marxism) have been exhausted and untenable (Hornborg 2015; Malm 2013; Malm and Hornborg 2014; Moore 2015, 2016a). It's the consequences of 'cheap nature' as Moore (2016b) succinctly puts it: the 'four cheaps' being labour, food, energy and raw materials. Adrian Parr (2013) has clearly presented the entwinement of neoliberalism and the politics of climate change, articulating links with capitalism in a broad range issues, from questioning its green camouflage, population arguments, access to food, and water as manipulated by corporations and animal pharma, the danger and devastation of oil spillages. As Parr says, 'Adaptability, modifications, and displacement ... constitute the very essence of capitalism' (146). By 2050 it will all be over!

Elmar Altvater (2016) is also especially elegant in showing how capital and the evidence of Earth scientists are intimately entangled. An alternative sustainable economy has been posited Herman Daly (2005, 2010), as has a set of nine planetary boundary conditions based on the projected Earth Science statistics that are not to be crossed if Holocene epoch is to maintained and the biosphere remains liveable: climate change, ozone depletion, atmospheric aerosol loading, ocean acidification, global freshwater use, chemical pollution land system change, rate of biodiversity loss, and biogeochemical loading-the global Nitrogen and Phosphorus (N&P) cycles (Rockström et al. 2009a, b).

This stark division, driven by capitalist interests, has also provided an understanding of how necropolitics (the politics of death) has supplanted the biopolitics of life within disciplinary societies (Mbembe 2003; Thacker 2011; López Petit 2011; Gržinić and Tatlić 2014). Some have called it a Necrocene. Justin McBrien (2016) writes powerfully how capitalism's end goal leads directly to a 'New Death.' Capitalist expansion can go nowhere except to a '*becoming extinction*' (116). The accumulation of capital leads to a form of autolysis as opposed to apoptosis: cells begin to destroy themselves through the action of their own enzymes in a form of self-digestion. The traumatic injury of such disease turns the body's drives against themselves, like an autoimmune disorder where a body attacks itself. Capitalism transmutes life into death, and then death into capital. In 'societies of control' the context of sovereignty and post-colonialism has been modified in a neoliberal era where terror, precarity, and insecurity persist globally. Death becomes the ultimate exercise in domination as well as resistance. Put bluntly, some lives are always worth more than others across the entire spectrum of the living, both human and non-human. Certain lives are always more disposable, both sexually, racially, and economically. The regime of necropower, as extended to the Anthropocene, brings with it a blurring of the lines between resistance and suicide, sacrifice and redemption, as well as martyrdom and freedom. For situations within sovereign war-states present a choice without a choice where death and terror paradoxically turn into terror and freedom. The Anthropocene in this sense presents the worry that 'life' can be disposed off or artificially manufactured at will.

NECROCENE

Related to necropolitics, or rather the ground upon which necropolitics emerges, is the biopolitics (or biopower) (or technology of power) of modernism as developed by Foucault's late work on governmentality: the power of the sovereign to take life or to let it live, is replaced by the regulating power of 'making live and letting die' (Foucault 2003, p. 247). This is a highly contested zone of theorization, concentrated (of late) largely in the Italian context, which includes the writings of Georgio Agambem, Hardt and Negri, Roberto Esposito, and various former members of Italian Autonomia Operaia movement of the 1970s and 1980s, the best-known members being Paolo Virno, Toni Negri, and Franco 'Bifo' Berardi; all of whom have their own nuances and directions in the way they (after Foucault) take up Spinoza's univocty and Deleuze's A Life of pure immanence. To this list can be added Rossi Braidotti (2017), who, from a deeply committed feminist viewpoint, gives short shrift to Esposito, Negri and Hardt in her own writings, although her own developments are deeply invested in the zoë|bios complex (formless life| formed life), and postanthropocentrc thinking (e.g., Braidotti 2013), which this body of theoretical work covers, especially in its importance to the Anthropocene.

Agamben's persuasive thesis states that the 'state of exception' of *homo sacer's* being is the political figure that embodies the originary political relation. This both questions and furthers Foucault's initial insights. It has, however, been severely criticized for its negative biopolitics as an aporia of Western politics, and for its deep commitment to the Greek logos wherein zoë and bios, as intertwined dichotomies, fortify and immunize the state (Dubreuil 2006). Zoë in this schema is equated to 'bare life' that is to be disposed; the Nazi social democracy exemplified this as the modernist norm. Esposito (2008) displaces Agamben's grasp of Nazi biopolitics. He replaces or bypasses Agamben's figure of the *musulman* as the embodiment of bare life, the *homo sacer* of the twentieth century, with that of the German *genos*. Here, the 'immunitary paradigm' that pervades the Nazi regime is understood as a biocracy. The immunization of the Aryan race is profoundly tied to the scientific and the medicalization of the body politic via genetic experimentation and physical exercise. The concentration camps are there to both exterminate the weak, the inferior Jews and to strengthen, protect and purify the Aryan race. The manipulation and modification of life becomes the Nazi's transcendental achievement that pervades all aspects of its social order. In a biocracy, bios is a category under juridical control, whilst the law (*nomos*) is biologized along blood lines to establish a racial 'norm of life,' as Esposito puts it. 'Biology and law, and life and norm, hold each other in a doubly linked presupposition' (Esposito 2008, p. 183).

In contradistinction to Agamben's negative biopoltics, Roberto Esposito develops a positive biopolitics wherein the political immunization, as enacted through the necropolitics of death, is overcome by dispelling the existence of any fundamental norm that involves the making of a genetic individual via bioengineering. Drawing on Deleuze's 'A Life' that deemphasizes individual life, as characterized by an active vitalism of a person striving towards predetermined goals and norms, Esposito turns to the absolute 'singularity' of an impersonal life and Gilbert Simondon's development of 'individuation' and the 'transindividual' to develop his positive biopolitics. A Life here is not some extrinsic normative value to be strived towards, which is the usual notion of vitalism, a becoming that achieves preset goals. A Life, or immanent life is *not* bound by principles of organization. Rather, it is a *style* of becoming, constantly transforming itself through encounters from without; encounters with what is other than itself. It is the acknowledgement of inorganic life, life that remains contingent as well as virtual; it is the recognition of forces that are alien to human. The potential of zoë, as Deleuze develops it, is virtual pure immanent life yet to be actualized. Paradoxically it is 'non-living,' yet to take form.

Esposito intertwines nature and culture in what has become a standard tenet by those interrogating the Anthropocene, sometimes written as natureculture. *Bíos,* in his position, always already contains zöe as an open unfolding process of life. He follows ontogenesis as developed by Gilbert Simondon, where there is a series of 'births,' each birth actualizing a potential through an event that its effects remain as a trace. The condition of life is a perpetual birth: life and birth being contraries of death, the synchronicity of the first is supplemented by the diachronicity of the second. Death in this view is constantly deferred, what Deleuze and Guattari (1987) and Deleuze (1997) develop as 'becoming-child.' Such vitalism is not orientated towards death, rather it transverses determinate life forms that are always subject to change. 'Norm and life,' in this account are intrinsic to the organism, not an extrinsic norm as to what is allowed and prohibited. Life becomes the immanent expression of its own unrestrainable power to exist, and this 'power' is extended to each and every organism. Such an extended Spinozian insight presents the constitution of the 'norm' to be both singular as well as a plural, following Jean-Luc Nancy's (2000) being-with-many (others) that allows for the creation of the new; the plural constitution of the norm emerges in reciprocity with all other singular norms in what is understood as a metastable, normative and hence dynamic system 'at the edge of chaos' (e.g., Bell 2006).

Esposito (2012) (see, Campbell 2006) provides an important political position known as the 'third person' that enhances his own affirmative biopolitics as well as Deleuze and Guattari's two important concepts that address their developments of a virtual (or passive) vitalism: 'Body without Organs' (BwO) and 'minoritarian politics'; the former concept refers to a state of virtual potential, transposed, if you like, as a quantum vacuum, a zero-point vacuum energy; although it 'appears' empty it is fully active and real as virtual particles blink into existence and then annihilate in a timespan too short to observe; the latter concept refers to actualized elements as developed within a system's 'smooth spaces' that are antecedent and precursive to deterritorializing the existing seemingly stable system held together through desire by a persistent dominant image of thought. Esposito's two concepts, the 'impersonal' and the 'impolitical' are ways to open up non-representational space that is at odds with representational politics, viewed as 'idolatrous' as he says, with its admiration for norms as the submission to accepted forms of the 'person.' Impolitical and impersonal seem to be closer to Deleuze and Guattari's 'smooth spaces' where other forms of relationality can occur, not limited to the person: hence open to non-human life (animals, plants, bacteria, viruses, and so on). In the indeterminate space that Deleuze and Guattari develop, it is the 'ego' that subsides as 'identity' transforms. Again, a passive vitalism, and not one of mastery, is at work here; the 'impersonal' is a subjectivity that manifests itself in the gap between zoë and bios (birth|rebirth as discussed above). The emergent 'becoming' of an encounter or 'event' ends in a shared *bíos*, an always already impolitical situation as individuation advances. The identity (or person) through the event is overcome; its symbolic place falls away. The apperceptive form of the 'I' (as the transcendental form of the consciousness) no longer holds ground. We have the displacement of politics cum person cum individual with impolitics cum impersonal cum individuation; this latter development being a non-representational theorization of human to non-human relationships; much like Deleuze and Guattari's exploration of becoming-other beginning with becoming-woman, which Colebrook (2014b) will argue is the necessary first step to 'queer' the Anthropocene (more below). Esposito, on the other hand, is more in tune with becoming-animal (2012, pp. 19, 114, 149–150). The relationality of 'becoming-animal' has nothing to do with person or thing, it is the achievement of the impersonal, which he then paradoxically calls 'the living person' as that moment of becoming that ends his book, *Third Person* (2012).

This is precisely the same track taken by (post)Deleuzian Elizabeth Grosz (2008, 2011), also Yusoff et al. (2012), maintaining that the interval between the human and the animal is one of degree and not kind. Grosz, throughout her writings, stresses sexual difference as the engine of all variation and diversity; it raises the tensions of the nuances and disagreements whether this is 'queer' enough, as Colebrook (2014b) maintains (more below); or whether sexual difference and the dynamics of sexual selection as privileged by Grosz is *the* condition for the emergence of all existing difference—including those that are said not directly linked to sexual dimorphism as well as polymorphism. Does sexuality itself need to function aesthetically (artistically) to be adequately sexual? Grosz's position effectively equates difference *qua* difference as sexual, weighing down affirmative desire with residual representational human-animal interests. There are challenges to this, as in the explorations of Lucia Parisi (2003) on 'abstract sex' that addresses reproduction away from biology to the inorganic, much closer to Deleuzian (passive) vitalism of difference and the virtual. Sex becomes queer in this regard, and prior to Grosz's appropriations. It raises questions whether 'luminous' display by certain bacteria are to be seen as an indication of sexual aesthetic display? Is the 'poetry' of nature always already present to us, as in the well-known example of the orchid and wasp developed by Deleuze and Guattari (1987) where there is the crossing of interkingdoms? Is Graham Harman (2005) correct to claim aesthetics as the first philosophy, and that 'allure' is felt between objects? What then does one do with asexual reproduction (parthenogenesis) where no display seems to be identifiable, or with so-called virgin births, or terminal fusion automixis that produces 'half-clones' of the mother and sometimes full clones, in Burmese snakes for instance (Groot et al. 2003).

'Abstract sex' is actualized into a variety of potentialities that recognizes the multiplicity of asexuality, creatures that show less ostentatious display of coupling, which is the aesthetic that binds the animal with the human in Grosz's account. Sex now becomes an infinitely mobile and mutable form: microbial sex, bacterial sex, aquatic sex, meiotic sex, turbo sex, cybernetic sex, and so on. In this series, the multiplicity of expressions of human sex is simply another variation subject to change. In Brian Massumi (2014) case, also a (post)Deleuzian like Parisi, there is a way to further radicalize 'difference,' by theorizing play along Batesonian lines. Play takes on the role of 'difference' by being the explorative energy in the zone of indiscernibility between man and animal. It is play that breaks the norm, but in doing so the norm becomes a *supernormal variation*. There is no 'norm' to adapt to, no style of the animal, or mimicking the animal; rather the differential power from which the qualities emerge are 'supernormitized.' Claire Colebrook (2014b, pp. 109, 172) in the case of 'becoming-woman' gives the example of Madonna and Lady Gaga who are not Marilyn Monroe impersonators, which would be the aim and norm to strive for, but singularities of the qualities and tendencies of style, movement, display, performance of Monroe's hyper-femininity.

This above somewhat tangential discussion for the Anthropocene draws some weight when it comes to the feminist response to the Anthropocene: what position should it take? Should it rest with the eco-feminisms emphasis on the harmony of Nature? Should it take the tact of Rosi Braidotti's (1994, 2013) insistence on sexual difference: Luce Irigaray being *the* anchoring point for both her and Grosz, or should one follow Colebrook (2014a, b) in her Anthropocene-inflected explorations of sex (after life and extinction)? She has effectively questioned Judith Butler's position in relation to performativity as always being positioned against some norm, provocatively asking in one of her chapters: 'How Queer Can You Go?'

The racial and feminist critique of the Anthropocene are well established in the ecofeminist writings of Vandana Shiva (2016), the science and technology scholar Donna Haraway (2016a, b) and the cultural theorist and philosopher Claire Colebrook. Claire Colebrook, who has intensively concentrated on the Anthropocene through a consistent series of reflections since about 2002, provides yet another meditation on feminism and the Anthropocene by raising life as being *indifferent* (more below). By this, she seems to be signalling the radical sense of Spinoza's *conatus*: the striving of life in its rogue or anarchic ways where distinctions, identifications, boundaries are done away with, in a destructive manner as well as a symbiotic one. After all, we cannot attribute 'morality' to Nature, if nature is taken as 'pure immanence' (zoë). It points to the obvious realization that the Earth is constantly deterritorializing itself. How to cope with the Earth's constant dynamism? One way is offered through 'aboriginal cosmopolitanism.' Nigel Clark (2008), writing on the Australian Aborigines, identifies them as 'nomads' (in the Deleuze and Guattarian sense of the term). They have learned to 'nomadically' dwell in harsh climates by staying in the same 'place' and 'going with the flow' of the Earth's forces. Clark reminds us that when it comes to rapid environmental changes one suffers estrangement and becomes a stranger to the familiarity of one's own place; the world that was known is simply left behind. It becomes shattered and upended. Aboriginal cultures have learned to cope with such estrangement. They are capable of 'fire farming' (Pyne 1991), to control fire's energy and proclivities.

This 'aboriginal cosmopolitanism' raises the question of 'climate change' in a deconstructive way. It suggests that controlling and stabilizing 'Nature' via industrialized agriculture changes the Earth's biomass and the rhythm of the seasons. Large-scale agribusiness has now led to a further change, a new instability and intensity that must be faced. What is a stable or an unstable climate is simply dichotomous thinking; the case being that there is *only climate variation* that must be endured. Colebrook (2017), drawing on Deleuze and Agamben, makes the point that indifference and difference are entwined in such a way that when difference emerges from indifference the horizon of its disappearance is already in sight. If indifference is basically the instability of Nature, it's ebbs and flows, destructive as well as symbiotic, Nature does not discriminate as to where and how devastation or stability of climate will occur when it comes to our species, then the constant attempt to stabilize nature, for example through agricultural technologies that rely on genetic modification of crops, soil manipulation, irrigation, and so on, paradoxically leads to climate change that is of our own making. The invention of Nature becomes a necessity, a 'thing' to manipulate, modify and control. To do otherwise; that is, to think of an Anthropocene counterfactually where this is not necessary, seems equally a dead end, a Romanization of sorts—to simply harmonize with Nature's unpredictability, like anti-quake architecture that only prevents the risks of such a catastrophe just so far, i.e., the percentage of death will not be as high. Accepting Nature's *indifference* include all the sublime horror as well as the beauty, which is a road already well travelled. For Colebrook then, a 'feminist Anthropocene' requires that the question for 'whom' does the Anthropocene speak for? Anthropos as Man. Who is the reader of the 'non-future? The personal becomes the geological in her account— the relations between human and non-human—as the 'anthropological machine' (cf. Agamben) continues to valorize the exceptionality of our species: Man is capable of destruction as well as self-annihilation, capable of technological wonder and advancement. Yet, as Bronislaw Szerszynski (2016a, p. 16) put it: 'the very notion of the Anthropocene contains an element of indecision: is this the epoch of the apotheosis, or of the erasure, of the human as the master and end of nature?'

A Radical Hypothesis

Matter is life *itself; it is* energy as intensive dynamic differentiation, a continuous variation of becoming that is composed of infinite singular events that are taken to be positive differences. This is life as pure desire that enables connections and relations to take place. Desire is like dark energy; it exists but we have no clue just what it 'is.' Dark matter, it seems to me, is equivalent to the virtual singularities of difference that are actualized into visibility. The dynamics of this 'life' are little understood at the quantum levels. Differential forces ('singularities' or 'multiplicities') are not directed towards realizing definable forms, rather they signal a range of potential relations and affect producing mutations, the new, that 'deterritorializes' established organized life. Hence, there only ever is 'life,' meaning a particular becoming that is not generalizable only universalizable. Or, put another way, there is only 'mind' or micro-minds as multiple differentiations not bounded to an organism. The idea is that *passive vitalism* (Colebrook 2010) is basically open system's thought; it is another name for 'difference' in and of itself, what many Deleuzean influenced writers take as the primordial generative force that produces constant variation, diversification and mutation (e.g., Grosz 2011).

The above discussion is crucial to theorizing the Anthropocene, which must address issues about life, anthropocentrism, the non-human and the inhuman (technology) as the unfolding of a new conversation. Life, itself, remains non-representational and unknowable (e.g., Helmreich 2011), and therefore speculative as a world-without-us ushering in what can be called a post-ontological problematic, the post playing its usual dualistic position of a *before* as well as an *after; after* because the Anthropocene presents a point where the histories of the Earth and our species intersect, enabling the realization that they have always already intersected in various degrees, establishing a *before*, but now 'we' find ourselves at a limit point, a shift from degree to kind. The 'origins' of the Anthropocene, in this account, are always coming to present, since the origin is never present to itself; beginning and commencement remain in perpetual entanglement as Heidegger (1968, p. 34) developed this in his discussion between *Beginn* and *Anfang* (i.e., a point—then movement). Which is why, Deleuze and Guattari will argue that 'we' always begin *in media res*. Life, however, has become a centering problematic of the Anthropocene as questions of death and extinction surround it, along with non-life (inorganic and inanimate) in what has come under the umbrella term panpsychism. Elizabeth Povinelli (2017a) names such non-life as *geos* where the relationship to bios (human life) has three varying levels: the first to Gaia, as the living planetary organism; then to geology (as the non-living), and finally as 'no relationship,' if it has no place in liberal thought where *anthropos* reigns without question.

The most radical account of accepting the Anthropocene as an epoch, following this paradox of origins, would be to propose (after van der Pluijm 2014; Lewis and Maslin 2015, p. 177) that there is no justification in retaining the Holocene, which is a climatic marker that follows the Pleistocene. The Holocene does not provide any 'golden spikes' (GSSP) that show stratigraphic evidence of change. Homo sapiens, however, are a Late Pleistocene species where the evidence of human activity is clear. Here, it can be argued that through technology—the harnessing of fire and tool use (hand axes and so on) as the development of a persistent technological prosthetic—both modify the physiology of our species throughout human history, as well as modifying the ecology (Nature) that in turn shapes us not totally in predictable ways. Fire, in Stephen Payne's (1995) analysis, expanded the range of (inhospitable) climates that became available to our species. Homo sapiens by definition always already are in the process of becoming Other than what they are, even if that is happening in the naivety that this process is somehow controllable and predictable. Such a position is 'post' Anthropocene in the sense that modifying 'nature' from the advent of our species simply means there is no 'nature,' our species has always already transformed it. To take such a radical accounting would mean a rethinking of species becoming, given that the Anthropocene 'event' enables a global refleXive (jagodzinski 2008, pp. 29–46) encounter to take place as shaped by the history of Homo sapiens becoming from 'then' (the

Pleistocene) to 'now' (understood as the threshold of our species ending). Such an enfoldment enables openings for new imaginaries to emerge. Whilst this does lead to such speculations, it also sets up a wish fulfilment that simply says we can now 'start over,'—self-redemption and human exceptionalism—rather than grasping the forces that are constantly at work.

An even more radical and extremely controversial hypothesis can emerge from this proposition by maintaining that the Homo sapiens genome has undergone imperceptible changes over human history; changes that take place when mutations result in niches where the assemblages are radically different. There is already evidence that the Homo sapiens genome had modifications to it through genetic intermingling from so-called archaic-forms of the genus Homo outside of Africa (Hammer et al. 2011; Gibbons 2011; Clive 2011), and the widely circulated 'fact' that Homo sapiens interbred with Neanderthals (4% of the Eurasian genome), and there is genetic evidence that the Denisovan interbred as well (Stringer 2012). When one looks at the most recent diagrams in these reputable journals, such as *Science* and *Nature*, the 'one species' idea seems in jeopardy as 'human' origins seem to follow a rhizomatic dispersal of multiplicities, rather than the 'tree lineage' adhered to before. In this speculation then, Homo sapiens physiology and encephalization during the Paleolithic period would not be the 'same' as the Homo sapiens populations living during the Neolithic (understood as a contested 'cereal-centric model' that has many nuances to account for morphological plant and animal change as well as genetic drift, i.e., Denham and Peter 2007; Jones and Brown 2007) as these are incomparable and incompatible cultural formations with changes at all levels of socialization, physical comportment, physiology, brain changes—such as changes within the corpus callosum due to the technological invention of writing in what has be speculated as a 'bicameral mind' (Jaynes 1976)—as well as grasping the variety of relations to the non-human world across cultures that have been explored by anthropologists like Philippe Descola (2013) and Eduardo Viverios de Castro (2014) (see Latour 2009), Lynne Margulis and Dorian Sagan's (1995) endosymbiosis and the 'genetic trading' between bacteria, and so on. Such a hypothesis would support de Castro's more radical view that that there are only a multiplicity of 'natures' globally, not some underlying 'relativist' universality that is deemed as 'human.' Homo sapiens and the 'human' cannot be conflated. The mounting epigenetical evidence would be another clue to argue such a hypothesis with more vigour and substance (e.g., Landecker 2011; Niewöhner 2011). Ontogeny does not recapitulate phylogency. It changes at the deep levels of structure (human anatomy and brain functioning). Imperceptible changes occur over generations by way of invented and augmented technologies and the changing sociocultural norms. For example, the world 100m champion, Usain Bolt is a 'product' of a contemporary assemblage, one that could not have emerged in another era, say the Neolithic. The way the natureculture of our species changes is dimly understood.

The further issue is that philosophical roots of ontological thought in the western world are grounded in an anthropocentrism of a world-for-us as developed by the Greeks; as Alfred North Whitehead (1978, p. 39) once quipped, all of philosophy is a footnote to Plato, necessitating the recovering of a minoritarian trace within philosophy, following Deleuze and Guattari, that opens up other imaginaries. But even here the Deleuzian position is but an inversion of Plato; Platonic Ideas come down to Earth, so to speak, as potential multiplicities that are actualized in various ways when addressing never-ending unsolvable problematics in what is called transcendental empiricism. There are openings, of course, into the non-human through virtual potentialities. But, the issue is more than that. The dialectic of the Enlightenment scenario (in Adorno and Horkheimer's sense 1972) that spreads instrumental reason and rationalism from its Euro-centred context is based on the retrieval of the Greco-Roman world that now gives us claims as to what is 'human,' what is 'humanism,' what is a person, a civic subject, global universalism, slavery, and so on. Industrialized technology in this sense is *always already* based on human enslavement made possible by a Greek philosophical elite, who were not only able to have the free time to philosophize 'for the love of wisdom,'—otium—but also to harness the Earth (via animal power) to free up more leisure time, which then produced 'culture' in its highly narrow elitist sense, where architecture and art, in particular, are given high status of splendour and crafts*man*ship. Dipesh Chakrabarty (2015) maintains that slavery (of women, races, etc.) lessens over history with the rise of industrialized technology as labour power becomes cheaper via machines, a trend that continues today. Not much has changed: a violent cosmos based on some form of servitude leads to the necessity of human universal rights.

This radical position presents an understanding of the Anthropocene as a necessary (present) juncture *for the re-evaluation of our species-becoming*, especially now that it is known that technological modifications can lead to our species extinction, to the extinction of other species (the endangered species on the 'red list') through our own doing, all related to the widely discussed sixth extinction event of the present (Kolbert 2014). For Brumo Latour (2017) the awareness of the Anthropocene makes us 'Earthbound'; the infinite possibilities projected by scientific progress are suspended. Colebrook (2017) makes a similar meditation, questioning the *hyper*-modernism of posthuman, postfeminism, and postracism projections which attempt to overcome representational differences through constant strategies of inclusion whereby infinite possibilities still remain for those yet to be included; a 'being of pure inconsistent multiplicity' in Alain Badiou's (2005) ontology, which is then made accountable via a *'counting-as-one'* through set-theory. Colebrook contrasts this with a *hypo*-modern position: the claim of passive vitalism which she now identifies as *indifferent*, partly to distinguish it from differences that capitalism caters to through its commodity strategies. Life as indifference once again draws on contingency and passive vitalism, and to the complications that such life brings, with its ups and downs, as intensive and extensive forces are at work, which are only partially controllable. Life as 'indifference' is undifferentiated life. Colebrook's theoretical move for Anthropocene thought is to escape *either* the universalist position where differences are recognized but subsumed, *or* the earth bound modified Gaia position of an interdependent organism, and maintain the necessity of a passive vitalism that enables an open system where 'the difference that makes a difference' (as Bateson would say) is recognized as an aleatory moment of change. The Anthropocene marks a shift in kind rather than degree; a recognition that tipping thresholds have been surpassed, placing us into a stark realization of what anthropogenic activity can actually 'do.' The Anthropocene, in this sense, has to be 'readable,' but only *if* it constitutes a significant difference.

Colebrook goes on to explore such a thought experiment via two counterfactuals as to what this entails, raising the issue of thought and its outside in good Deleuzian fashion: the three thought positions being, first, the relationship of humanity to what is to come (posthumanity-inhumanity-nonhumanity); the second, the relationship of human time with Earth time (the question of temporality itself), and the last concerns sexual difference in relationship to gender, an area Colebrook (2012) has carefully examined before. The counterfactual of imaging a non-Anthropocene species, a species that is not exceptional in its self-destructive potential seems impossible. It is, however, following Deleuze, necessary for 'counteractualizing' the *event* of the Anthropocene to continually make 'sense' of it. This counterfactual imaginary raises the question: at what point in human history does one go back where such a possibility (meaning the Anthropocene is an accident, which is itself therefore possible) is mitigated or completely prevented? As a species, this would mean some sort of total harmonious integration and balance with Nature, a myth since all existing creatures face extinction as a 'normative' part of species being. The same myth persists in the 'opposite' end of such a thought experiment: what is the 'ultimate' horizon that our species is striving for? Here all teleologic ideals (freedom, liberal democracy, justice, and so on) end up as, what Elizabeth Povinelli (2012) calls, a 'surround' without an opening as all differences disappear into the 'same.' The loss of desire for a horizon ends up as something not human, the stuff of sci-fi. A limit point is reached when no more 'children' are born that promise the future as Oedipally explored in the film *Children of Men*. The end of species-becoming ends in either annihilation or the birth of a post-human, a Nietzschean *Überman*, like the projections of many transhumanist scenarios (e.g., Bostrom 2014). Finitude perhaps becomes earthbound once more. Colebrook's second counterfactual addresses the non-renewal fossil fuels; here, there is an invention of technologies that succeed working with Nature, something that has already taken place in the fields of biosynthesis and biomimicry. However, as she says, this fantasy scenario ends up in positing something like a pure ecology where there is no measure of cost as there is only perfect efficiency.

RESPONSIBILITY AND COEXISTENCE

It is perhaps with Jean-Luc Nancy's (1997) meditations on justice and responsibility that may provide some insight when it comes to the global worries of cohabitation that besiege the Anthropocene. Nancy addresses this sharing of the world or coexistence as the law of the world; being is shared and divided at once (the nuanced meaning of '*partage*' in the French). Nancy maintains that globalization requires a new understanding of the notion of a 'just measure.' Humanity is confronted by a lack of criteria; the relationship between measure and the immeasurable ends up in terms of scale and greatness where the world has become its own measure. For Nancy, this requires an absolute, limitless responsibility, an archi-responsibility that precedes all measure. How, asks Nancy, can one think of justice from a global perspective in a world that has become its own measure, so that a 'just measure' might be found? Although, a full answer is impossible, Nancy's coexistence within the Anthropocene requires that the justice of dividing–sharing is always thought though the problematic of the singular–plural of coexistence, where the evidence of an unbearable injustice is all around (natural disasters, illness, viral diseases, unspeakable crimes). No theological excuses can be made, no ontotheological escape is possible, and no theodicy to generate a good|evil divide, only a multiplicity of instances traversed by interventions, passions, suffering, compassions, and struggle. Nancy identifies *comperance*, as the existential condition of every appearance before 'justice,' where the surround of the 'com' with 'appearance' provides a context for summoning archi-responsibility. Existence here accepts those contingencies of life, its actualizations that are paradoxically shaped by 'the law without law.' The 'law without law' would then command the justice of a coexistence. Here, the encounters of becoming offer no direct criteria: only a promise shaped by responsibility.

Kathryn Yusoff et al. (2012), picks up and further extends Nancy's struggles to relate to the coexistence of others: how to, as Nancy (1997) develops it, untie the knot that is sense. As Nancy puts it, 'How to think the tie as always *still to be tied*? … [I]nstead of conferring sense on the presupposed knot, to make the tying of the (k)not into sense itself' (103, original emphasis). Yusoff extends this difficulty to non-human subjectivity, especially given the loss of biodiversity, which is lost to sense; there are aspects of interaction that simply do not representationally register for us: for instance, endocrine-disruptive work of PCBs and POPs in the fatty tissues of seals (218). Yusoff's unique approach is an engagement with the 'insensible,' which is in-between sense and nonsense, an agitative or contagious force or motivation that oscillates between material and the virtual, organic and inorganic, time and the untimely, inhuman and the human. The difficulty of articulating such a position is evident throughout Yusoff's writing, which draws on Karen Barad, Isabelle Stengers, and Nancy, who plunge the problem into deeper recesses of unknowability. This question of the Anthropocene imaginary: What is our responsibility to the future, from the future? How do we co-inhabit with nonhuman; how to live with nonrelation, as the indifference or dejection of the world-without-us? How does one cohabit 'insensible alterity'—like mercury poisoning in the high arctic? How, in short, is one able 'to become more attuned to *no*thing, rather than *some*thing' (2013, p. 215). Yusoff answer is to think 'between natures' as the space between sensible entities to push into these insensible spaces. Easier said than done, of course, as she admits that patience with the insensible is a mode of reception that certainly questions how thought discloses and encloses itself, yet it risks falling into nonsense. Whether this engagement with insensibility is a 'willed' event or one of contingency, like the indeterminate space-time of becoming Deleuze and Guattari are concerned with, is not clear. Nancy's politics of knots seems to point in the same direction as Deleuze and Guattari's zones of indeterminacy. 'The (k)not: that which involves neither interiority nor exteriority but which in being tied, ceaselessly makes the inside pass outside, each into (or by way of) the other, the outside inside, turning endlessly back on itself without returning *to* itself' (1997, p. 111, original emphasis). Yusoff equates insensibility to the interiority of relations with things; she understands it as a force of alterity (the force of the (k)not in Nancy's terms) that has within it the paradox of being both unbounded and binding at once, crucial to how 'we' are called to justice. Such a discussion points more to the pedagogy of the Anthropocene, than it does to any forms of politics as is usually understood, for it suggests a reorientation of ourselves to the world-without-us, a minoritarian position at best; to 'inhabit that which is strange, nonintuitive, insensible—that which is remote from human comprehension or intelligibility—like phytopankton seeds, fungi, geological epochs, or multicelled organisms at the beginning of time' (225).

Anthropocene Imaginaries

The question of Anthropocene imaginaries within the US context has been reviewed by Levy and Spicer (2013) who provide four contested positions along with three historical developments over the past two decades within the country's political economy: 'fossil fuels forever,' 'climatic apocalypse,' 'techno-market' and 'sustainable lifestyles.' It should come as no surprise that the framing of each imaginary along with its values is dictated by capitalist interests as presented by the three periodizations in which they occur: the carbon wars of the 1990s, the carbon compromise between 1998 until 2008 when the financial crisis emerged, resulting in a climate impasse from 2009 on. The 'catastrophism' that characterizes the second imaginary of 'peak oil' has been carefully explored by Imre Szeman (2007, 2017, Truscello, Chapter 11 in this volume), who refers to such an imaginary as being shaped by 'eco-apoc-alypse discourses' that point to the dire sociopolitical and environmental consequences if fossil fuel use is not curbed. The Obama administration attempted to jump start the climate conversation by supporting the 2016 Paris Agreement, and stalling on pipeline legislation as well as protecting large tracks of land as protected national parks. The Trump Whitehouse, beginning in 2017, has been busy overturning Obama's legacy. Trump has placed his authority in restoring the 'fossil fuels forever' imaginary, pulling out of the Paris Agreement on Climate Change, cutting staff at the Environmental Protection Association (EPA), discouraging and sus-pending its scientific work, and promoting the coal industry, and allow-ing offshore oil drilling that President Obama had suspended. Despite these setbacks, it is in the interests of many US based companies to con-tinue the 'techno-market' imaginary via Green Capitalism to promote 'clean energy' that the Obama administration had promoted.

Against these more conventional imaginaries of viable business mod-els is Elizabeth Povinelli's (2017a) three figures of the Anthropocene. The first figure is a 'geontology,' which supplements the redisclosing of biontology. The desert, with its central imaginary of Carbon (Oil)—the percepts and affects are not difficult to imagine here as the desert seems (mistakenly) denuded of life, with oil underneath it as the trapped 'ancient sunlight' (cf. Tom Hartmann). From the International Space Station Dubai's simmering human-made aqua waterways highlight the surrounding bleak landscape, as if Dubai was some elaborate waterhole found in the dystopian film series *Mad Max* (Arabian Business 2014). Animism, as a second figure, has its central imaginary of the indi-gene. Here vitalism is the key issue in complete opposition to the (life-less) desert, and the question that should be of concern: what sort of vitalism? Indigenous and native peoples are presented with the core of biontological knowledge in a particular understanding of Nature. How to show gratitude to what is generally understood as the harmo-nious working of an interconnected planet (e.g., Jenkins 2012), which can be quickly grasped as animistic and transcendentalist in reference to a Great Creator, or Great Spirit; or, like the figure of Temple Grandin, whose activism presents a (post)humanist orientation, offers yet another direction. It is possible to see aspects of this orientation resulting in becoming more effective engineers and designers as clearly evidenced in biosynectics and biomimicry that support adventure capitalism. Povinelli's last figure is that of the terrorist with the central imaginary of the virus. Here, biosecurity becomes an issue in terms of geosecurity related to the economic crisis, energy shortages, the devastation occurring from catastrophic events, as well as meterosecurity: climate change means breaching borders into more liveable habitants. The terrorist is the emergent disrupter of the prevailing social order; this includes the virus that is liable to start an epidemic—in the popular imagination the Zombie becomes the terrorist that destroys the species. There are also the drug-resistant bacteria and even the worry of what may be happening in waste sites that may well be the breeding ground of as yet unknown strains of microbes and bacteria that could begin to spread without any way to stop them (Hird and Yusoff 2016).

The imaginary of 'geologic life' in the Anthropocene is developed throughout the writings of Kathryn Yusoff (2013), who perceives the 'geo' as a collaborator on particular formations of subjectivity, yet appearing as a benign entity that is taken as neural ground that can be manipulated to human will alone. Yusoff turns to two notions of the fossil to show how the geoformation of subjectivity is at stake in the Anthropocene. The first formation is 'liquid' in the form of fossil fuels. These fuels create a new geologic subject as they form the material conditions that subtend contemporary geopolitical life. The second formation is solid in its form: bone turning to stone, becoming a mineralized fossil, a becoming-mineral when our species becomes extinct, to be contemplated by some 'geologist-to-come.' Here, Yusoff reviews the hominid fossil evidence that has dispelled the out of Africa hypothesis of 'one species.' Her 'fossil theory' based on this view, presents a difficult picture as to just how the agency of fossils as forms of geologic thought shape and are in turn shaped by the geopolitics of life. Fossil fuels present new forms of 'sacrificial responsibility' (should the 'gift' that they offer to be distributed differently for the future?), whilst hominid fossils are a reminder that the *geo* is not pregiven as a common ground under the sign of anthropos, a point made by many commentators.

There are also 'alter-Anthropocene' imaginaries that highlight the problems with categories of *the* species and *the* human (Neimanis and Walker 2014). To imagine the Anthropocene 'otherwise' has been the challenge. Stacy Alaimo (2008) provides once such path through her conception of 'transcorporeality,' which is basically another way of recognizing the imbrication, or 'intra-action' as it has now become popularized via Karen Barad's (2007) seminal claims concerning particles and observation at the quantum level, between the 'human' and 'non-human.' Alaimo (2017) has sought to mobilize this affective force through her explorations of artists who address an aesthetic sensibility that cultivates a 'fleshy posthumanist vulnerability that denies the possibility of any living creature existing in a state of separation from its environs' (113). Alaimo dwells on the exchanges that take place between the human and the non-human, supporting a 'geological turn,' after Kathryn Yusoff (2013) who investigates 'geological life,' a lithic sensibility as a missing mineralogical dimension of human composition as discussed above. An example of the transcorporeal subject, the subject as a material being entangled with its environment, Alaimo points to people with multiple chemical sensitivity (MCS), who have to find living spaces that are less toxic. Environmental health and justice are at the forefront in such cases.

Alaimo's direction follows what might be a broader imaginary to find a new relationship to the Earth. Astrid Neimans and Rachael Loewen Walker (2014) develop the imaginary of 'weathering' and 'thick time' to further this concept of transcorporeality. Humans and nonhuman climate and weather phenomena are mutually emergent and co-constitutive and coextensive with one another within a space-time. Following Nancy Tuana (2008, p. 194), the encounter is marked by a membrane logic of 'viscous porosity' where a certain 'resistance and opposition' is emphasized to the changes a form undergoes: hence the infinitive 'weathering.' Whilst Neimans and Walker call on Karen Barad's (2007) 'differential patterns of mattering' and Irigaray's 'elemental passions' that inform this 'weathering' imaginary to forward their feminist agenda, Deleuze and Guattari present the same conceptual basis via 'haecceity,' the recognition that there are multiple relations of co-creation, each retaining their discreteness as bodies that produce the qualities of 'things' and desire as to what holds the assemblage in a metastate for a particular temporality. Transcorporeality is synonymous with Deleuze and Guattari's assemblages and haecceities. In a somewhat different yet similar context Bronislaw Szerszynski (2010a) deconstructs the scientific practices of reading the climate-weather from a biosemiotic analysis in relation to the environment. The exchange of the organism with its environment are metabolic processes that include both endosemiotic and exosemiotic relations, the first referring to the inside of the organism, the second to its outside. However, such an approach is limiting when it comes to the Earth (Gaia) where the internal dynamical forces within its 'body' are quite different than an organism's exchanges between its inside and outside. Following Bernard Stiegler and André Leroi-Gourhan,

Szerszynski maintains that the further developments of technology and language provide different metabolic exchanges with the Earth, consonant with the radical thesis presented above where the symbolic plays into the physiology of the body as well. Szerszynski, after Derrida, calls on 'writing' the weather, which is both one of responsibility and charged with ethics. 'Weather-wising' is no longer in play (the idea of grasping seasonal variations and changes), and grasping the weather kairologically to feel the quality of particular periods of time, such as changing seasons and local conditions is also passé. The weather has become almost unruly for this sensibility to persist. Nevertheless, the demand on us is still to recover this sort of *sensibility of time* that weather bestows on us; it is a reminder that all of us must face this constant 'weathering,' which the logocentrism of technological forecasting simply glosses over through the constant repetitions of televised weather channels. In brief, 'writing' the weather, in Derridean terms, should bring us back to the sensibility of the inequalities and irresponsibilities that surround 'climate change.' With hurricanes like Katrina (2005), Harvey (2017) and especially Irma (2017), the most powerful Atlantic hurricane yet recorded, raises the reporting of weather to new levels in terms of the devastation, loss of life, and a sampling of the social hysteria that will only multiply in intensity in the near future.

Richard M. Dolye (2011) coins the word 'ecodelic,' which Alaimo (2016a) acknowledges, to shift the rhetoric and the affect away from disparaging discourses that surround entheogenic, psychedelic hallucinogenic plants, and chemical use to argue that Homo sapiens have always historically had a 'spiritual' mystical, and symbiotic relationship with nature's pharmacopoeia so as to erase, suspend, or modify the barrier between self and the biosphere. Dwelling on Cannabis, LSD (Doyle 2002), and especially ayahuasca and its use during a ceremony, Doyle (2005) claims it cured his asthma. Doyle's mission is to promote the experimentations of ecodelic plants and drugs to cultivate 'psychonauts' (those who are capable of navigating their psyches) for a 'closer' and more profound relationship with the energy flows of the Earth; in Deleuzian terms, perhaps to 'feel' the immanence of life and become 'imperceptible.' The (post)Deleuzian Manuel Delanda hinted as much in one of his YouTube lectures regarding the use of hallucinogens. In a particularly interesting passage, when recognizing that a 'pornography' of Cannabis plants has arisen as a way to sell them as a product, Doyle says the potential buyer is closer to 'becoming-insect' (see Parikka 2010) when inspecting their qualities (seeds, leafs) than the sexuality that resonates with flowers and vaginas—for the heterosexual male gaze, that is (Doyle 2011, p. 238). With cannabis becoming legal in certain states in the US and in Canada, one can quip that these two country's population will become 'closer' to nature (to a certain 'degree' at least) like the '68 'flower power' to start Doyle's 'transgenic involution.'

This ecodelic road to the Anthopocene imaginary seems to take two routes; the first can dangerously drift into an over-romanticized view of Nature advocated by many eco-feminists, a danger spelled out by Colebrook (2014a) as well. Here the idea is to get in touch and feel Nature's flow, a biophiliac sensibility, which is said to be totally interrelated at every level; there is little talk or exploration about the deterritorialization that is part of this process. 'Catastrophism' in all its forms takes a back seat. There are, however, interesting aspects of such a participatory view that can lead to another route. Despite the harmony between eros and Nature advocated pedagogically by Kerry Brady and Brian Swimme (2012, p. 119), there is a remarkable description as to how a particle of light might be understood as being expressive of Nature at the quantum level. They write that light at the quantum level is absorbed by the tree and disappears into the energetics of the leaves which interact with it; the tree's leaves are said to transform the quantum of light creating a new photon that can be absorbed into the energetics of a person's eye. The new quantum, as an expression of the tree, enters into the tissue of the human who is perceiving the tree; here an exchange of expressions takes place. Such a description would follow Deleuze and Guattari's explorations of becoming other—in this case 'becoming-tree.'

This other imaginary route seems to lead to such claims as the exchange of information between entities through an 'ecology of mind' in Gregory Bateson's terms (see Bateson 1979; jagodzinski 2017). It also suggests that minds are *extended*, as first developed by Andy Clark and David Chalmers (1998), who maintain that the contents of our thoughts are determined by features of the environment, and that 'the proper faculties of cognition are not limited to the brain and neuronal system, or even to the body as a whole, but that they are spread out into the environment' (18) as 'becoming-tree' would suggest. Philosophers such as Andy Clark (2005, 2010) and Alva Noë (2010) have developed this hypothesis further. Such a position is certainly controversial. However, if this hypothesis is related to the most recent work in epigenesis, it becomes more convincing than less. Epigenetics refers to the long-term functional change in gene expression; a process that is 'beyond' and 'below' the genes, through methylation and histone modification that does not involve the DNA nucleotide sequence itself. The continuity of the germline and the one-way flow of information from DNA to RNA to protein as usually understood from neo-Darwinian evolutionary biology is challenged, and the dismissed Lamarkian idea (the passing on characteristics to its offspring during its lifetime) become tenable. 'Junk DNA' has been recognized for its 'potential' as a change agent (Bardini 2011).

The historian of science, Hannah Landecker's (2011) research in nutritional epigenetics explains the effects of nutrition on gene expression as the general 'molecularisaton of the environment': the health-determining environment of the physical body in relation to the molecular changes to the body's metabolism when food (engineered in all its variants) is understood molecularly as it is ingested, for example, by mothers during pregnancy, effecting their newborn's physical and mental growth. Molecules enter the body during 'critical periods' of development, which then shape the very metabolic system that the body will be able to process food in the future. Layers of human intervention go all the way down to the molecular levels when it comes to a body's physiology. 'The new metabolism is no longer the interface between Man and Nature, as it was in the nineteenth and twentieth centuries, but a metabolism for the human condition in technical society, where the food is manufactured and designed at the molecular level, the air and the water are full of the by-products of human endeavour and manufactured environments *beget different physiologies*' (190, added emphasis). Jörg Niewöhner's (2011) review and understanding of environmental epigenetics is more dramatic in its implications. Environmental epigenetics basically questions how both individual and social behaviour impact on (patho)physiology through epigenic pathways; it seeks questions that have to do with changes in the social and material environment as they impact individuals and forms of social physiologically, which raises the further question as how these changes are passed on to subsequent generations. These questions produce an 'embedded body' imprinted by evolutionary and transgenerational time. The anthropologist Tim Ingold (2013) has explored such epigenetic modifications to show how dwelling in a particular environment provides an entanglement between body and its 'context.' Niewöhner (2011) calls this the '*molecularization of biography and milieu*' (291, original emphasis).

In her analysis of human-non-human relationships, the anthropologist, Henrietta Moore (2011, pp. 170–205; 2012, 2014), ceaselessly it seems, reminds those who have deeply embraced the 'affective turn,' that affect is not autonomous in the way it has been suggested by Brian Massumi (2002) (see Leys 2011). The work on animals shows that affect and cognition are not separate realms; in humans, affect is developmentally connected to language, the norms, and values that are in play through the earliest interactions. Moore wishes to remind us that human sociality remains a key factor in the capacity for virtuality; that is, human agency endows things of the imagination and mind with meaning and significance. This means that human sociality depends on things that are *not* of this world, a point that Deleuze makes as well; this dimension of incorporeality is most often overlooked in human sciences where embodiment is constantly stressed. Incorporeality (Grosz 2017) might best be described as 'thought-becoming' when concepts are created. The invisible forces that collide, clash, symbiotically coexist and interact belong to this virtual realm (dark matter?). They are 'material' in the sense that they are a dimension of the senses, ideas and, conceptuality. Such material carries with it an excess over corporeality that enables it to intensify itself so as to maximize sensations, what Deleuze and Guattari deemed 'microbrains' as forms and degrees of conceptuality, a 'thinking' that is orientated to the materiality of terrestrial and cosmic forces. Moore's argument is that affect (however defined) is not the same 'entity' (intensity, energy, potentiality) before it enters human relations and once it enters human relations. Affect's singularity has to be recognized through the events the organism undergoes.

Passive Vitalism and the Anthropocene, Non-human Interactions

In her two volumes on the Anthropocene, Colebrook (2014a, b) develops her understanding of the two forms of vitality that Deleuze and Guattari (as stated in *What is Philosophy?* 1994, p. 214) develop throughout their work, pointing out, again and again, that the *usual* notion of vitalism that is forwarded is a Bergsonian one where becoming is a process as it shapes an aesthetic where goals are aimed for and reached; creativity is understood as an unfolding according to the talents that are made manifest in the child. In various religious contexts, each essential being moves towards the expression and fulfillment of the divine, each in its own capacity—like in the Great Chain of Being. This form of vitalism is well in sync with animism and spiritualism that plays itself out, not only in various philosophies of idealism, by identifying a force outside the organism that provides the necessary energy, or 'life' to animate matter. God, in any of its multiple names, and the One, in any of its philosophical forms, is the originating source that transfers this invisible energy presenting all sorts of ethical issues to the medical body: when is a body officially 'dead'; when is an abortion to be is allowed, and so on. The second form, or passive vitalism is what Colebrook spends time discussing as this vitalism is incorporeal, virtual, and maintains a hesitation or gap between contemplative condensed thought and action. This is the emanating force of life in difference. Colebrook wants to emphasize this passive vitalism as being 'queer' as it is identified with immanent force, which always destabilizes being. So, whilst positive vitalism provides the claims that the subject makes decisions, gives meaning to life through various signifiers, categorizations, autobiographies, and historical recording to provide social constructions, passive vitalism deals with events and encounters that radically affect identity, the potential for variation is always there.

Queer vitalism dispenses with universal humanity, where differentiated gender, racial, class categories, and so on can eventually be overcome at some ideal telos. Queering in this thinking, pushed to nth degree, means that an individuated actualization is a genetic potential for (sexual) differentiation where the chromosomal and hormonal composition is singularized, to which is added body comportment as well as sartorial style. Passive vitality exceeds bodies and their actions: this means that a re-singularization or counter-actualization is always a potential in the making. It's not a question of achieving an ideal that one has of oneself; rather this counter-actualization refers to taking what characteristics and features there are and expanding on them. What is smooth can become striated, what is molar can come minoritarian; in brief, the potential for 'queering' is what is vital. Life is queer in this sense, as it takes in all those potential differences that both exceed and also infinitely divide each body. Each body is then negotiated by multiple affections and attachments, the conjunction 'and' is always at play in a conjunctive logic as Deleuze points out. Desire in this case, is pre-individual; it's force is genetic, political, social, metabolic, fantastic. Identity in this Deleuzian sense is distinct yet unclear, the paradox of a formation that is always about to change; each being is distinctly individuated, being composed of only encounters and affections. Certain affections are felt with certainty, yet felt only dimly in relation to how others feel those same affections.

Colebrook's queer vitality has nothing to do with the destabilization of norms, or their solicitation; rather it is the creation of differences that are not grounded in the subject, or generating life in the way positive vitalism presents it—the striving to reach certain goals. This radical position requires thought without an image, the creation of aleatory point via the qualities and potentials that are there; theory is given the task liberated from any normative image by opening itself up to problems (Ideas), hence hybrid and incorporeal. Colebrook in a particularly lucid paragraph says:

> The vitalism is passive because it is not the case that there is something like life that surges from itself in order to arrive at itself; [as in emergence theory] instead there are quantities of force [intensity] entering into undecided relations *from which* being emerges that would then provide some basis for active decisions. (2014b, p. 191, author's italic)

Colebrook is searching for an imaginary that moves away from any norm; she finds that potential in the passive vitalism where a force encountering another force becomes the posing of a problem. Such are queer encounters that are neither determined according to recognition (identity) or reproduction (sexuality, sexual difference). The mode of relations remains to be determined (relations being external to their terms—forces for structuration that exceed human thinking).

The question for the Anthopocene imaginary is how far do you go when it comes to an openness to the material world that defines the concept of transcorporeality as Stacey Alaimo and others embrace it? Or, in Colebrook's question 'how Queer do you go?' Jeffrey J. Cohen (2015) for example, presents a 'human-lithic enmeshment' in his meditation on stone. 'Geophilia: The Love of Stone,' and 'Queering the Inorganic,' titles of previous published articles are explored in his book on stone, subtitled 'An Ecology of the Inhuman.' Cohen raises fundamental questions as to our relations with the mineral world given the long history with stone tools. The artist Ilana Halperin, combines a deep interest in physical geography with studiously written descriptions that draw her and her audience closer to observing the details of landforms, perhaps furthering the exchange of molecules of light mentioned earlier. In the case of the 'weathering' imaginary that Neimanis and Walker (2014) theorize, two artists are advocated on their behalf who exemplify such exchange with the environment. The first is US-based Basia Irland, who carves books out of ice that are studded with seedpods; she releases these 'books' into rivers to be 'read' by the waters. A reseeding ceremony is then performed as witnessed by local indigenous community members. The idea being that such a performance curbs the localized effects of climate change and effects the watershed. The further claim is that such performative acts of material empathy initiate a rethinking of the (presumed) distance between human embodied selves and meterological phenomena. Their second example is Karolina Sobecka's 'Thinking like a Cloud' (Project Sky+ 2013), which is more interactive. A Cloud Collector contraption is sent into the troposphere attached to a weather balloon to collect condensed cloud samples. The cloud's microbiomes are then consumed by participants. No mention is made of the potential pollutants that are equally consumed. Perhaps the hesitation on the faces of participants before the 'cloud' is drunk *is* the encounter of 'becoming-cloud'? Alaimo's (2016b, 2017) artistic examples explore a sense of precarious and corporeal openness to the material world as an environmental stance to perform an 'insurgent vulnerability' (see also Papadopoulous 2010), much like Danish artist Kirsten Justesen does (Alaimo 2016b, p. 91); her naked exposure to ice reverberates with the broader politics of exposure to carcinogens, endocrine disruptors, radiation and so on. Here Rhona Zwillinger's photographs under the series title: *The Dispossessed—Living with Multiple Chemical Sensitivities* (1998) is a prime example. Adrian Parr (2009), in a similar vein, presents the 'naked' installations of Spencer Turnick against the backdrop of a Swiss glacier on the front cover of her book in the way sustainability discourse has been usurped by green capitalism as part of an 'ethics of exposure.' Given this array, what does one do with the experimental performance 'May the Horse Live in Me,' by Art Orienté Object (the team of Marion Laval-Jeantet and Benoît Mangin) where 'becoming-animal' is taken literally as a step towards hybridity: the change of blood between a horse and Marion Laval-Jeantet through a series of carefully monitored blood transfusions. Is such bioart then queer enough when it comes to transcorporeality?

The relationship between climate change and violent conflicts are emerging as a result of limited resources, especially the dependence of oil by nation states (Szeman 2007; Scheffran et al. 2012). Oil marks the current end point of the path to a fossil economy that was laid down by Homo Erectus, our hominid ancestor who learnt to control fire—the energy reserve stored in detrital biotic carbon (dead leaves, peat, and wood), and then the detrital fossil carbon stored in, at first coal, then oil and gas that was subsequently trapped in undergroud pockets (Malm 2016). The argument here is that the emergence and development of the Anthropocene as a planetary phenomenon was triggered as a result of this first event of energy extraction. What followed provided a multiplier effect with each new technological discovery dependent on the so-called exosomatic energy flows that do not pass through the body of a life form. Our species have modified the carbon cycle ever since until it has now reached a critical threshold (Raupach and Canadell 2010). The Anthropocene, like life, is carbon-based.

While the question of economic sustainability has already been hijacked by capitalism (Parr 2009), there are many economists who bluntly point to the 'uneconomic growth' of capitalism that leads to environmental catastrophe and our species death (Daly 2005, 2010; Oosthoek and Gills 2005). In terms of the laws of thermodynamics, our planet is headed to a world-without-us (Newman 2008). The Second Law of Thermodynamics states that entropy (disorder) will increase in an isolated system; while the Earth is not an isolated system, and at the quantum levels there has been a complete rethinking of Thermodynamic states when it comes to dissipative structures, the biosphere that sustains us, which is increasing in CO_2 emissions, will eventually reconstitute itself (self-organize). The Earth's system will 'dissipate,' reversing the second law, and enter into a new phase space, to achieve a new equilibrium or state of order, which will not support our species as it exists today. Whether there is enough political will to stop this seems bleak. Various discourses of 'resilience' have been developed to at least address this scenario (Scott Powell et al. 2014). A crucial point made by Derek Woods (2014) in terms of the scale of the Anthropocene is simply this: while most of the discourses are about the transformative 'force' the species has on the Earth; that it to say, it is a 'geological force,' it is perhaps more accurate to think of assemblage theory (following Deleuze and Guattari) where the terraforming is composed of the assemblages of humans, non-humans, and technics (what I would call the inhuman). Such as position would be another piece of the puzzle for what may seem as a ridiculous hypothesis mentioned earlier: that there is no Homo sapiens species 'per se' that can be lumped together, but only variations as the epigenetic evidence seems to suggest where social, political, and environmental factors alter bodies and brains.

WHAT IS THE HUMAN?

If 'Man' is the zenith point, the transcendent idealization that is constantly coveted and captured by a select percentage of the world's population to claim the status for setting standards of moral conduct of behaviour: the laws of God and Allah made manifest through patriarchal leaders as prescribed by the Great religions that arose during the Axial Age (Jaspers 1953); or maintaining values that are said to be 'democratic': the string of ideals as outlined by the American Constitution, each of which is subject to agonistic debate. What is the common Good? What is Justice, Equality, Liberty, the pursuit of the Good Life and Diversity in relation to Sameness, and so on? These are certainly contentious issues when determining what is 'human' in relation to such transcendental aspirations. The Anthropocene, therefore, raises the nadir point of debate as to the fate of the Earth as we know it as we are but one species of many.

Many authors writing on the Anthropocene narrative mention, time and time again, that the categories 'species' and 'humanity' are problematic terms in their generalization (Szerszynski 2010b; Read 2017; Povinelli 2017; Chernilo 2017a). Bronislaw Szerszynski (2010b, p. 181) draws on both Derrida's grammatology and Foucault's archaeology to question any form of classification. 'Man' becomes simply a 'denouement'; the Anthropocene is both the apotheosis as well as an eclipse of Man. As the *Aufhebung* of 'Man,' the Anthropocene presents the heights of technological elevation; at the same time, this very development is also its negation. The coming geological layer of the Anthropocene that is being built up around the globe is a *memento mori*, a reminder that the fossils we will become point to our becoming-mineral, like the stones of the Earth.

The most common claim is that humanity or human nature remains undefined, open to change, there is no 'human essence' as the species is constantly changing its ecology. Bluntly put, ecology is necessarily an empty signifier to be filled out ideologically. The field of philosophical anthropology has debated this for some time. Homo sapiens lack instinctual determination and organ specialization; it is dispossessed by a lack of adaptation and specialization like other higher order mammals (e.g., Gehlen 1980). The defining feature of humanity becomes the externalization of memory, knowledge, and habits in the symbolic forms of tools, signs, and images (e.g., Leroi-Gourhan 1993). Technology is especially stressed; Derrida's (1995) view is that our species lives in a state of 'originary technicity.' It is impossible to define *the* 'human' as a biological entity (hence the body as a species called Homo sapiens), nor as a philosophical state (soul, mind, or consciousness); rather, Derrida argues, human 'nature' is constituted in relation to technological prosthesis. Bernard Stiegler's trilogy on technics and technology provides an articulation of Derrida and Leroi-Gourhan's positions via epiphylogenesis (Stiegler 1998). He points to three identifiable forms of memory: genetic memory that is programmed in our DNA, epigenetic memory that is acquired during a lifetime and stored in the central nervous system (CNS), and finally the epiphylogenetic memory that is embodied in the technical systems or artefacts. It is this last form of memory that is unique to our species via tools, language, writing, computer use, and so on.

To maintain that our species has no 'nature' per se, or that our species is entirely a part of nature leads to interesting meta scenarios for the Anthropocene. As a species that is entirely a part of nature, let say for argument, not claiming any exceptionalism, or 'speciesism,' has been the trend by a number of commentators. As Andrew Kipnis (2015) put it, an extreme position would be a posthuman one, those 'analytic stances that grant agency to nonhuman entities and that downplay the differences between human and nonhuman agency' (44). Perhaps the most ardent being Donna Haraway's (2016a, b) position articulated in 'Staying with the Trouble,' where the poetic imaginary of her Chthulucene overflows with multispecies interactions, symbiotic and contestatory alike. Haraway's Chthulucene has a strong biophillic resonance to it. Her feminism sides with the Gynecene thesis developed by Alexandra Pirici and Raluca Voinea (2015) that calls for new forms of eco-feminist stewardship, reverence for the Mother Earth with links to Indigenous spiritual beliefs concerning land claims. Perhaps more intense is the biophillia espoused in the 'Ecosexual Manifesto' as part of the SexEcology movement spearheaded by Elizabeth Stephens and Annie Sprinkle (n.d.). They write, 'Ecosexuals can be GLBTQI, heterosexual, asexual, and/or Other. We will save the mountains, waters, and skies by any means necessary, especially through love, joy and our powers of seduction (n.p.).'

Given the early discussion, we can ask after Colebrook (2014b) whether this direction is indeed 'queer' enough?

Maintaining we are but *one* species of many that inhabit the Earth has the usual consequence: the Anthropocene is the result of our human nature. This is to say, we have no nature or determined behaviour. We have come to the point of recognition that this 'is' indeed our nature— namely, to modify our environment to the point of realizing that we are paradoxically either helpless in controlling it or, just the opposite: our species faces its greatest challenge to technological modify the Earth for our benefit. This last view flips the argument over from: we are part of nature to we are all culture; technologically culture is our 'true' nature that shapes nature. Both grand narratives of the Anthropocene provide heroic progressive change via an active vitalism based on the notion that our species must, on all fronts, overcome the forces that are said to plunge us towards death and extinction.

This either/or logic is best overcome by what Jason Moore (2015) refers to as 'double internality.' The overcoming of humanity-in-nature and nature-in-humanity (the classic nature-culture divide) is understood through the recognition that our species' nature is simultaneously being produced in and through nature as we modify it. So, there is no nature per se, just different historically produced natures as argued earlier in this introduction. For Moore, global capitalism is a specific production of 'human nature' that in turn also exploits, alienates, and extracts labour from non-human resources. Capitalist exploitation of labour is dependent on 'cheap nature,' and cheap labour power—made even cheaper through unpaid housework, cheap food (via agricultural revolutions), and cheap fuel. The Anthropocene points to the very limit of capital as the 'natures' it produces cannot be sustained.

Such an imaginary is always challenged by the claims of technological fixes as the future is never fixed. There are projections for the possibility of fusion energy predicted for mid-twenty-first century that would introduce a multiplier effect as to the energy produced. Again, capitalism would be 'saved.' There are also the developments of biomimicry and biosensing as expanding the more-than-human sensorium attributable to plants, animals, and bioengineered organisms via biotechnology (Johnson 2017). The engineering of cells capable of registering the presence of certain molecules, and the creation of synthetic life—organisms with an 'expiry date' built into their DNA's—are projected to be introduced into ailing ecologies to help 'green' them, as technology is on the cusp of yet another 'revolution' to manipulate the environment. Both human as well as 'machinic labour' is now supplemented with all these forms of non-human labour to further capitalist exploitation by the manufacturing and reproducing 'life.' Genetic capitalism is well underway (Bardini 2011).

The Marxist distinction between 'living' and 'dead' labour seems to collapse. What exactly is creative human activity (living labour) in a post-Fordian economy? Seems all living labour becomes 'dead' labour very quickly, like the technology and machinery Marx was referring to. The so-called 'general intellect' is now being shaped, not only by real subsumption, where human immaterial labour via social media is being captured for profit, but we are now apprehending the world through the tools and prosthetic technologies of big data that are leading to 'algorithmic governmentality' (Rouvroy and Stiegler 2016). Knowledge becomes machine intelligence, which is constantly exteriorized via patterns and tendencies that are revealed in big data that then shape the managerial direction that corporations in leagues with nation states need to take in order to maintain their profit margins.

Paolo Virno (2015) presents yet another variation of philosophical anthropology in unravelling the 'mystery' of what is our 'species' nature. In evolutionary terms, encephalization of the brain has given our species language centres as the formation of tongue and lips became modified, externalizing a symbolic coded system. The loss of hair and any instinctual determination for shelter resulted in the development of clothing and architecture. For Virno, and other notable philosophical anthropologists (Helmuth Plessner, Arnold Gehlen) there is no 'natural' basis for such activities; there is no 'one' language, one style of clothing, one set of customs, and so on. Natural potentials are all actualized through history, with the added feature that language, habits, customs, fashions are all generic and transindividual; they all require a constant exchange of relations in order to exist, like the aboriginal Thangedl elder, Warwick 'Chook' David, who at 80, is the sole survivor of an indigenous language that echoes the 48,000 years of tribal lore (Michael 2016). Such attributes are not faculties of the mind; they are the potential of humanity as a species—our species being. Virno (2009) calls this development 'historico-natural' as diagrams (graphs, maps etc.) of human nature. They are pre-individual conditions that need to be given transindividual articulation, otherwise the feral child is unable to fully integrate into the symbolic. Of course, such talk raises issues of neuro-normativity. Yet, the point here is that human nature is its history; it's something that can

only be thought in its singularity, relationality and in its historical specificity. Language, habits, and culture—constitutes a particular anthropogenesis through a historical ensemble of social relations. Virno stresses
that capitalist *formal subsumption* has turned into *real subsumption* where
communication, intelligence, and affects have all been harnessed for
profit. '[T]he biological prerogatives of the human animal have acquired
undeniable historical relevance in the current productive process' (2009,
n.p.). It seems then, that capitalist mode of production appears to be
identical with 'human nature' as all possible aspects of mind/body
are exploited for profit. It puts anthropogenesis to work on all levels,
exploiting human potential *as* potential.

Virno appears to make a strong case for our species being. Yet, when
the 'human' is a representational category, the qualities that define it
become the criteria against which those who are said to be 'human' are
measured; they are then ethically and politically recognized. The category is not open to 'everyone.' The human becomes necessarily exclusionary, thereby closed. Construed as a universal category, the human is
a member of a moral community where everyone is treated as an equal.
Under this ideal, to be human is to possess all the rights and responsibilities that follow from such equality. Along with this comes the responsibility of moral behaviour and worth, and ethical obligations to the
Other in relation to established constituted laws. Parr (2017), for
instance, raises the significance of human dignity as it relates to stewardship of nature following the thought of George Kareb. She maintains that dignity is 'essentially secular and existential' (129). However,
Parr's rethinking of dignity from its humanist roots of status as an individual property and morality heads in a different direction and conclusion. Environmental degradation demonstrates assemblages of desire that
bring together the complexities of the human (e.g., gender, status, inter-
generational relations) with the inhuman (e.g., chemicals, air pollution,
industrial waste) and the non-human (e.g., animals, excrement). Dignity
within this context is an emergent quality that comes from the relational
operation and cooperation between the nonhuman (light, air, food, animals) with the inhuman (the material power that is available) along with
the human dimensions of expression (music, laughter, joy). 'This is a
power that cannot be located in the international development agency
or state actor' (134). From a Deleuze–Guattarian–Hjelmslevian inflected
point of view, Parr shifts the 'common' grounding of human dignity to
articulate how the complex relationship between form of content and its

form of expression (Nairobi slums in this case) is shaped and manipulated through the design aesthetic of a neoliberal conception of the 'good life' where high-rise housing is said to impart dignity to displaced slum-dwellers. The notion of a distinctive human nature, as so many critics have shown, what is 'properly' human is shaped by religious, social, political, cultural mores, and ideologies. The exclusions are legion: those who are mentally disabled, lacking the clarity of mind to be responsible as the gift of reason and rationality, which has been historically levied against women, labouring classes, untouchables, comatose patients, foreigners, immigrants, LBGTQ+, and so on.

The essential 'human' emerges from the nation-building and imperial colonial expansion. As a representational category, the essential human maintains itself through a constant battle of 'inclusion' by the 'excluded' to be counted as fully human—to overcome the signifiers of being less-than-human or not human, based on the mythic utopian ideal that there indeed is the possibility of an equal and just society where all are counted. It is the dream of liberal plural democracies, the striving to achieve justice and equality based on a state of being or attribute, however, a status that is achieved already based on a predefined definition that is hierarchically politically, socially, economically, and mentally structured, available to only a small percentage of 'humans.' Such western thought is a perversion of Christianity and Capitalism: those who eventually reach Paradise, the perfect Utopia are the moral humans. They have, as espoused in its extreme variant by the Weberian Protestant ethic, aligned themselves with capitalist values of hard work and frugality, in distinction to those who don't. Those who are condemned to toil on Earth unable to live off the fruits of their labour. The three C's have now all come together in their own complex assemblage: Capitalism, Christianity, and Climate Change.

Human exceptionalism as differentiated from the animal is based on the power of rational thought (reason), the ability of speech and writing that enables some form of self-reflection to assert agency, and the moral capacity to experience both guilt and shame in wrongdoing. Whilst animals are certainly capable of such attributes, the claim is made that this is a difference in kind and not degree, although an argument can be made for a reverse scenario. It is true that evolutionary history shows that we share many genes with animals such as mice, dogs, chickens and so on, but especially with other highly sentient creatures, such as the great apes of the family Hominidae: gorillas, orangutans, chimpanzees, and bonobos, the last two being our closest relatives, sharing between 96 and 98.8% of their DNA depending on calculations. Yet, there is no *one* gene (or pair of genes) that specifically marks 'human' difference (Best 2014). This, however, means very 'little' when it comes to differentiating us from 'them' as a gap is always generated through what Giorgio Agamben (2003) calls an anthropogenic or 'anthropological machine' (26) that generates the 'human.' To be human then is to recognize a differentiation from a non-human Other (animal, monster, alien). Although the Homo species is certainly a hybrid, an amalgamation of Neanderthal and Denisovans, with early Homo types (Homo Floresiensis and Dmanisi) appearing outside Africa (Aiello 2010; Callaway 2014; Lordkpanidze et al. 2013), it is the species Homo sapiens who remain the highest encephalized species on Earth. As discussed earlier, responsibility cannot be feigned.

NEW MATERIALISM

The Anthropocene has heightened an ecological consciousness where the recognition of non-human life has become acute in decentering the question of the 'human.' A 'distributive subjectivity' where life in all its unknowable forms, life as energy, has become a key concern under the larger rubric of a 'new materialism.' Agency, as defined in 'human' terms, namely being synonymous with self-consciousness, cognition, or rationality, becomes decentered. It is 'oddly' a post-ontological position in that the limitations of ontology as inherited by the western embracement of Greek-Egyptian-African philosophies fall short. As in the anthropological philosophy of Eduardo Viveiros de Castro (2014), what does one do with Amerindian cosmology where there are only 'human beings.' Humans and animals are not distinct. For Ameridian peoples (Achur and Runa), plants and animals as non-humans are *not* simply other species belonging to nature, but *'persons,'* human persons who are distinct from 'human' humans, *not* from their lack of consciousness, language and culture, but by the simple observation that they have different 'bodies,' endowing them with specific subjective and cultural perspectives. Cast into Deleuze-Guattarian terms, all actualized Life are bodies that relate through different states of becoming between them; bodies are always emergent properties that have different speeds—some faster, others slower. The ethological genealogy of Jacob von Uexküll is well-trodden ground (see Buchanan 2008). What this means is that the entwinement or entanglement, to use a popular term, between what are heterogeneous opposites, most often in the grammar of speaking, sub-ject–object, nature–culture, materialism–idealism, human–non-human, natural–social, and so on, breaks down when these dualisms are recog-nized in their recognition as being part and parcel of various processes of emergent becomings. This decentering of the human agency produces a flat ontology (e.g., Bruno Laour), where there is a state of constant becoming via dissipative and creative assemblages held together through various forms of desire (symbiotic, parasitical, and antagonistic). Human embodiment, as an ineluctable animality, weds our 'species becoming' to the non-humanness of nature, both inorganic (mineral, geological) and organic life (biosphere). Mind, following such early pioneers as Gregory Bateson, is suffused throughout an assemblage where human and non-human worlds are ecologically interrelated, and animal consciousness spans the effects of suffering, pain, joy, along with various 'degrees' of cognitive abilities (for a review see Allen and Bekoff 2007).

The point that needs to be made is the degree to which an ecological niche (*Umwelt*) can be modified by a species varies and depends on the degree of evolutionary encephalization. Put bluntly, the 'agentic capac-ity' of a stone depends on the specific conditions: molten lava's material-ism is quite different than a sedimentary stone. Both terms, *Umwelt* and *Lebenswelt* seem too limited in grasping the changed understanding of a distributed subjectivity. At its most radical, this post-ontological position posits immanent life (e.g., or A Life in Deleuze's oeuvre, or pure imma-nence) wherein the emergence or production of 'things' as matter is due to the mysteries at the quantum levels such as its fractures, relationality at a distance, gravitational forces at certain limits, and so on that endow matter with a panpsychism, some form of internal effulgence that matter seems to possess. Affect theory presents the various difficulties and con-fusions (as emotions) that surround this line of thought, as 'affect-itself' can easily be thought as this imperceptible energy, 'life' (which is not definable), pure immanence leading us into the question of dark matter and the impenetrable thought of François Laruelle's (2013) non-philos-ophy of the 'radical Real.'

But, there is also the added and crucial difference that *does* make a difference in terms of the forces harnessed by our species, and that is technicity; the ability to utilize the encephalized ability of our brains to exteriorize a technology that has a multiplier effect, both in terms of pro-ducing energy as well as changing the physiology of the very embodi-ment. Throughout my work (e.g., jagodzinski 2010), I call this the inhuman as opposed to the non-human shaping of our species becoming, requiring a new understanding of self-refleXion and Xpression that moves towards a post-ontological position (jagodzinski 2008, pp. 29–46).

THE COMMONS

There has been a sustained attempt to rally the Anthropocene around the imaginary of the 'commons,' and cosmopolitanism as a way of per-haps forcing some sort of unified sense of the Planet to be shared, a way to live together (Gilroy 2004; Derrida 1997; Stengers 2010, 2011; Parr 2017). For Izabelle Stenger's this question is more to ask about the cos-mos as the root of cosmopolitics as she presents her own questioning of the sciences. Ulrich Beck's (1998) *Cosmopolitan Manifesto* maintains that new forms of solidarity and community may arise because of the shared risk to human survival. Yet, most of this literature presents a Eurocentric view, where difference is theorized in representational terms. Tariq Jazeel (2011) provides a succinct critique of cosmopolitan 'universal-ity,' an Apollonian gaze as he calls it, attempting to change the discourse by calling on Gyatri Chakravorty Spivak's (2003) notion of 'planetarity' where there is a constant deconstruction of knowing, a perpetual uncer-tainty and process of de- and re-inscribing the whole earth image. Virno (2015), as one of many, draws on Gilbert Simondon's development of ontogenesis as individuation to identify the 'commons' of our species, language being a primary common. The shared faculty of language results in a multitude of singularities, as such there is no dividing line per se between and amongst various cultures, only the potential to per-form language differently in each and every case. Language belongs to the deep, deep structure of our species being, hardwired so to speak; it is also the slowest code to change.

Virno, like Deleuze, draws on Simondon's 'pre-individual reality' as the generator of the common, which is always in a metastable state that undergoes transformative change as levels of potential energy are triggered and fluctuate within a system that define the process of indi-viduation. (As developed earlier, this is precisely in tune with Roberto Esposito's own account of the 'third person.') These dynamics of dif-ferentiation at the pre-individual reality, what Simondon calls 'trans-duction,' are physical, biological, and psychic processes, manifested as social operations between heterogeneous structures that are in relation with each other. The pre-individual is the biological–physical basis of our species, the *generic capacity* of the human rather than any specific individual. As such, sensory perceptions open up the larger domain of the impersonal and hence the common. Language, which is shared within a community, belongs to everyone and yet to no 'one.' Perception, language, memory and feelings are common, belonging to collective individuation, what Simondon termed as transindividuation (the collective). Collective individuation furthers individuation, rather than diminishes differences. It is at the pre-individual level that mutation and variation take place. For Simondon, there is no human nature per se, and hence his orientation has the advantage of being post-anthropological in decentering Man. Individuation refers to transitions and thresholds that define the 'human' as a particularly unstable field of actualization, in Deleuzian terms. Moreover, it refers to the inorganic and organic alike. There are only degrees of individuation as inorganic (or 'physical individuation') is marked by a more stable energy than living organic being ('vital individuation'). Individuation is not 'human,' rather it emerges from an inhuman milieu and unfolds in innumerable directions. The implications are evolutionary as Keith Ansell Pearson has explored (1999).

Given Virno and other views of the common, the politics for the Anthropocene are difficult to sort out as Virno, following Simondon and Deleuze's own theorizations, presents an affirmative political position that insists on the virtual (real) dimension of the pre-individual as the source for change. In short, the metastablity of any system at the pre-individual level provides a 'disparation' minimally between two heterogeneous orders, two disparate scales of reality, at which point there is no interactive communication. However, within this initially incompatible series, communication within a social field can be established, and hence the invention of a new 'common' can emerge that was not imagined or governed in advance. The 'new' common emerges from the difference (and not diversity) as the ontological background is itself unequal (not a cohesive organic whole). It waits for a metastable turning point, a moment of the conditioned contingency of actualized political invention, an *event* must occur. This, then becomes a question of human 'energetics' rather than dialectics, the latter already is an actualized politics; the former is one of an emergent subject, immanent to the metasabilty of various social groups (see Toscano 2007).

Given the *event* of the Anthropocene, what sort of 'subject' is manifesting itself remains in balance. The Anthropocene forces a deindividuation of one kind or other. This emergent subject, however, has to belong to a transindividual collective, and that itself is paradoxical as a gap is created between, what Simondon sees as a paradigmatic subject who is an inventor, and someone who necessarily separates or breaks away from a communitarian bond. A deindividuation must take place, or a deterritorialization into a becoming 'other' in Deleuze and Guattarian terms, when the metasability of the earth is addressed so that a new subject position becomes possible. The question of the common in relation to the Earth has opened up a wide ranging debate (Harvey 2011). One major direction mapped by the trilogy of Michael Hardt and Toni Negri (*Empire, Multitude, Commonwealth*), replaces the worker with the multitude—receiving its fair share of criticism. There is also the proliferation of rethinking the common within new radical discourses that reimagine communism from its failed historical attempts (e.g., Badiou 2010; Landa 2013; Douzinas and Žižek 2010; Dean 2012). As Isabelle Stengers (2015) and Wark (2015) argue, the telos of the commons has no guarantees; its fascist elements are equally as probable. The election of Trump as the 45th president of the US seems to highlight this worry. The political dividing line still remains between a politics of preindividuated 'energetics' and that of post-Hegelian inflected dialectics. The Anthropocene problematic has yet to emerge a 'new' common that is actualized beyond the grips of capitalism. Adrian Parr (2017, Chapter 5), however, has forcefully made the case for a 'new earth' and a 'new land' by articulating the 'smooth' space of *commonism* that emerges as a 'radical alterity' within the privatized landscape of neoliberal planetary urbanization. [Commonism] mediates between the natural common of innovative ideas and collective practices, distributing the benefits, resources, and opportunities that arise from the urban common in support of collective well-being' (117–118). Drawing on the urban farming and land-use policies of Cuba and Venezuela as well as Detroit Black Community Food Security Network, Parr maintains that that these exemplars of commonism create alternative collaborative values to the surplus value that drives the large-scale corporate model of industrialized monoagriculture that sells its goods to the global markets. 'Commonism struggles to reclaim the common' (118), a common that is constantly captured by neoliberal forms of capitalist exploitation. Commoning, as a collaborative sharing of resources that are neither publically (via collectives or the state) nor privately owned *must* be accessible to all (like air, water, gifts from the soil), as a fundamental human right, which *must* be protected. Commonism has the potential to deterritorialize neoliberal planetary urbanization.

ARTISTIC RESPONSES

The artistic response to the Anthropocene has been far 'ra(n)ging.' *The Art of the Anthropocene* (Davis and Turpin 2015) provides a broad array of aesthetical, political and environmental positions by artists who echo many of the concerns raised in this introduction. In a comprehensive article, Kathryn Yusoff and Jennifer Gabrys (2011), for instance, define three approaches to the artistic imaginary that this introduction has already noted: the first imaginary explores scenarios that surround climatic disasters; dystopian scenarios that address the range of Anthropocene problematic: resource shortages, population explosions, peak oil, environmental collapse, and technoscientific accidents that are likely to result from experimentation. A second approach is to bring the Anthropocene 'home,' to one's backyard—so to speak, to publicize, and relocate the imaginary in a personal way so as to generate adaptive, usually strategic survival strategies that are sustainable. Lastly, what they refer to as a 'cultural turn,' is the engagement of climate science with art and art-public collaborations to generate new forms of knowledge. Yusoff and Gabrys fill out these three imaginaries through various examples that are viewed as being paradigmatic to each: the usual disaster films and literature for the first; the second (adaptation), focuses on the art, design, architecture, and engineering aimed at the city, community, and the home, being the key sites for reconsidering scale for sustainability with utopic resonances. The 'other side' of this development is simply the folly of the sustainability itself and the adaptive mentality of 'survival' in the broader Anthropocene context. Here The Yes Men's *Survivaball* (2006) is certainly iconic of such an ironic stance. The last category, which weds climate science with art and art-public collaborations, offers a wide complexity that can heighten sensibilities towards climate change and make them 'less abstract.' Weather becomes one such interface, the best-known examples staged through the installations of Olafur Eliasson, who cooperates with scientists in his studio as mutual collaborators and advisors (e.g., *Weather Project* 2003–2004; *The Glacierhouse Effect versus the Greenhouse Effect* 2015).

The visualization and the 'sounding' of the Earth patterns of weather, its atmospheric changes, the melting of the glaciers through time lapse photography, the quality of air and water, plant photosynthesis, sounds of marine life, animal habit and movement, and so on; in short, all aspects of the environment provide a rich source of experimentation by artists to bring public awareness of the nonhuman world, which remains non-representational and 'invisible' to human perception unaided by various forms of technologies (Kanngieser and Beuret 2017). Imaginative narratives elaborate, intervene, interrupt, and reimagine historical narratives as stories open up history, both real or imagined (e.g., as iconically presented in Peter Huyghe's *A Journey that Wasn't*, 2005). The play with Earth time, or deep time, is explored as well. Iconic are artists like Katie Paterson. Her *Fossil Necklace* (2013) is composed of 170 fossils carved into spherical beads, each fossil representing a major event in the evolution of life over the *longue durée* of geological time. Bioart, which has grown in profile and prominence in the past decade, offers yet other questions for the Anthropocene as the development and growing of synthetic cells and tissue engineering introduces more complications ethically and politically (jagodzinski 2014). The coming of science with art is especially impressive in cases like Rachel Armstrong's (2015) 'living architecture,' where the synthetic creation of cells are produced to bond with rotting wood underwater in cities like Venice. Bionics is a science-design-art-technology multidisciplinary field that is said to rescue ailing ecologies and present the future of 'smart' design for the Anthropocene problematic. Synthetic biology is perceived as the new technological fix to what ails us today, it is a way of genetically programing organisms, disassembling and classifying DNA sequences, and reorganizing them to meet the emerging needs.

The installation work of Natalie Jeremijenko (2017) is perhaps one of the most well-known and well-respected attempts at bringing art and engineering science together to highlight the Anthropocene problematic. Jeremijenko's Anthropocene art are experiments in 'lifestyle,' where forms of biomemesis in her design technology are in play to explore ideas of nature's intelligence. For instance, her MUSSEL CHOIR utilizes a 'computer-model-based approach.' Mussels are fitted with sensors that measure the opening and closing of their shells in their reaction with the freshness and health of the water they are placed in. The data collected is then converted into sound for public display, which makes legible 'mussel' behaviour that is otherwise imperceptible. Mussels, in this case, are but another example of harnessing the biosensing abilities of the nonhuman (as mentioned earlier in this introduction, see Johnson 2017). Jeremijenko perceives this as a way of making the public aware of the transformative changes that take place all around them, which are now visualized or sonically made hearable via these small-scale experiments. Jeremijenko's installations provide a pedagogical function to those spectators who are willing to be involved in the processes that she uncovers at work in nature. Her 'model' for embracing nature through technology, engineering and design is one of 'mutualism,' a profound grasp that there is symbiosis at work wherever you look. This is consonant with Haraway's (2016b, p. 57) approach where interspecies collaboration is the 'sympoiesis' and 'symbiogenesis' of co-becoming that determine the very material conditions of existence. Jeremijenko's *Signs of (intelligent) Life* (2015) is an example this as commercial signs become displays of perennial flowers to support pollination and increase human health benefits.

The artistic response to the nonhuman is an ongoing work in progress. The bioartistic responses, spearheaded by Eduardo Kac's own genetically modified installations (2007) has reinvented a new meaning to Duchamp's 'Readymades.' His transgenic GFP Bunny disturbs the technogenetic experimentations of biosynthesis, raising ethico-political concerns for the creation of such creatures. There are artists like Terike Haapoja, a Finn, who collaborates with Laura Gustafsson in attempting to give 'voice' or a 'place at the table' for nonhuman life through his conceptual performative practices. The project, *History of Others* (2013–ongoing), consists of interspecies cosmopolitical mediated performances, images, imaginary institutions, and various social agents, consonant with what Rosi Braidotti (2013, p. 60) calls a 'zoe-centred egalitarianism': a forum where a legal-political equality amongst species might be developed (see also Chakrabarty 2015; Demos 2016a). Stagings such as *The Trail* (2014), where hunters are charged for murder; *Museum of the History of Cattle* (2013), which, like Temple Grandin, tries to present a 'cow's view' of the world are paradigmatic examples. Haapoja even initiated a *Party for Other* (2011), which was supposed to be an interspecies political organization that ran in the Helsinki parliamentary elections! Daniel Falb (2015) maintains that such artistic projects are post-Conceptual in their form (cf. Peter Osborne 2013, 2014) as Anthropocene art is distinct from Ecological or Environmental art in the sense that the challenge becomes how to address this event that escapes easy visualization—like greenhouse emissions, for instance, that remain invisible, or natural processes that escape the radar of human consciousness; or, how would one 'represent' the ephemerality of cyber-data? (see Carruth 2014). Environmental and ecological art forms are usually able to do this representationally, drawing reverberations with romantic strategies of landscape depiction of one sort or another.

In somewhat of a contrasting tone to Daniel Falb's post-Conceptualist position, T.J. Demos (2016b), in *Against the Anthropocene: Visual Culture and Environment Today*, interrogates the representational visuality of the Anthropocene made possible by the digitalization of processed quantities of data that are collected from satellite-based sensors, invisible to human perception, but presented as stunning aestheticized visuals of the Earth as well as beautiful geo-sculpted landscapes; the proliferation of photography of the corporate-made devastation of the Earth (e.g., especially the photography of Edward Burtynsky and Louis Helbig) provide mixed-messages, a question mark concerning how they are to be ethically and politically grasped. Whilst Falb maintains that Land Art should be considered 'early' Anthropocene Art, distinctions would need to be made between, say, Michael Heizer's *Double Negative* (1969–1970), which leaves a trace of two enormous trenches on the Earth as photographed from satellite pictures, and the landscapes of human activity that Burtynsky and Helbig show us, such as Burtynsky's *Oil Fields* and Helbig's tar sands photography like *Effluent Steam*.

Google Earth, Global Positioning Satellites (GPS), data and imagery from the World Wide Web have 'visualized' the Anthropocene through the machinery of this vast sociotechnical system. Despite their seeming 'legibility,' these images seem to join the incomprehensibility of 'hyperobjects' that Tim Morton (2013) has popularized. But these technoscientific, militarized, objective images, argues T.J. Demos, are hardly indecipherable from a military-state-corporate mindset, which needs them for surveillance. The Earth, as an object of contemplation, becomes an ideal image to be manipulated geostrategically via geoengineering as well as for geopolitical strategizing. Elizabeth DeLoughrey (2014) provides a succinct review of how the globe has been visualized via NASA's space program, noting specifically, following Paul Virilio, that 'cinematic derealization'—the experience of detachment—perfectly plays into the technologies of militarism. Being both connected to the Earth and yet apart is what Gayatri Chakravorty Spivak (2003) refers to this distancing as 'planetarity.' Such a mode interprets the world in such a way that it is not simply reduced to a homogenization via globalization, rather it becomes a form of ecological theorizing that presents the planet as a form of 'alterity.' Lisa Park's (2005) *Cultures in Orbit*, provides an overview of just how satellite imagery shapes the mastery of Earth's visualization via television images that are beamed globally in such a way that it appears that the Earth itself, as an 'alterity,' is controllable and subject to seemingly easy manipulation as its scale is miniaturized.

T.J. Demos position is to further the visualization of the Anthropocene as the Capitalocene, pointing to the photography of Richard Misrach's *Petrochemical America* as providing a strong contrast to any forms of hyper-aestheticization of sublime destruction. Along with the landscape architect, Kate Orff, who integrates her 'Ecological Atlas' with the photographic series of *Petrochemical America* (2012). She documents—through informational maps, diagrams, and flowcharts—the chemical and ecological devastation along the 150-mile Mississippi River corridor between Baton Rouge and New Orleans. Throughout his book, Demos provides many key examples of how activist artists stage ways to push back against capitalism, with the strong hope that such cultural activism will make a difference.

Aesthetics and its imaginary are indispensable for providing new potentials to be collectively actualized in order to keep the future open in what is quickly approaching as a phase-change to the Earth's biosphere (Ostberg et al. 2018). The creative dimension of human activity must generate new forces of desire that can open up cracks in capitalist global expansion. Such an aesthetic imaginary, 'a redistribution of the senses' in Jacques Ranciére's (2004) terms, is desperately needed. Adrian Parr (2017, pp. 191–195) provides a number of artistic examples as part of the *Artists 4 Climate* exhibition, as coordinated with the Paris COP21 talks (November to December of 2015) meant to stir such an imaginary, as was the 'thousand empty shoes protest,' a public art installation in front of Place de République (not far from the site of terror attacks on November 13) by climate change activists as an 'silent' protest against the government's ban on public demonstrations.

DIVINE AND THE DARK

I end this long introduction with one last important section, which is a response that addresses the species as a 'divine animal,' not a God of the 'new Prometheans' of geoengineering, as some have explored to present the options, dangers and possibilities (e.g., Lynas 2011), but a 'spiritual' species capable, as Joanna Zylinska (2014) maintains, *a minimal ethics* for the Anthropocene. The 'divine' raises questions that address religion, theology, spirituality, issues surrounding animism, and panpsychism that feeds into the new materialism in ways yet to be fully understood, which hold the promise as to what might, pray tell, be the next shift our species must make to generate a planetary spiritualism—a geo-spiritual formation of the Anthropocene. Included in this spiritual direction are approaches to understanding Earth's memory (e.g., Szerszynski 2015). In which way does the Earth remember and forget, as anthropomorph-ically speculative as this may sound? The search is essentially for a new cosmology that situates Earth in a decentered multi-universe. It is an attempt to desecularize the Anthropocene. This shift towards 'spiritual-ism' turns its direction away from transcendentalist theologies, especially established patriarchal world religions that still grip the Anthropocene globally in various ways, despite Pope Francis' second encyclical *Laudato Si'* (2015) on climate change, and despite in some church congregations (e.g., United Church) the acceptance of gender equality, homosexuality, and same-sex marriage. The shift in postmodernity is that religions have become more cerebral, dropping their mimetic and mythic forms (see Turner 2016). Ritual embodiment and religious habituation have faded as the 'letter' of the sacred texts, as the pure (yet interpreted) message of the Qur'an, the New Testament, Buddhist *dharma*, the *Analects* of Confusion are what matter. One thinks here of the dystopian post-apoca-lyptic film, *The Book of Eli* (2010), where its main character, who is blind, has memorized the entire King James version of the Bible given that books have become rare, and he has the last copy, in Braille! He carries the Word in him like some prophet. Oliver Roy (2011) makes the point that Islam in particular has become deterritorialized and de-cultured as Muslims are uprooted from their original spatial and cultural origins. Given the urban contexts, religions have become mobile, cerebral, indi-vidualized, in some cases promoted on the Internet, with Facebook posts to advertise Easter services, marriages, and upcoming events; perhaps the apotheosis of this development is the Hack Temple in San Francisco that has now branched out to include The Silicon Valley where 'technology' has taken on the mantle of religion.

The Axial-age religions (800–200 BCE) (Jaspers 1953; Bellah and Jonas 2012) that spawned the Hebrew Prophets, Buddha, Confucius, Socrates, Plato, Zoroaster, Lao-Tse developed a moral code to live by that provided for a transcendent world to come that was meant for everyone, perhaps not a stretch given that the empirical reality was deeply unsatisfactory (disease, disabilty, ageing, the hierarchy of power, and so on.) There is a remarkable inversion taking place in relation to this historical development as the shift from transcendent to immanence is taking place via new materialisms, which have become prominent and co-current with the Anthropocene. Once again there are existential questions of survival and extinction on the horizon. Those scholars speculating a geo-spiritual formation of the Anthropocene cast their eyes into matter-energy, 'downcast eyes' to recall Martin Jay's (1993) study of antiocularism that eventually led into the contemporary flourishing of non-representational thought. Clayton Crockett (2016) turns to 'insurrection theology' distancing himself from forms of transcendentalist theology and ecotheology alike. His most radical thesis is that 'theology for itself is energy. Theology is energy when thought according to an insurrectionist theology. ... Energy is the best name we have for the real as real' (57). For Crockett, the new materialism is a theological materialism based on energy transformation, because 'theology for itself is energy' (28). When it is not self-serving ideology, theology concerns the Real. Crockett draws heavily on Deleuze and Guattari, and throughout his writing, he attempts to make a bridge between their work and the Hegelian inflected positions of Žižek, Badiou, Lacan and Malabou (see Crockett and Robbins 2013; Crockett 2014). Whether such a bridge is indeed tenable remains open to debate since his attempt is to avoid the antagonism that Hegelian negativity generally brings, but then again, the authors mentioned would disagree as they rewrite the Hegelian dialectic for their own purposes. The point is that this theoretical base enables Crockett to draw concepts from each so that he might continue to speculate what such an Anthropocene theology might 'do.'

Bronislaw Szerszynski (2017) has also thought deeply about a geo-spiritual formation of the Anthropocene by identifying two developments of the 'Planetary Spirit' along with their respective issues and concerns. He, too, is working with the new materialisms. For the first level of planetary spirit, Szerszynski identifies four key components: the first is the need to draw on contemporary thermodynamics to grasp energy and entropy (e.g. dissipative structures); the second, is a grasp of immanence and emergence. The third is to distinguish geo-spiritual formations historically; fourth is to understand how these different geo-spiritual formations address Georges Bataille's 'accursed share' of excess energy. Against this formation, Szerszynski presents six alternative coding, recodings and decoding of the flows of energy and matter, which he calls the high gods of the Earth's new epoch. In order, they are: *Anthropos* or abstract human (Man); capital; sun (as the dominant source of 'free' energy); Earth (e.g., Latour's Gaia, Haraway's Chthulucene, Reza Negarestani's 'tellurian rebellion'); Yahweh, Allah, the God of Abraham, and the 'final god' or non-god is the cosmos—as in cosmic nihilism (e.g., Ray Brassier 2007) and cosmic contingency (e.g., Reza Negarestani and Eugene Thacker). As a final reflection, Szerszynski explores a second level of planetary spirit. Here, he develops four more aspects of geo-spirituality. He supplements Thermodynamics with the notion of *turbulence*; he extends the vocabulary of fluid dynamics to include a distinction between '*diffusion* (local, uncoordinated, multidirectional movement) and *advection* (mass parallel or laminar flow)' (262). He dwells on 'enchantment' when it comes to making things *motile* (potentiality in motion) and *mobile* (actually in motion). He identifies primitive accumulation as a key factor when it comes to 'motilization' and the coupling of diffusion and advection. Lastly, the concept of *spiritual entropy* he maintains, needs more articulation (see also Szerszynski 2016a).

Both Crockett (2012) and Szerszynski draw on Deleuze and Guattari as a way to develop their independent understanding of planetary spirit in terms of flows of energy, power, value, entropy, and negentropy. For Szerszynski, thermodynamics and fluid dynamics are crucial as they are for Crockett. Szerszynski, for his part, seems to draw on the Deleuze and Guattarian distinction of striated and smooth space and cast them in fluid terms: the 'high gods' shaping the Earth's ongoing molar transformation, whilst the 'low spirits' are in the role of smooth spaces where local gradients change realizing counter energies. Both recognize the significance of the quantum field. The quantum level of particle physics points to the deep entanglement between our species and 'nature.' The split between wave and particle at the subatomic level cannot be overcome, in the same way species becoming is shaped by a both-and logic; humanity is *both* alienated from nature *and* belongs to it, an enfoldment that always produces an excess of life. This excess, as some have noted, can be attributed to the thermodynamics of nonequilibriun systems and the phenomenon of self-organization (autopoiessis). This is the cosmological direction insurgent theology is heading, with new forms of ethical possibilities as are being explored by post-Deleuzian like Brian Massumi's (2017) development of Guattari's 'virtual ecology,' where the question of value (axiology) is reworked to provide a possible anticapitalist embrace of new values to live by. That said, there is also a 'dark side' that presents a highly dystopian narrative, an imaginary that is a constant accompaniment with the Anthropocentric imaginary captured admirably by Emily Apter (2013) in her discussion of 'planetary dysphoria.' Against, for instance, Bruno Latour's (2017, p. 117) call for the recognition of the Earthbound, '"bound" as if bound by a spell, as well as "bound" in the sense of heading somewhere,' a geopolitical commitment to Gaia against Humans who appear oblivious to the Anthropocene, and against, for instance, Bronislaw Szerszynski's (2014, 2016a) careful and meditative questioning of Earth's memory and deep time in relation to Earth's 'monuments' (that reverberates with art as a monument developed by Deleuze and Guattari in *What is Philosophy?*), there is the 'tellurian' ranting of Reza Negarestani (2008) and Ben Woodward (2013).

It is the Earth's dynamic unstable geology that is at stake, going back to the debates initiated by the Volcanists. The solidity of the Earth falls away with worms, volcanoes, earthquakes, erosion, crevices, and human athropogenic 'ungrounding,' as do claims to solidity via the 'grounding' of anthropological theories. This is extreme deterritorialization (in Deleuze and Guattari's terms), constantly 'ungrounding' itself as the 'New Earth' takes on a more horrific meaning than the one usually called upon by (post)Deleuzians (e.g., Parr 2017), where toxic human waste become land marks that are permanent crypts, forever uninhabitable. The geologic character of the Earth, for Woodward (2013, p. 86) becomes 'a storm of forces, as darkly productive monster.' Reza Negarestani (2008) is darker still, as developed in his *Cyclonopedia* (2008), which presents a 'telluriam rebellion.' Briefly, it is a theory fiction that locates a living being within the Earth that carries out a rebellion against the 'Solar Empire.' Humanity is but a side-effect in this rebellion, manipulated by the sentient agency of oil seen as a 'autonomous terrestrial conspirator' that consumes the Sun's energy, as it is the Sun, which paradoxically gives us life and marks our death in the future. It is the radical Outsider. The metaphysics suggest that the Sun represents hierarchical system of control and repressive wholeness; the Earth is always attempting to escape its system to become 'otherwise.' This process is taking place underground as it is constantly deterritorializing the coherency of the Earth's surface. Negarestani's '()hole complex' of the Earth (237), is perhaps a way to puncture or to further understand the smooth spaces Deleuze and Guattari write about, but now it is holes that do the rupturing. It is a 'poromechanical earth' (53), an obvious reference to 'pourous,' as surface and depth are breached. This is a political move, called 'polytics' where strategies operate in local events of disruption—'riddle the Earth with holes (of plot, of function, of structure and wholeness)' (243). Negarestani (2013a, b) has a Cthulhuoid Ethics, which is the general 'labour of the inhuman.' The human is a becoming shaped by non-human and inhuman forces that then open up a 'what else' (Negarestani 2013a, p. 5). The vision here is that the connections always need to be made with the non-human and inhuman as a form of liberation, as 'the continuous unlearning of slavery' (Negarestani 2013b, p. 13). For Negarestani then, the recognition of the tellurian dynamism of the Earth requires an embracement of inhuman technologies and techniques, a 'dark imaginary' that drags the Anthropocene into the depths of the Earth, and no longer on its surface.

In contrast to the bleak 'science fictioning' of Negarestani is the more levelled, thoughtful, and sociopolitical work of Adrian Parr as forcefully presented in her latest work at the time of this writing, *Birth of a New Earth* (2017). Parr has been at the forefront of thinking through Anthropocene problematic through her political commitment to water and urban renewal projects in Nairobi, Kenya. She bears the official title of a UNESCO Water Chair and has a Kenyan name–Wanjiru. She is quite aware of her outside privilege as a 'white' female university professor who does her level best mediating capitalist institutions, NGO's and the local self-help collaborative organizations in Nairobi. It is a nomadic schizo-position of an indeterminate in-between that she is constantly walking and negotiating. Her situation raises all those difficult complexities between love and hate. As Jacques Lacan (1981) once famously said, '*I love you, but, because inexplicably I love in you something more than you – the* objet petit a – I *mutilate you* (p. 268, original emphasis).' Parr's thesis raises the question concerning the paradoxical position of environmentalism as a 'realist' fetish: as an ideological supplement it speaks to its appropriation by neoliberal capitalism, yet its interrogation and re-examination is nevertheless crucially required. As she writes in her introduction, 'The axiomatic of environmentalist politics, one that also serves as the source for the birth of a *new* earth, is simple: the refusal to surrender life to the violence of global capitalism, corporate governance, and militarism' (xv, original emphasis). Environmental degradation, she argues, is directly tied to capitalism and militarization within an inequitable world order. Yet, it is an 'opportunity to strategically interrupt the march of global capitalism' (xvi).

Parr provides a new roadmap, a way out from the morass of issues that surround green environmentalism, sovereign exceptionality, and the dangers of fascistic environmentalism. Parr also shows us how the 'clear-cutting' of cities, like Aleppo, Gaza, Baghdad, and Pervomaisk, through militarization of the world order, removes and levels entire populations, turning them into refugees and nomadic immigrants. In these collapsed urban ecosystems, life becomes 'an endless repetition of the same' (162); its inhabitants live in a 'pure present' with no future; all that is left is rubble as traditions, history and heritage are erased, any forms of urban infrastructure are simply gone and the 'collective imagination and memory' or Idea that held the city together becomes only a haunt—its ghost. It is the loss of 'belonging' that 'facilitates daydreaming and emancipatory imagining, without which hope sours and turns to revenge' (165). Parr identifies this devastation by war as a 'blind spot in environmental politics and the collective imagination' (xix). How can a critical politics turn violence into solidarities between various assemblages of ethnicity, race, gender, class, generations, and species so as to provide an alternative imaginary? How can apocalyptic scenarios of the end of the world be avoided 'when suffering collapses into ecstasy' (xx)?

Part of Parr's rethinking the roadmap to a new Earth is to explore an 'urban commons' as a way of subverting the relations of power so that the basics of water and affordable food are met. She forcefully argues for a *communism* that is able to generate alternatives to the law of surplus values through collaboration, cooperation, trust and generosity. Commonism's radical alterity requires three processes, as she puts it (118–119): the need to construct coalitions involving individual, local, regional, national international levels to ground a transformative politics; to initiate urbanization processes that challenge and construct alternative to the profit taking of surplus value. Lastly, to institute those collaborative activities that transform the 'common [as] uncommon,'; that is, to subvert was is usually either a public or private hold of resources that should be available to all. In this way, Parr maintains, a 'new land' is created.

Equally, Parr thinks through the discourse of human dignity, which is treated as a commodity of status and moralism, an inalienable human right within a neoliberal capitalist discourse. Throughout her work in the slums of Nairobi (primarily Kibera), Parr recognizes dignity as a social relation, 'as a combination of actions that enhance the ability of all to partake in collective actions (decision making, ecological flourishing, financial autonomy, food and water sovereignty)' (140). Parr follows a Deleuze-Guattrian understanding of 'utopia' where the 'nowhere' of utopia becomes the 'now here' with the potential of the future being open. Dignity is a 'transformative élan, a utopian form, that activates activist practices throughout the slum' (141). She identifies three public interest design groups: Kounkey Design Initiative, Solidarités Internationale and Human Needs Project who work with Kibera communities in ways contra to the neoliberal aesthetic that the state has instituted through a series of high-rise developments for slum-dwellers. The sustainable urban farms that are created by these organizations are (posthuman) assemblages of human, inhuman and nonhuman inter and intra-relationships where the dignity that emerges directly engenders productive connections of life enabling activist agency and well-being. '[T]he demand for dignity is a calling into existence through an act of communing' (144).

Parr ends her journey with the importance of a political imaginary that overcomes the dead end of the apocalyptic spectacle that saturates the entertainment industries (notably Hollywood). Here, she also includes the effective imaginary of ISIS, equally pernicious in its abilities to draw in supporters for what is a utopian ideal that appeals to Islamic fundamentalism. '[A]n apocalyptic imagination resolves the feelings of anxiety and powerlessness over the future that living under a violent socioeconomic system produces into an awe-inspiring image of natural disaster and redemption' (180). Drawing on Gilles Deleuze, Jacques Rancière and Rosi Braidotti's vision of posthumanism, Parr makes the case as to why 'a promise' addresses a 'people yet to come' in that it needs to be constantly revitalized through repetition to ensure a becoming-other. Parr stresses the need of such a 'emancipatory imaginary' is crucial for a renewed sensibility that addresses the Anthropocene, which moves away from the devastating global effects of 'disaster' capitalism that has caused so much environmental degradation, and continues to do so by latching on to new land deals as the climate changes. Parr concludes her thesis ('So to Speak') with the realization that these are bleak times where the populism of The Right and Alt-right is on the rise and felt globally in its effects especially on immigration, yet Parr sees hope in an *emancipatory imaginary*. It is not a question of faith, but a battle 'at the level of imagination and ideas' (208).

To end my introduction is perhaps to reiterate the task that is before this generation, especially those privileged enough to think, write, protest, and be active against a vile system of capitalism that, on all counts, shows the growing division between the have and have nots, such a division resting on capitalism's imaginative ability to exploit the 'commons' for profit ends, leaving a devastated environment behind, eventually a deserted Earth as our time as a species, an experiment of Nature, will be over. These are the competing narratives for a rethinking of a cosmology and geo-spiritualism of the Earth to question global capitalism and an emergent new set of values in the Anthropocene. Both take their leave from the conceptual toolkit of Deleuze and Guattari but with opposing positions: a dark and light Deleuze, ironically as played out in the saga of *Star Wars*.

* * *

This collection is divided into four sections, the first being *Capitalist Framings*, a strong current that runs throughout the Anthropocene, often dubbed the Capitalocene. The lead chapter is by Nick Dyer-Witheford entitled *Struggles in the Planet Factory: Class Composition and Global Warming*, provides an insightful expansive narrative of class struggles and their dynamics, covering capital's mining and oil industries, Fordist factory production, leading up to our contemporary situation where cognitive labour of digitized capitalism has led to identifying the dangers of global warming and has fostered new environmental protests. Dyer-Witheford continues to describe and explore the anti-extractivist movements originating in indigenous, nomadic, and peasant communities, raising questions concerning global migration as the emergence of new proletarians who are besieged by unbearable environmental conditions. He ends his chapter by speculating as to what new struggles will emerge from the conflations of class and climate within global capitalism. Chapter two, by Alexander M. Stoner & Andony Melathopoulos is entitled: *Stuck in the Anthropocene: The Problem of History, Theory and Practice in Jason W. Moore and John Bellamy Foster's Eco-Marxism*. They provide a critique of what they see as a failure in the theoretical work of Jason W. Moore and John Bellamy Foster, both of whom have developed eco-Marxist analysis of the Anthropocene in their respective ways. Stoner & Melathopoulos, however, maintain that there still is a failure to address history adequately, both in theory and practice. It is their attempt to provide such a missing dimension. The closing chapter, *Making Our Way in a World of Our Making: The Anthropocene, Debt-Money, and the Pre-Emptive Production of Our Future* by Matthew Tiessen, continues the exploration of the Anthropocene within a capitalist economy. Tiessen succinctly develops the role of money in relation to the private banking system within a capitalist system wedded to the Anthropocene. Tiessen raises difficult questions regarding the financial system that operates only in relation to profit. How is ecological change even possible given such a framing?

Part II is entitled *Planetary Projections.* The imaginary is an important consideration when thinking through the Anthropocene narratives. The lead chapter by Jason Wallin, *Catch 'Em All' and Let Man Sort 'Em Out':* *Animals and Extinction in the World of Pokémon,* is an indicator of how that imaginary can be put to use to articulate a thoughtful problematic. Human-non-human relationships form a large part of the Anthropocene imaginary, and in this chapter, Wallin explores our relation to animals in ways few 'could' imagine. He speculates that the disappearance of animals, especially in their 'wild' state, is being revived in the popular imagination by the worldwide phenomenon of Pokémon GO, discussing the consequences of such a trajectory. Janae Scholtz's *Intervals of Resistance: Being True to the Earth in Light of the Anthropocene* is a tour de force of the imaginary addressing Deleuze and Guattari's call for 'a belief in the world' and the question of a New Earth. The Anthropocene, she argues, calls for an ontological shift in awareness, which she entitles moving 'from the earth to the cosmic.' This pedagogically involves attuning (both conceptually and materially) to the level of the imperceptible forces, intensities, and affects which populate the earth—the cosmic level of being. Scholtz then outlines what such a shift would need to consider. This section closes with a chapter by Mickey Vallee, entitled *Sounding the Anthropocene.* Again, the imaginary is of a crucial issue. His chapter is about 'sounding' the Anthropocene in contemporary scientific research of bioacoustics. Vallee argues that a 'technological ear' is needed to extract imperceptible vibrations that constitute movement within an ecological environment. How to make these imperceptible vibrations resonate against a dystopian Anthropocene, is the crux of his problematic.

Part III, *Media and Artistic Responses,* perhaps unsurprisingly, forms the largest section of this collection. Like the imaginary in Part II, media and artistic responses are indispensable for exploring the range of affective responses to the various narratives in play. The lead chapter is by David Fancy. *Geoartistry: Invoking the Postanthropocene via Other-Than-Human Art* is his attempt to elaborate Deleuze and Guattari's insights into the aesthetic and artistic expressions of other-than-human entities as they relate to human artistic expression. Fancy fleshes out these connections when it comes to animals, birds, and insects (fireflies), and speculates on what these natureculture relations might mean for a postanthropocene future. Bradley Necyk & David Harvey's chapter, *'Like Watching a Movie': Notes on the Possibilities of Art in the Anthropocene* addresses the post-conceptual problematic of representing the Anthropocene where many of the changes are subtle, under the radar and invisible to the human eye. Necyk & Harvey draw on the work of Rob Nixon in exploring one example of Anthropocene violence: the forest fires in the northern Alberta city of Fort McMurray, Canada. In contrast to the way this climatic disaster was spectacularized, Necyk and Harvey offer a counter-narrative by drawing on contemporary video and installation art that offer alternative anti-narrative artistic discourses when addressing the Anthropocene.

Throughout this collection, a number of watercolors are presented at the end of each chapter, along with each section break. These are the artistic watercolors of Mia Feuer as is the front cover, a sculpture that speaks to the Anthropocene: *Totems of the Anthropocene 2017 (found sleds, found vessels formerly containing petroleum products, individually cast and assembled polyester resin crystals)*. Chapter 10, *FOAMA or ... You Make Me Feel the Way Gasoline Looks on Water*, is her artistic statement that covers her relation to organic and synthetic materials she uses in her sculpture studio to make her environmental installations. Feuer's works spans the sites of Suez, Fort McMurray, Svalbard, Bayou Point au Chien and Tehran. Her relation to these sites is briefly addressed in her artist's statement. Michael Truscello's *Catastrophism and Its Critics: On the New Genre of Environmental Documentary*, Chapter 11 in this collection, carefully examines four recent eco-opportunist documentaries: *Racing Extinction, This Changes Everything, Chasing Ice*, and *Chasing Coral*, carefully analyzing each of their structures for their affective forces to present a particular ideological narrative when it comes to 'saving the planet.' Truscello argues cogently that these documentaries pay little attention to the root cause of the Anthropocene—namely capitalism, and concludes what this change of orientation would require. Patti Pente, an artist and colleague, contributes to this section with *Slow Motion Electric Chiaroscuro: An Experiment in Post-Anthropo-Scenic Landscape Art*. Chapter 12 examines the posthuman shift in ontology. Pente addresses the digital—analogue enfolded world of glitch art within the concepts of 'fail and fix,' keying on the concept of disjunctive synthesis as developed by Deleuze and Guattari. Pente ends her chapter with a remarkable proposal, an artistic experiment where landscape art is virtually redefined by its affective force. Closing this section is Leslie Sharpe's *Situations for Empathic Movement* (Chapter 13). An ecological installation artist, Sharpe sets up an interactive artwork where the audience can make a human connection with real and online tracking of animals to begin to grasp the impact of climate change on their behaviour, health, migration and habitat in the Canadian North.

Part IV, *Pedagogical Considerations* are responses to just how are we to react to the Anthropocene and what can we do about it in face of such an overwhelming sublime 'hyperobject.' The lead chapter, *Against Climate Stoicism: Learning to Fight the Anthropocene* (Chapter 14) Ted Stolze confronts this issue squarely, avoiding the usual 'all is lost' narrative. Stolze's thoughtful chapter challenges the climate of stoicism that maintains climate change must be accepted as it remains beyond human control. Against Roy Scranton, who maintains humanity must 'learn to die in the Anthropocene,' Stolze marshals a rigorous contra-argument where a militant philosophy is proposed as an alternative strategy to learn to fight in the Anthropocene. In the following chapter, Nathan Snaza provides us with a meditative essay that is a challenge to educators. In Chapter 15, *The Earth Is Not 'Ours' to Save*, Snaza questions the notion of human agency in three recent works: *Occupy Education* by Tina Evans, Bruno Latour's *The Ecological Thought* and Tim Morton's *Hyperobjects*. Turning to feminist and queer posthumanisms and the new materialism, Snaza develops what he calls 'bewildering education' where animate and inanimate agencies are recognized. Paradoxically, he argues, such an education requires that we learn to 'fail' at being geological agents. Our final Chapter 16 is by Jessie Beier. It is an experimental form of writing to develop a fabulated 'dispatch from the future.' Beier investigates the role of 'science fictioning' as a conceptual weapon so as to shake up pedagogical thinking. Calling on Deleuze and Guattari as well as Vilém Flusser as speculative philosophers, it is her attempt to actualize alternative imaginaries to the Anthropocene and possibly break with the habituated repetitions that are now in play.

Image 1.1 Construction (Mia Feuer, watercolor)

REFERENCES

Adorno, T. W., & Horkheimer, M. (1972). *Dialectic of the Enlightenment*. New York: Herder and Herder.

Agamben, G. (2003). *The Open: Man and Animal* (K. Attell, Trans.). Stanford, CA: Stanford University Press.

Aiello, L. C. (2010). Five Years of Homo Floresiensis. *American Journal of Physical Anthropology, 142*(2), 167–169.

Alaimo, S. (2008). Trans-Corporeal Feminisms and the Ethical Space of Nature. In S. Alaimo & S. Hekman (Eds.), *Material Feminisms* (pp. 237–264). Bloomington: Indiana University Press.

Alaimo, S. (2016a). *Exposed: Environmental Politics and Pleasures in Posthuman Times*. Minneapolis: Minnesota University Press.

Alaimo, S. (2016b). Climate System, Carbon-Heavy Masculinity, and Feminist Exposure. In S. Alaimo (Ed.), *Exposed: Environmental Politics and Pleasures in Posthuman Times*. Minneapolis: Minnesota University Press.

Alaimo, S. (2017). Your Shell on Acid: Material Immersion, Anthropocene Dissolves. In R. Grusin (Ed.), *Anthropocene Feminism* (pp. 89–120). London and Minneapolis: University of Minnesota Press.

Allen, C., & Bekoff, M. (2007). Animal Minds, Cognitive Ethology, and Ethics. *The Journal of Ethics, 11*(3), 299–317.

Altvater, E. (2016). The Capitalocene, or, Geoengineering Against Capitalism's Planetary Boundaries. In J. W. Moore (Ed.), *Anthropocene or Capitalocene? Nature, History, and the Crisis of Capitalism* (pp. 138–152). Oakland, CA: PM Press.

Ansell Pearson, K. (1999). *Germinal Life: The Difference and Repetition of Deleuze*. New York: Routledge.

Apter, E. (2013). Planetary Dysphoria. *Third Text, 27*(1), 131–140.

Arabian Business. (2014). *NASA Astronaut Tweets Images of Dubai from International Space Station*. http://www.arabianbusiness.com/nasa-astronaut-tweets-images-of-dubai-from-international-space-station-572147.html.

Armstrong, R. (2015). *Vibrant Architecture: Matter as a Co-designer of Living Systems*. Warsaw and Berlin: De Gruyter Open.

Badiou, A. (2005). *Being and Event* (O. Feltham, Trans.). London: Continuum.

Badiou, A. (2010). *The Communist Hypothesis*. London: Verso.

Barad, K. (2007). *Meeting the Universe Halfway: Quantum Physics and the Entanglement of Matter and Meaning*. Durham, NC and London: Duke University Press.

Bardini, T. (2011). *Junkware*. Minneapolis: University of Minnesota Press.

Bateson, G. (1979). *Mind and Nature: A Necessary Unity*. New York: E.P. Dutton.

Beck, U. (1998, March 20). The Cosmopolitan Manifesto. *New Statement*, pp. 28–30.

Bell, J. (2006). *Philosophy at the Edge of Chaos Gilles Deleuze and the Philosophy of Difference*. Toronto: University of Toronto Press.

Bellah, R. N., & Joas, H. (Eds.). (2012). *The Axial Age and Its Consequences*. Cambridge, MA: Belknap Press of Harvard University Press.

Best, S. (2014). Minding the Animals: Ethology and the Obsolescence of Left Humanism. In S. Best (Ed.), *The Politics of Total Liberation: Revolution for the 21st Century* (pp. 107–136). New York, NY: Palgrave Macmillan.

Bostrom, N. (2014). *Superintelligence: Path, Dangers, Strategies*. Oxford: Oxford University Press.

Brady, K., & Swimme, B. (2012). Nature and Eros: An Educational Process for Engaging with a Living Universe. *World Futures, 68*(2), 112–121.

Braidotti, R. (1994). *Nomadic Subjects: Embodiment and Sexual Difference in Contemporary Feminist Theory*. New York: Columbia University Press.

Braidotti, R. (2013). *The Posthuman*. Cambridge, UK and Malden: Polity Press.
Braidotti, R. (2017). Four Theses on Posthuman Feminism. In R. Grusin (Ed.), *Anthropocene Feminism* (pp. 21–48). London and Minneapolis: University of Minnesota Press.
Brassier, R. (2007). *Nihil Unbound: Enlightenment and Extinction*. New York and London: Palgrave Macmillan.
Buchanan, B. (2008). *Onto-Ethologies: The Animal Environments of Uexküll, Heidegger, Merleau-Ponty, and Deleuze*. Albany, NY: SUNY Press.
Callaway, E. (2014, January 29). Modern Human Genomes Reveal Our Inner Neanderthal. *Nature*, p. 14615.
Campbell, T. (2006). Bios, Immunity, Life: The Thought of Roberto Esposito. *Diacritics, 36*(2), 2–22.
Carruth, A. (2014). The Digital Cloud and the Micropolitics of Energy. *Public Culture, 26*(2), 339–364.
Chakrabarty, D. (2015, February 18–19). *The Human Condition in the Anthropocene: The Tanner Lectures in Human Values*. New Haven: Yale University. http://tannerlectures.utah.edu/Chakrabarty%20manuscript.pdf.
Chernilo, D. (2017a). The Question of the Human in the Anthropocene Debate. *European Journal of Social Theory, 20*(1), 44–60.
Chernilo, D. (2017b). *Debating Humanity: Towards a Philosophical Sociology*. Cambridge: Cambridge University Press.
Clark, A. (2005). Intrinsic Content, Active Memory and the Extended Mind. *Analysis, 65*(1), 1–11.
Clark, A. (2010). *Supersizing the Mind: Embodiment, Action, and Cognitive Extension*. New York: Oxford University Press.
Clark, N. (2008). Aboriginal Cosmopolitanism. *International Journal of Urban and Regional Research, 32*(3), 737–744.
Clark, A., & Chalmers, D. J. (1998). The Extended Mind. *Analysis, 58*, 7–19.
Clive, F. (2011, December 30). Viewpoint: Has 'One Species' Idea Been Put to Bed? *BBC News*. http://www.bbc.com/news/science-environment-16339313.
Colebrook, C. (2010). *Deleuze and the Meaning of Life*. London, Oxford, New York, New Delhi, and Sydney: Bloomsbury.
Colebrook, C. (2012). Sexual Indifference. In T. Cohen (Ed.), *Telemorphosis Theory in the Era of Climate Change* (Vol. 1, pp. 167–182). Ann Arbor, MI: Open Humanities Press.
Colebrook, C. (2014a). *Death of the Posthuman: Essays in Extinction* (Vol. 1). Ann Arbor, MI: Open Humanities Press.
Colebrook, C. (2014b). *Sex After Life: Essays on Extinction* (Vol. 2). Ann Arbor, MI: Open Humanities Press.
Colebrook, C. (2017). We Have Always Been Post-Anthropocene: The Anthropocene Counterfactual. In R. Grusin (Ed.), *Anthropocene Feminism* (pp. 1–20). London and Minneapolis: University of Minnesota Library.

Cohen, J. J. (2015). *Stone: An Ecology of the Inhuman*. London and Minneapolis: University of Minnesota Press.

Corcorn, P. L., Moore, C. J., & Jazvac, K. (2014). An Anthropogenic Marker Horizon in the Future Rock Record. *GSA Today, 24*(6), 4–8.

Crockett, C. (2012). Beyond Heat: Energy for Life. In D. Bowman & C. Crockett (Eds.), *Cosmology, Ecology, and the Energy of God* (pp. 59–69). New York: Fordham University Press.

Crockett, C. (2014). The Triumph of Theology. In C. Davis, M. Pound, & C. Crockett (Eds.), *Theology After Lacan: The Passion for the Real* (pp. 250–266). Cambridge, UK: James Clark.

Crockett, C. (2016). Earth: What Can a Planet Do? In W. Blanton, C. Crockett, J. W. Robbins, & N. Vahanian (Eds.), *An Insurrectionist Manifest: Four New Gospels for a Radical Politics* (pp. 21–59). New York: Columbia University Press.

Crockett, C., & Robbins, J. (2013). *Religion, Earth & Politics: The New Materialism*. London and New York: Palgrave Macmillan.

Daly, H. E. (2005). Economics in a Full World. *Scientific American, 293*, 100–107. https://doi.org/10.1038/scientificamerican0905-10.

Daly, H. E. (2010). From a Failed-Growth Economy to a Steady State Economy. *Solutions, 1*(2), 37–43. https://www.thesolutionsjournal.com/article/from-a-failed-growth-economy-to-a-steady-state-economy/.

Davis, H. (2015). Life & Death in the Anthropocene: A Short History of Plastic. In H. Davis & E. Turpin (Eds.), *The Art of the Anthropocene: Encounters Among Aesthetics, Politics, Environments and Epistemologies* (pp. 347–358). London: Open Humanities Press.

Davis, H., & Turpin, E. (Eds.). (2015). *The Art of the Anthropocene: Encounters Among Aesthetics, Politics, Environments and Epistemologies*. London: Open Humanities Press.

Dean, J. (2012). *The Communist Horizon*. London: Verso.

de Castro, E. V. (2014). *Cannibal Metaphysics: For a Post-structural Anthropology* (P. Skafish, Ed. and Trans.). Minneapolis, MN: Univocal Publishing.

Deleuze, G. (1997). What Children Say. In *Essays Critical and Clinical* (D. W. Smith & M. A. Greco, Trans.) (pp. 61–67). Minneapolis: University of Minnesota Press.

Deleuze, G., & Guattari, F. (1987). *A Thousand Plateaus. Vol. 1 Capitalism and Schizophrenia* (B. Massumi, Trans.). Minneapolis: University of Minnesota Press.

Deleuze, G., & Guattari, F. (1994). *What Is Philosophy?* (H. Tomlinson, Trans.). New York, NY: Columbia University Press.

DeLoughrey, E. (2014). Satellite Planetarity and the Ends of the Earth. *Public Culture, 26*(2), 257–280.

Demos, T. J. (2016a). *Animal Cosmopolitics: The Art of Terike Haapoja*. The Center for Creative Ecologies. https://creativeecologies.ucsc.edu/demos-haapoja/.

Demos, T. J. (2016b). *Against the Anthropocene: Visual Culture and Environment Today.* Berlin: Sternberg Press.

Denham, T., & Peter, W. (Eds.). (2007). *The Emergence of Agriculture: A Global View.* London: Routledge.

Derrida, J. (1995). *The Rhetoric of Drugs* (P. Kamuf, Trans.). In E. Weber (Ed.), *Jacques Derrida: Points...Interviews, 1974–1994* (pp. 228–254). Stanford, CA: Stanford University Press.

Derrida, J. (1997). *On Cosmopolitanism and Forgiveness* (M. Dooley & M. Hughes, Trans.). London and New York: Routledge.

Descola, P. (2013). *Beyond Nature and Culture* (J. Lloyd, Trans. and M. Sahlins, Foreword). Chicago: University of Chicago Press.

Douzinas, C., & Žižek, S. (2010). *The Communist Hypothesis.* London: Verso.

Doyle, R. M. (2002). LSDNA: Rhetoric, Consciousness Expansion, and the Emergence of Biotechnology. *Philosophy and Rhetoric, 35*(2), 153–174.

Doyle, R. M. (2005). Hyperbolic: Divining Ayahuasca. *Discourse, 27*(1), 6–33.

Doyle, R. M. (2011). *Darwin's Pharmacy: Sex, Plants, and the Evolution of the Noösphere.* Seattle and London: University of Washington Press.

Dubreuil, L. (2006). Leaving Politics: Bios, Zoë, Life. *Diacritics, 36*(2), 83–98.

Esposito, R. (2008). *Bíos: Biopolitics and Philosophy* (T. Campbell, Trans. with introduction). London and Minneapolis: University of Minnesota Press.

Esposito, R. (2012). *Third Person: Politics of Life and the Philosophy of the Impersonal* (Z. Hanafi, Trans.). Cambridge, UK: Polity Press.

Falb, D. (2015). Epstemologies of Art in the Anthropocene. In C. Behnke, C. Kastelan, V. Knoll, & U. Wuggenig (Eds.), *Art in the Periphery of the Center* (pp. 302–317). Berlin: Sternberg Press.

Foucault, M. (2003). *Society Must Be Defended: Lectures at the Collége de France, 1975–1976* (M. Bertani & A. Fontana, Eds. and D. Macey, Trans.). New York: Picador.

Francis, P. (2015). *Laudato Si': On Care for Our Common Home.* https://laudatosi.com/watch.

Galloway, T. S., Cole, M., & Lewis, C. (2017, April 20). Interactions of Microplastic Debris Throughout the Marine Ecosystem. *Nature, Ecology & Evolution, 1*, 116.

Gehlen, A. (1980). *Man in the Age of Technology.* New York, NY: Columbia University Press.

Gibbons, A. (2011). A New View of the Birth of Homo Sapiens. *Science, 331*, 392–394.

Gilroy, P. (2004). *After Empire: Multiculture or Postcolonial Melancholia.* London: Routledge.

Groot, T. V. M., Bruins, E., & Breeuwer, J. A. J. (2003). Molecular Genetic Evidence for Parthenogenesis in the Burmese Python, *Python molurus bivittatus. Heredity, 90*, 130–135. https://doi.org/10.1038/sj.hdy.6800210.

Grosz, E. (2008). *Chaos, Territory, Art: Deleuze and the Framing of the Earth.* New York: Columbia University Press.

Grosz, E. (2011). *Becoming Undone: Darwinian Reflections on Life, Politics, and Art.* Durham: Duke University Press.

Grosz, E. (2017). *The Incorporeal: Ontology, Ethics, and the Limits of Materialism.* New York, NY: Columbia University Press.

Gržinić, M., & Tatlić, Š. (2014). *Necropolitics, Racialization, and Global Capitalism: Historialciation of Biopolitics and Forensics of Politics, Art, and Life.* Lanham, Boulder, New York, and London: Lexington Books.

Hammer, M. F., Woerner, A. E., Mendez, F. L., Watkins, J. C., & Wall, J. D. (2011, September 13). Genetic Evidence for Archaic Admixture in Africa. *Proceeding of the National Academy of Sciences in the United States, 108*(37), 15123–15128.

Hannah, D., & Krajewski, S. (2015). *Placing the Golden Spike: Landscapes of the Anthropocene.* Milwaukee, WI: Publication Studio—Jank Editions.

Haraway, D. (2016a). *Staying with the Trouble: Making Kin in the Chthulucene.* Durham, NC: Duke University Press.

Haraway, D. (2016b). Staying with the Trouble: Anthropocene, Capitalocene, Chthulucene. In J. Moore (Ed.), *Anthropocene or Capitalocene? Nature, History, and the Crisis of Capitalism* (pp. 34–76). San Francisco, CA: PM Press.

Harman, G. (2005). *Guerilla Metaphysics: Phenomenology and the Carpentry of Things.* Chicago, IL: Open Court.

Harvey, D. (2011). The Future of the Commons. *Radical History Review, 109*(Winter), 101–107.

Heidegger, M. (1968). *What Is Called Thinking?* (F. D. Wieck & J. G. Gray, Trans.). New York, Evanston and London: Harper and Row.

Helmreich, S. (2011). What Was Life? Answers from Three Limit Biologies. *Critical Inquiry, 37*(4), 671–696.

Hird, J. M., & Yusoff, K. (2016). Subtending Relations: Bacteria, Geology and the Possible. In A. Avanessian & S. Malik (Eds.), *Genealogies of Speculation: Materialism and Subjectivity Since Structuralism* (pp. 319–342). London and New York: Bloomsbury.

Hornborg, A. (2015). The Political Ecology of the Technocene: Uncovering Ecologically Unequal Exchange in the World-System. In C. B. C. Clive Hamilton & F. Gemmene (Eds.), *The Anthropocene and the Global Environmental Crisis* (pp. 57–69). London: Routledge.

Hornborg, A. (2017). Artifacts Have Consequences, Not Agency: Toward a Critical Theory of Global Environment History. *European Journal of Social Theory, 29*(1), 95–110.

Ingold, T. (2013). Prospect. In T. Ingold & G. Palsson (Eds.), *Biosocial Becomings: Integrating Social and Biological Anthropology* (pp. 1–20). Cambridge: The Edinburgh Building, Cambridge University Press.

International Geosphere-Biosphere Programme. (2015). *The Great Acceleration.* http://www.igbp.net/globalchange/greatacceleration.4.1b8ae20512d-b692f2a680001630.html.

jagodzinski, j. (2008). *Television and Youth Culture: Televised Paranoia*. London and New York: Palgrave Macmillan.

jagodzinski, j. (2010). *Visual Art and Education in an Era of Designer Capitalism*. London and New York: Palgrave Macmillan.

jagodzinski, j. (2014). Life in Art|Art in Life: Bioart Ethics within the Anthropocene. In H. Sederholm (Ed.), *Synnyt/Origins: Finnish Studies in Art Education Special Issue (Bio/Art/Education)* (Vol. 3, pp. 13–25). https://wiki.aalto.fi/display/Synnyt.

jagodzinski, j. (2017). Bateson with Guattari: Ethico-Aesthetic Responsibility in the Anthropocene; or, What Is an Artisan to Do? In N. Bateson & M. Witkowska-Jaworska (Eds.), *Towards an Ecology of Mind: Batesonian Legacy Continued* (pp. 187–209). Dabrowa Górnicza: USB, Scientific Publishing, University of Dabrowa Górnicza.

Jaspers, K. (1953). *The Origin and Goal of History* (M. Bullock & Michael, Trans.) (1st English ed.). London: Routledge and Keegan Paul.

Jay, M. (1993). *The Denigration of Vision in Twentieth-Century French Thought*. Berkeley: University of California Press.

Jaynes, J. (1976). *The Origin of Consciousness in the Breakdown of the Bicameral Mind*. Toronto, ON: University of Toronto Press.

Jazeel, T. (2011). Spatializing Difference Beyond Cosmopolitanism: Rethinking Planetary Futures. *Theory, Culture & Society, 28*(5), 75–97.

Jenkins, J. (2012). Nature Awareness and Panpsychic Ritual Gratitude: Revitalizing our Ancestral Heritage. *World Futures, 68*, 104–111.

Jeremijenko, N. (2015). Signs of (intelligent) Life. In D. Hannah & S. Krajewski (Eds.), *Placing the Golden Spike: Landscapes of the Anthropocene*. Milwaukee, WI: INOVA (Institute of Visual Arts).

Jeremijenko, N. (2017). New Experimentalism: Dehlia Hannah in Conversation with Natalie Jeremijenko. In R. Grusin (Ed.), *Anthropocene Feminism* (pp. 197–219). London and Minneapolis: University of Minnesota Library.

Johnson, E. R. (2017). At the Limits of Species Being: Sensing the Anthropocene. *South Atlantic Quarterly, 16*(2), 275–292.

Jones, M., & Brown, T. (2007). Selection, Cultivation and Reproductive Isolation: A Reconsideration of the Morphological and Molecular Signals of Domestication. In T. Denham, J. Iriarte, & L. Vrydaghs (Eds.), *Rethinking Agriculture: Archaeological and Ethnoarchaeological Perspectives* (pp. 36–49). Walnut Creek, CA: Left Coast Press.

Kac, E. (Ed.). (2007). *Signs of Life: Bio Art and Beyond*. Cambridge, MA: MIT Press.

Kanngieser, A., & Beuret, N. (2017). Refusing the World: Silence, Commoning and the Anthropocene. *The South Atlantic Quarterly, 116*(2), 363–380.

Kipnis, A. B. (2015). Agency Between Humanism and Posthumanism: Latour and His Opponents. *HAU: Journal of Ethnographic Theory, 5*(2), 43–58.

Kolbert, E. (2014). *The Sixth Extinction: An Unnatural History.* New York: Henry Holt.

Lacan, J. (1981). *The Seminar of Jacques Lacan. The Four Fundamental Concepts of Psychoanalysis.* Book XI (A. Sheridan, Trans.). J.-A. Miller (Ed.). London and New York: W.W. Norton & Company.

Landa, I. (2013). True Requirements or the Requirements of Truth? The Nietzchean Communism of Alain Badiou. *International Critical Thought,* 3(4), 424–443.

Landecker, H. (2011). Food as Exposure: Nutritional Epigenesis and the New Metabolism. *BioSocieties,* 6(2), 167–194.

Latour, B. (2009). Perspectivism: 'Type' or 'Bomb'? *Anthropology Today,* 25(2), 1–2.

Latour, B. (2017). *Facing Gaia: Eight Lectures on the New Climatic Regime.* Hoboken, NJ: Wiley.

Laurelle, F. (2013). *Philosophy and Non-philosophy* (T. Adkins, Trans.). Minneapolis: University of Minnesota Press.

Leroi-Gourhan, A. (1993). *Gesture and Speech* (A. B. Berger, Trans.). Cambridge, MA: MIT Press.

Levy, D. L., & Spicer, A. (2013). Contested Imaginaries and the Cultural Political Economy of Climate Change. *Organization,* 20(5), 659–678.

Lewis, S., & Maslin, M. A. (2015, March 12). Defining the Anthropocene. *Nature, 519,* 171–180. https://doi.org/10.1038/nature14258.

Leys, R. (2011). The Turn to Affect. *Critical Inquiry, 37*(3), 434–472.

López Petit, S. (2011). What If We Refuse to Be Citizens? A Manifesto for Vacating Civic Order (S. Touza, Trans.). *Borderlands e-Journal.* http://www.borderlands.net.au/vol10no3_2011/lopez-petit_refuse.pdf.

Lordkpanidze, D., Ponce de León, M. S., Margvelashvili, A., Rak, Y., Rightmire, G. P., Vekua, A., & Zollikofer, C. P. (2013, October 18). A Complete Skull from Dmanski, Georgia, and the Evolutionary Biology of Early Homo. *Science, 342,* 326–331.

Lynas, M. (2011). *The God Species: Saving the Planet in the Age of Humans.* London: Fourth Estate.

Malm, A. (2013). Sea Wall Politics: Uneven and Combined Protection of the Nile Delta Coastline in the Face of Sea Level Rise. *Critical Sociology, 39,* 803–832.

Malm, A. (2016). *Fossil Capital: The Rise of Steam Power and the Roots of Global Warming.* London: Verso.

Malm, A., & Hornborg, A. (2014). The Geology of Mankind? A Critique of the Anthropocene Narrative. *The Anthropocene Review, 1,* 62–69.

Mann, C. (2006). *1491: New Revelations of the Americas Before Columbus.* New York: Vintage Books.

Mann, C. (2011). *1493: How the Ecological Collision of Europe and the Americas Gave Rise to the Modern World.* London: Granta Books.

Margulis, L., & Sagan, D. (1995). *What Is Life*. Berkeley and Los Angeles: University of California Press.

Massumi, B. (2002). *Parables of the Virtual: Movement, Affect, Sensation*. Durham, NC: Duke University Press.

Massumi, B. (2014). *What Animals Teach Us About Politics*. Durham, NC and London: Duke University Press.

Massumi, B. (2017). Virtual Ecology and the Question of Value. In E. Hörl & J. Burton (Eds.), *General Ecology: The New Ecological Paradigm* (pp. 345–373). London, Oxford, New York, New Delhi, and Sydney: Bloomsbury.

Mbembe, A. (2003). Necropolitics. *Public Culture, 15*(10), 11–40.

McBrien, J. (2016). Accumulating Extinction: Planetary Catastrophism in the Necrocene. In J. Moore (Ed.), *Anthropocene or Capitalocene? Nature, History, and the Crisis of Capitalism* (pp. 116–137). San Francisco, CA: PM Press.

Michael, P. (2016, November 25). Queensland: Thangedl Elder Warwick 'Chook' Davis Is the Last of His Kind. *The Courier Mail.com.au*. http://www.couriermail.com.au/news/queensland/thangedl-elder-warwick-chook-david-is-the-last-of-his-kind/news-story/0f00c7ea1b2f836f80f0a6d49f7c7815.

Moore, H. L. (2011). *Still Life: Hopes, Desires and Satisfactions*. Cambridge, UK and Malden, MA: Polity Press.

Moore, H. L. (2012). Avatars and Robots: The Imaginary Present and the Socialities of the Inorganic. *The Cambridge Journal of Anthropology, 30*(1), 18–63.

Moore, H. L. (2014). Living in Molecular Times. In B. Blaagaard & I. Van de Tuin (Eds.), *The Subject of Rosi Braidotti: Politics and Concepts* (pp. 47–55). London, New Delhi, New York, and Sydney: Bloomsbury.

Moore, J. W. (2015). *Capitalism in the Web of Life: Ecology and the Accumulation of Capital*. London and New York: Verso.

Moore, J. W. (Ed.). (2016a). *Anthropocene or Capitalocene? Nature, History, and the Crisis of Capitalism*. Oakland, CA: PM Press.

Moore, J. W. (2016b). The Rise of Cheap Nature. In J. Moore (Ed.), *Anthropocene or Capitalocene? Nature, History, and the Crisis of Capitalism* (pp. 78–115). San Francisco, CA: PM Press.

Mora, C., Tittensor, D. P., Adi, S., Simpson, A. G. B., & Worm, B. (2011). How Many Species Are There on Earth and in the Ocean? *PLoS|Biology, 9*(8), e1001127. https://doi.org/10.1371/journal.pbio.1001127.

Morton, T. (2013). *Hyperobjects: Philosophy and Ecology After the End of the World*. Minneapolis: University of Minnesota Press.

Nancy, J.-L. (1997). *The Sense of the World* (J. S. Librett, Trans.). Minneapolis: University of Minnesota Press.

Nancy, J.-L. (2000). *Being Singular-Plural* (J. S. Librett, R. Richardson & A. O'Byrne, Trans.). Stanford, CA: Stanford University Press.

Negarestani, R. (2008). *Cyclopedia: Complicity with Anonymous Materials*. Melbourne: re.press.

Negarestani, R. (2013a). The Labor of the Inhuman, Part 1: Human. *e-flux Journal, 52,* 1–10.

Negarestani, R. (2013b). The Labor of the Inhuman, Part 2: The Inhuman. *e-flux Journal, 53,* 1–13.

Neimanis, A., & Walker, R. L. (2014). Weathering: Climate Change and the "Thick Time" of Transcorporeality. *Hypatia, 29*(3), 558–575.

Newman, S. (2008). In the End: Thermodynamics and the Necessity of Protecting the Natural World. In S. Newman (Ed.), *The Final Energy Crisis* (pp. 309–312). London: Pluto Press.

Niewöhner, J. (2011). Epigenesis: Embedded Bodies and the Molecularization of Biography and Milieu. *BioSocieties, 6*(3), 279–298.

Noë, A. (2010). *Out of Our Heads: Why You Are Not Your Brain, and Other Lessons from the Biology of Consciousness.* New York: Hill and Wang.

Osborne, P. (2013). *Anywhere or Not at All: Philosophy of Contemporary Art.* London: Verso.

Osborne, P. (2014, March/April). The Postconceptual Condition or, the Cultural Logic of High Capitalism Today. *Radical Philosophy, 184,* 19–27.

Oosthoek, J., & Gillis, B. K. (2005). Humanity at the Crossroads: The Globalization of Environmental Crisis. *Globalizations, 2*(3), 283–291.

Ostberg S., Boysen, L., Schaphoff, S., Lucht, W., & Gerten, D. (2018). The Bioshpere Under Potential Paris Outcomes. *Earth's Future, 6*(1): 23–39.

Papadopoulous, D. (2010). Insurgent Posthumanism. *Ephemera: Theory & Politics in Organization, 10*(2), 134–151.

Parikka, J. (2010). *Insect Media: An Archaeology of Animals and Technology.* Minneapolis: Minnesota University Press.

Parisi, L. (2003). *Abstract Sex: Philosophy, Biotechnology and the Mutations of Desire.* London: Continuum Books.

Parks, L. (2005). *Cultures in Orbit: Satellites and the Televisual.* Durham and London: Duke University Press.

Parks, B. C., & Timmons, R. J. (2007). *A Climate of Injustice: Global Inequality, North-South Politics and Climate Policy.* Cambridge, MA: MIT Press.

Parr, A. (2009). *Hijacking Sustainability.* Cambridge, MA: MIT Press.

Parr, A. (2013). *The Wrath of Capital: Neoliberalism and Climate Change Politics.* New York: Columbia University Press.

Parr, A. (2017). *Birth of a New Earth: The Radical Politics of the Environment.* New York: Columbia University Press.

Payne, S. J. (1991). *Burning Bush: A History of Australia.* New York: Holt.

Payne, S. J. (1995). *World Fire: The Culture of Fire on Earth.* New York: Holt.

Pirici, A., & Voinea, R. (2015). *Manifesto for the Gynecene – Sketch of a New Geological Era.* https://www.digitalmanifesto.net/manifestos/205/.

Povinelli, E. A. (2012). After the Last Man: Images and Ethics of Becoming Otherwise. *e-flux Journal, 35.* http://www.e-flux.com/journal/35/.

Povinelli, E. A. (2017a). The Three Figures of Geontology. In R. Grusin (Ed.), *Anthropocene Feminism* (pp. 49–64). London and Minneapolis: University of Minnesota Library.

Povinelli, E. A. (2017b). The Ends of Humans: Anthropocene, Autonomism, Antagonism, and the Illusions of Our Epoch. *The South Atlantic Quarterly, 116*(2), 293–310.

Rancière, J. (2004). *The Politics of Aesthetics: The Distribution of the Sensible* (G. Rockhill, Trans. and Introduced). London and New York: Continuum.

Raupach, M. R., & Canadell, J. G. (2010). Carbon and the Anthropocene. *Current Opinion in Environmental Sustainability, 2*, 210–218.

Read, J. (2017). Anthropocene and Anthropogenesis: Philosophical Anthropology and the Ends of Man. *The South Atlantic Quarterly, 116*(2), 257–273.

Rockström, J., Steffen, W., Noone, K., Persson, Å., Stuart III Chapin, F., Lambin, E., & Lenton, T. M. (2009a). Planetary Boundaries. *Ecology and Society, 14*(2), 32. https://www.ecologyandsociety.org/vol14/iss2/art32/.

Rockström, J., Steffen, W., Noone, K., Persson, Å., Stuart III Chapin, F., Lambin, E., & Lenton, T. M. (2009b). A Safe Operating Space for Humanity. *Nature, 461*(24), 472–475.

Rouvroy, A., & Stiegler, B. (2016). The Digital Regime of Truth: From the Algorithmic Governmentality to a New Rule of Law (A. Nony & B. Dillet, Trans.). *La Deleuziana, 6*(3), 6–27. http://www.ladeleuziana.org/wp-content/uploads/2016/12/Rouvroy-Stiegler_eng.pdf.

Roy, Oliver (2011). *Holy Ignorance: When Religion and Culture Part Ways* (R. Schwartz, Trans.). New York: Columbia University Press.

Satterthwaite, D. (2009). The Implications of Population Growth and Urbanization for Climate Change. *Environment & Urbanization, 21*, 545–567.

Scheffran, J., Brzoska, M., Kominek, J., Link, P. M., & Schilling, J. (2012, May 18). Climate Change and Violent Conflict. *Science, 336*(6083), 869–871. https://doi.org/10.1126/science.1221339.

Scott Powell, N., Larsen, R. K., & van Bommel, S. (2014). Meeting the 'Anthropocene' in the Context of Intractability and Complexity: Infusing Resilience Narratives with Intersubjectivity. *Resilience: International Policies, Practices and Discourses, 2*(3), 135–150.

Shiva, V. (2016). *Who Really Feeds the World: The Failures of Agribusiness and the Promise of Agroecology*. Berkeley, CA: North Atlantic Books.

Spivak, G. C. (2003). *Death of a Discipline*. New York: Columbia University Press.

Steffen, W., Broadgate, W., Deutsch, L., Gaffney, O., & Ludwig, C. (2015). The Trajectory of the Anthropocene: The Great Acceleration. *The Anthropocene Review, 2*(1), 81–98.

Stengers, I. (2010). *Cosmopolitics 1* (R. Bononno, Trans.). Minneapolis: University of Minnesota Press.

Stengers, I, (2011). *Cosmopolitics 11* (R. Bononno, Trans.). Minneapolis: University of Minnesota Press.

Stengers, I. (2015). *In Catastrophic Times: Resisting the Coming Barbarism* (A. Goffey, Trans.). London: Open Humanities Press.

Stephens, E., & Sprinkle, A. (n.d.). Ecosex Manifesto. *SexEcology.* http://sexecology.org/research-writing/ecosex-manifesto/.

Stiegler, B. (1998). *Technics and Time, 1: The Fault of Epimetheus.* Stanford, CA: Stanford University Press.

Stringer, C. (2012, May 3). What Makes a Modern Human? *Nature, 485,* 33–35.

Szeman, I. (2007). System Failure: Oil, Futurity, and the Anticipation of Disaster. *The South Atlantic Quarterly, 106*(4), 805–823.

Szeman, I. (2017). Introduction: Pipeline Politics. *The South Atlantic Quarterly, 116*(2), 402–407.

Szerszynski, B. (2010a). Reading and Writing the Weather: Climate Technics and the Moment of Responsibility. *Theory, Culture & Society, 27*(2–3), 9–30.

Szerszynski, B. (2010b). The End of the End of Nature: The Anthropocene and the Fate of the Human. *The Oxford Literary Review, 34*(2), 165–184.

Szerszynski, B. (2014). *The Anthropocene and the Memory of the Earth.* https://osmilnomesdegaia.files.wordpress.com/2014/11/bronislaw-szerszynski.pdf.

Szerszynski, B. (2015). *The Anthropocene and the Memory of the Earth.* https://www.youtube.com/watch?v=oqr6bH_U0-o.

Szerszynski, B. (2016a). Planetary Mobilities: Movement, Memory and Emergence in the Body of the Earth. *Mobilities, 11*(4), 614–628.

Szerszynski, B. (2016b). The Anthropocene Monument: On Relating Geological and Human Time. *European Journal of Social Theory, 20*(1), 111–131.

Szerszynski, B. (2017). Gods of the Anthropocene: Geo-Spiritual Formations in the Earth's New Epoch. *Theory, Culture & Society, 34*(2–3), 253–275.

Tauna, N. (2008). Viscous Porosity: Witnessing Katarina. In S. Alaimo & S. Hekman (Eds.), *Material Feminisms* (pp. 188–213). Bloomington: Indiana University Press.

Thacker, E. (2011). Necrologies; or, the Death of the Body Politic. In P. Ticineto Clough & C. Willse (Eds.), *Beyond Biopolitics. Essays on the Governance of Life and Death* (pp. 139–162). Durham and London: Duke University Press.

Toscano, A. (2007, September 9). *The Disparate: Ontology and the Politics of Simondon.* Paper delivered at the Society for European Philosophy/Forum for European Philosophy annual Conference, University of Sussex. http://www.after1968.org/app/webroot/uploads/Toscano_Ontology_Politics_Simondon.pdf.

Turner, B. S. (2016). Ritual, Belief and Habituation: Religion and Religions from the Axial Age to the Anthropocene. *European Journal of Social Theory, 20*(1), 132–145.

van de Pluijm, B. (2014). Hello Anthropocene, Goodbye Holocene. *Earth's, 2*, 566–568. https://doi.org/10.1002/2014ef000268.

Virno, P. (2009). Natural-Historical Diagrams: The 'New Global' Movement and the Biological Invariant (A. Toscano, Trans.). *Cosmos and History: The Journal of Natural and Social Philosophy, 5*(1). http://www.cosmosandhistory.org/index.php/journal/article/view/129/238.

Virno, P. (2015). *When the Word Becomes Flesh: Language and Human Nature* (G. Mecchia, Trans.). Los Angeles: Semiotext(e).

Wallerstein, I. M. (1976). *The Modern World-System 1: Capitalist Agriculture and the Origins of the European World-Economy in the Sixteenth Century*. New York: Academic Press.

Wark, M. (2015, December 28). Barbaris or Barbarism? *Public Seminar*. http://www.publicseminar.org/2015/12/stengers/#.WZcLMmQjFkY.

Whitehead, A. N. (1978). D. R. Griffin & D. W. Sherburne (Eds.), *Process and Reality: An Essay in Cosmology – Corrected Edition*. New York: Free Press.

Woods, D. (2014). Scale Critique for the Anthropocene. *Minnesota Review (New Series), 83*, 133–142.

Woodward, B. (2013). *On an Undergrounded Earth: Towards a New Geophilosophy*. Brooklyn, NY: Punctum Books.

Yoshida, S., Hiraga, K., Takehana, T., Taniguchi, I., Yamaji, H., Maeda, Y., ... Ocla, K. (2016, March 11). A Bacterium That Degrades and Assimilates Poly(ethylene terephthalate). *Science, 351*(6278), 1196–1199. https://doi.org/10.1126/science.aad6359.

Yusoff, K. (2013). Geologic Life: Prehistory, Climate, Futures in the Anthropocene. *Environmental and Planning D: Society and Space, 31*(5), 779–795.

Yusoff, K., & Gabrys, J. (2011). Climate Change and the Imagination. *Wiley Interdisciplinary Review: Climate Change, 2*(4), 516–534.

Yusoff, K., Grosz, E. A., Clark, N., Soldanha, A., & Nash, C. (2012). Geopower: A Panel on Elizabeth Grosz's Chaos, Territory, Art: Deleuze and the Framing of the Earth. *Environment and Planning D Society and Space, 30*(5), 971–988.

Zwillinger, R. (1998). *The Dispossessed: Living with Multiple Chemical Sensitivities*. Paulden, AZ: Dispossessed Project.

Zylinska, J. (2014). *Minimal Ethics for the Anthropocene*. London: Open Humanities Press.

PART I

Capitalist Framing

Image I.1   Gas pump (Mia Feuer, watercolor)

CHAPTER 2

Struggles in the Planet Factory: Class Composition and Global Warming

Nick Dyer-Witheford

Introduction: The Planet Factory

The idea of the Anthropocene has been booming for over a decade now, with atmospheric carbon emissions posited as the signature of humanity's new, epochal, earth-changing status. And burgeoning more recently is a critical counter-inscription of climate change as the creation, not of the species as whole, but rather of a specific mode of production with a compulsive drive to accumulation, outcome of what some term a Capitalocene (Moore 2016). With deep fidelity to its Marxist roots, this eco-socialist current has generated intense, not to say sectarian, debates, pitting the dialecticians of capitalism's "metabolic rift" (Foster et al. 2010; Angus 2016) against proponents of "world ecology" (Moore 2015). Yet relatively neglected, though with some stellar exceptions (Abramsky 2010; Mitchell 2011; Malm 2016), is the class composition of struggles in and against global warming.

Class composition seems an unlikely, even inauspicious, concept to bring to climate crisis. The invention of *operaismo*, a militant Marxism that flared briefly through the giant automobile factories of

N. Dyer-Witheford (✉)
University of Western Ontario, London, ON, Canada

© The Author(s) 2018
j. jagodzinski (ed.), *Interrogating the Anthropocene*, Palgrave Studies in
Educational Futures, https://doi.org/10.1007/978-3-319-78747-3_2

mid-twentieth-century Italy, it is a heterodox complement to orthodox Marxism's claim that capital tends to deepen its "organic composition," the ratio of machines and raw materials in production relative to human labor. In counterpoint, *operaismo* theorists analyzed the changing capacities of workers to oppose capital, capacities determined both by their technical composition (i.e., the structuring of the labor process) and political composition (i.e., the level of class unity and organization). For *operaismo*, to "start with the struggles" was a partisan political methodology, aimed to assist in the interconnection and circulation of worker revolts through cycles of composition, decomposition, and recomposition.

How remote this sounds from the urgent universalisms and plangent fatalisms of the Anthropocene. The distance appears yet greater because, as Sara Nelson and Bruce Braun (2017) observe, both *operaismo* itself and then the *autonomia, autonomist* or *post-operaismo* tendencies that succeeded to its theoretical legacy has generally shown little interest in, and sometimes outright hostility to, environmentalism. And yet … the factory is still with us. Not just because, despite all talk of postindustrialism, installations of industrial production continue to abound, even if today they are more likely to be found in the Pearl River Delta, Mexican border zones or Delhi suburbs than in previous northern locations. Beyond that, if we take "factory" as denoting not a specific manufacturing apparatus, but rather the domain in which capital exercises command over its laboring subjects by means of advanced technologies, then, far from shrinking and fading away, the factory has since the mid-twentieth century got bigger; much bigger.

For, in the thinking of *operaismo* and its successors, emphasis on the factory as the epicenter of class struggle was soon superseded by the idea of the "social factory." To Marx's famous account of the circuit of capital, moving from the purchase of raw materials, machines, and labor production, to commodity exchange, to finance, from whence money is thrown back into more production, the social factory model added an additional circuit, partially exterior to the first, but feeding into it, that of the reproduction of labor power. In this circuit, the supply of the vital human component of capital, the worker, is renewed in households, schools, welfare offices, and media, institutions which then become sites for compositions of class very different from those of the heavy industry, and for new struggles against capital. Elements of this social factory thought, particularly the emphasis on the role of unwaged female labor in the reproduction of labor power (Federici 2012; Caffentzis 2013; Midnight Notes 1992) are drawn on by Jason Moore (2015) in his discussions of capital's appropriation of the free gifts of nature—but stripped of the crucial *operaismo* emphasis on class conflict.

Today we are in a "planet factory." To conceptualize this, we must add to idea of the social factory another circuit of capital, well described by eco-socialist theorists for over a decade, that of the reproduction of nature as capitalism's "tap and sink" (O'Connor 1998)—that is, a source of both free resources and free waste disposal. The issue of the Anthropocene is then how far, and in what ways capital, subsumes nature within its circuit of valorization, and by both calculated design and catastrophic accident, creates its own ecology, a "second nature," more, or less, habitable for humanity. We can refer to a "planet factory" in a double sense, both as naming the operation of a world-market crisscrossed by supply chains feeding in and out of the ever-escalating industrial production of commodities, and as designating the process by which what is produced (and destroyed) in these operations is nothing less than the biosphere itself—with global warming as, of course, exhibit one. What *operaismo* and *post-operaismo* thought distinctively bring to such analysis is a focus on how the planet factory arises from, and in turn transforms, global cycles of class struggle.

If we are to investigate the class composition of climate change, the first and central proposition must be that capitalism drives climate change, and the class that drives capitalism is the class of capitalists. The key role of capitalists is as the personified representatives of capital, that is to say, as the owners, directors, and rulers a system committed to continuous profitability, and hence to infinite growth. This should be distinguished from the related but subsidiary issue of the carbon consumption of the wealthy. A number of recent studies, focused on consumers direct and indirect greenhouse gas emissions (car purchases, driving habits, domestic heat, and light, etc.) clearly established that consumption of the rich accounts for far more global warming than that of the poor. To give only one example, Oxfam's (2015) "Extreme Carbon Inequality" report estimates that the "average footprint of someone in the richest 1% could be 175 times that of someone in the poorest 10%" (see Chancel and Piketty 2015; Jorgenson et al. 2017). In class composition analysis, however, this issue is subordinate to that of production emissions and the crucial role of the rich as the owners, financiers, and managers of a production system destructive of global ecology. In the conclusion to this paper we will return to the class of capitalists and their strategies in the face of the global climate chaos. But first we will look at the implication and imbrication within climate chaos of the exploited and alienated populations on whose employment and unemployment capital depends.

This chapter therefore sketches a conceptual diagram of class composition and climate warming that includes both capital's ongoing burning of fossil fuels and the rebound of solar radiation, the ascent of emissions and the descent of heat, displaying the conflicts igniting, and ignited by each.[1] On the ascending emissions side of the cycle, we begin with the class dynamics of capital's mining and oil industries; go on to the class clashes and armistices that created the Fordist factory and its mass production of automobiles; and then to the cognitive labor of a digitized capitalism whose computerized infrastructures and scientific labors at once identified global warming, aggravated its industrial effects, and fostered new movements of environmental protest. On the descending side of the cycle, where the trapped greenhouse heat descends to land and ocean with increasingly calamitous result, we then encounter the revolts of anti-extractivist movements originating in indigenous, nomadic and peasant communities; the migrations of the world's most desperate "surplus populations" fleeing climate-driven war and famine; and the travails of new proletarians beset with smog and rising sea waters in China's urban centers.

What this diagram charts in a series of snapshots is thus a series of contested moments in the interplay of carbon's climate cycle and capital's valorization circuit.[2] These moments are variably distributed across the globe. Like all factories, the planet factory has a division of labor among various industrial bays, machines, and installations, conducting different but connected operations. As theorist of unequal ecological exchange and uneven ecological footprints make clear, the populations most imperiled by global warming are often those least immediately involved in the emissions that cause it. Looking at the class composition of the planet factory, a factory owned and managed by capital, destroys narratives of climate change as the common problem of undifferentiated humanity, but also shows the contradictions among proletarians who are in, of and only sometimes against capital, and hence also often divided against each other. Our tour therefore concludes with speculation as to what new compositions of struggle might counter capital's double convulsion of class and climate.

MINERS AND ROUGHNECKS

In Nova Scotia, on Canada's Atlantic Coast, memorials to mining disasters are everywhere; the Drummond Colliery Disaster 1873 (60–70 deaths); the Foord Pit Explosion, 1880 (50 deaths); the Springhill Mine Disaster, 1891 (125 deaths); the Dominion No. 12 Colliery Explosion, 1917 (65 deaths); the Albion Mine Explosion, 1918 (88 deaths); the cable break at Sydney Mines, 1938 (20 deaths); another Springhill Explosion, 1956 (39 deaths); the Springhill Bump, 1958 (74 or 75 deaths); the Westray Coal Mine Explosion, 1992 (26 deaths). Such commemorative landscapes are common in coal country. Early in the twentieth-century deaths of miners in accidents in the UK and United States were annually numbered in the thousands; today, this scale of deaths persists in China, Russia, India, and elsewhere (McNeill and Engelke 2014, p. 11). On May 30, 2014 three hundred and one workers died in a massive explosion at Turkey's Soma coal mine caused by cost cutting and neglect of safety equipment, their charred and choked bodies pulled to the surface from two miles underground. Turkish trades unions declared a one-day general strike that, in the context of wider protests against an authoritarian government, attracted wide popular support.

Fossil fuels are the major source of carbon emissions. Their extraction starkly demonstrates the class relations, at once brutal and contradictory, at the root of global warming. For the industries that supply these fuels have been, and are, the site of some of capital's fiercest class conflicts, but also those in which the life-and-death dependence of the working class on jobs—even on jobs that kill—is most apparent. From the start of capital's industrial revolution, mines were the necessary complement and precondition to the mechanized factory. Early versions of Watt's steam engines, the machine that arguably initiated the Anthropocene, were so inefficient that they could only be used in the mouths of mines where coal was literally lying around, for pumping and lifting—so that mines could be considered the first machine-driven factories.

They are also sites of some of capital's most tangible ecological despoliation. Mine tailings, ground subsidence and slag disfigure and poison coal vein landscapes; the surface or strip mining that now supplies some 40% of the world's coal can now, with advanced excavation equipment, gouge away the surface of the earth to a depth of fifty meters, dumping it into valleys and rivers (McNeill and Engelke 2014, p. 11). The extreme is "mountaintop removal"—blasting off the summits of ranges such as the Appalachians. Miners, and the women and children of the families that must absorb the hardness of their labor, live near mines, and bear the brunt of these ravages, but are dependent on mines for livelihood. Subterranean accidents make deep coal mining one of the most hazardous of capital's labor processes, but black lung disease (medically, pneumoconiosis, and silicosis) from years spent underground inhaling coal dust, a peculiarly vicious form of atmospheric pollution, kills far more, the numbers actually increasing with intensified mechanization. In the 1960s the fight waged by American miners against indifferent employers and corrupt unions for medical care and compensation for black lung victims precipitated groundbreaking campaigns for workplace health and safety legislation (Gottleib 2005, pp. 350–352), an example of industrial working class environmentalism.

Historically the dangers of mine work, its literally abysmal conditions, single-industry communities, and mine-capital's difficulties in supervising its underground workforce, made mining an archetypal scene of class conflict: The technical conditions of mine work gave mine workers a strong political composition. To this was added the all-important factor that, as coal became the carboniferous elixir of life for industrial production, capital as a whole was a rendered vulnerable to any blockage of supply. In his study of "carbon democracy" Timothy Mitchell argues that in the early twentieth-century miners' strike power did not just raise their wages, but also supported popular movements for issues such as the electoral franchise and an incipient welfare state; the "ability to interrupt the flow of energy had given organized labour the power to demand the improvements to collective life that had democratized Europe" (2011, p. 29). Before 1945 it was the mine, more even than the factory, that gave the industrial working class a capacity to quite literally turn off capital's lights.

However, mining also demonstrates how capital escapes struggles by what Beverly Silver terms technological and spatial "fixes"—resorting to new machines and new materials, and/or by shifting the location of production. Mitchell and others suggest advanced capital's mid-twentieth century shift from coal to oil as its major fossil fuel source, and specifically to the oil fields of Middle East, was driven not only by the increasing importance of the automobile, but also by industrial conflicts: Winston Churchill reportedly switched the British navy from coal to oil power from fear of strikes. Pumping oil had its own tough "roughneck" labor, but it required a smaller workforce than mining, and one that could be kept under close managerial supervision; liquid oil was more cheaply transported by pipeline than bulk coal by train or ship, and pipelines could be designed on a "dendritic" model evading choke points, such as ports or rail termini that might be blocked by strikers (Mitchell 2011, p. 30).

Eventually, the oil fix to the coal miner problem would itself be undone as Middle Eastern fields were disturbed by wars, insurgencies, populist oil nationalizations, and even by oil worker strikes. Oil corporations renewed the search for "energy security" in extreme locations—the seafloor, tropical forests, the Arctic, sites again supposedly immune from political disruptions. A yet more machinic apparatus of carbon fuel extraction took form, typified by vast post-human edifices of offshore drilling platforms, the largest of which stand "over 600 meters above the water, rivaling the tallest skyscrapers [and] dotting the North Sea, the Gulf of Mexico and the coasts of Brazil, Nigeria, Angola, Indonesia, and Russia, among others," a site of "massive pollutant leaks, accidents and blowouts," and simultaneously of great danger for workers: the disaster of Deepwater Horizon in 2010 exemplified both aspects (McNeill and Engelke 2014, p. 14).

The adoption of oil, and later of natural gas, did not, however, lead to any decline in the demand for coal, which as Andreas Malm (2016, p. 357) reminds us throughout the twentieth century "never disappeared from the calculus": Indeed, coal production grew relentlessly, especially in newly industrializing countries. Rather, miner militancy "provided one more incentive for the diversification, multiplication and expansion" of fossil fuel (Malm 2016, p. 357). But oil and gas did disempower miners. By a bitter irony, some of their last great struggles were to preserve the jobs that are in so many ways deadly and devastating. The British miner strike of 1984, fought by entire coal communities, women, and men, against the paramilitary state power of Margaret Thatcher (whose clarity about global warming seems to have partly derived from class hatred of coal workers) is a case in point.

What does this tell us about the class basis of the Anthropocene? Reliance on fossil energy is driven by capital's need to substitute machines for workers. Yet the very sites at which these fuels are extracted have themselves been sites of hard labor and class conflict. Miners and oil rig workers are makers of the Anthropocene, in so far as their labor power has been essential to fossil capital. They have also been among its first victims. Fossil capital creates (and then discards) workforces that are in, of, and against it: in it, as mining, oil, and fracking produce the devastated landscapes they inhabit; against, as they fight the exploitation of their labor and destruction of their lives; of it, insofar as they depend on the wages capital pays, identify with the work they do, and are often without other good options. No movement against global warming can afford to ignore these contradictions.

MASS WORKER

On October 7, 1913 the Highland Park Ford Plant, near downtown Detroit, became the first automobile production facility in the world to implement the moving assembly line. The conveyor belt production system was propelled by electricity generated from coal dust and gas-driven turbines; later, it would draw power from Ford's even more massive plant on Rouge River, fuelled with coal transported on his privately owned railways from his mines in Kentucky. On July 18, 2013 the city of Detroit filed bankruptcy, its factory complexes gutted and decaying, many of them disassembled and shipped to Mexico where labor was cheaper; what had become the industrial capital of the United States was now a site of "ruin-porn," supplying apocalyptic background for artistic vampire movies. Between these two events, Detroit was the hub of a century of class wars, deals, and broken truces that would not only make and then unmake the power of the "mass worker" but also determine the future of earth's atmosphere.

According to Malm (2016) fossil fuel was essential to capital in establishing the factory as an "abstract space," set apart from its surroundings, dedicated to the extraction of surplus value, and operating according to an "abstract time" removed from all organic rhythms. He argues that for early British industrialists, the key consideration bonding capital to coal was not simply that it drove the steam engines that were factories' prime mover. It was rather that the factory, as site of rigorously, disciplined labor required an energy source whose availability could be counted on and that could be transported to the urban centers where cheap labor congregated. Coal met these criteria in a way that alternatives—notably water power, for a time its serious competitor—could not guarantee. Class struggle thus lies at the root of climate change: it is capital's need to control labor, impose work, and replace workers at its own optimal rhythm that drives capital to coal.

This is the story Malm traces from the factory's genesis to its consolidation in the smoke-belching mills of the nineteenth century, before then leaping to the current global warming crisis. His narrative, however, jumps over a crucial part of the story. For if the coal mine, steam engine, and early industrial factory initiate the Anthropocene, what launches it into a rocket-like takeoff is what (well after Ford's death) became known as Fordism. This is not just because Ford's factories, by applying electricity to the moving assembly line and the precision machining of interchangeable parts, raised the carbon-propelled mechanization of production to a new level of intensity. Nor was it only because the commodity rolling off those assembly lines—the automobile—would become the centerpiece of a massive new cycle of carbon energy consumption. It was rather because Fordism cemented these two elements in a fateful pact between capital and its workforce.

This truce was preceded by class war. Ford's assembly line created a deskilled, routinized labor process whose pace, monotony, danger, and clamor was immediately felt by workers as inhuman, and resulted in huge labor turnover. It was as a solution to this problem of worker defection, and not from prescient wisdom about turning workers into consumers, that Ford introduced his famous $5 dollar day wage. But class conflict continued to rage in Ford's factories. By concentrating semiautomated, homogenized labor these factories created prime conditions for worker solidarity. The same continuity and systematized rationalization of work made possible by assembly lines meant that, if those lines were halted by strikes or sabotage, towering quantities of fixed capital would be paralyzed. It was these discoveries, famously put into action in the 1936 "sit-down strikes" in the auto-plants of Flint Michigan, that created the power of what *operaismo* would dub "the mass worker."

The postwar settlements of autoworkers unions with the big three Detroit automakers, in which the latter traded wage increases for, if not labor peace, at least orderly collective bargaining, became the benchmark for a new era of capitalism in which regularly rising worker incomes would be translated into consumption power to purchase the commodities streaming off assembly lines, in a virtuous circle that seemed a formula for perpetual economic growth. Fordism was primarily a white, male pact. Women were cast in the role of homemakers spending the bread-winner's regular pay packet. In 1967 sections of Detroit burned in mainly black uprisings. Nonetheless, the Fordist deal yielded a historically unprecedented increase in material living standards of the working class, however unevenly distributed.

It was also a massively CO_2 generating deal, even if the participants neither noticed nor cared. Not only did it consolidate the model of largely fossil-powered, Fordist manufacture. It also inaugurated, as the fruits on working-class wage victories, mass fossil-fuel powered consumption. This had two tidily gendered pillars. One was the personal, petrol-powered automobile, the ur-product of Fordist assembly line, its hegemony in the United States secured by careful corporate sabotage of public transport and electric car options. The other was the electrified home, now for the first time in reach of working-class families. The marriage of appliance-equipped suburban housing with highway construction and automobile commuting would make the United States, and then the European, economies into a gargantuan oil-sucking, coal-heaping, carbon-emitting machines. It is the start of Fordism's "thirty glorious years," the Golden Age of capital from 1945 to 1975, that see the graph of carbon emissions change from a slowly rising line to sudden hockey stick curve lancing into sky, the "Great Acceleration" (McNeill and Engelke 2014), which some declare the true advent of the Anthropocene.

That upward trajectory of emissions would continue; Fordism, however, did not. In the 1970s, as increased international competition cut into profits, major corporations chaffed at wage increases and welfare state taxation. With the governments of Reagan and Thatcher, this translated to the full-scale political offensive against labor known as neoliberalism. Of this offensive's many components—deflation, automation, anti-trade union legislation—of critical importance was the relocation of industrial plant to the developing world. Supply chains snaked out from the United States and Europe to seek sites for cheap offshore industrial production with low wages and little environmental regulation. Contrary to easy talk of deindustrialization, the overall share of industrial work in global employment remained relatively steady; indeed, total manufacture output more than tripled over the four decades from 1970 to 2011, while world population did not even double (United Nations 2013). But industrial work *was* trans-nationally reorganized, shifting from the former core of the capitalist system to its one-time periphery: Auto-production, shipyards, textile factories, electronics plants, and chemical processing went, first to special export zones in Central America, Eastern Europe, and Asia, then in China (Silver 2003).

This process, the real founding of the planet factory, had several dramatic consequences for global warming. One was the creation in new industrial, and heavily carbon-emitting, regions such as China, of new industrial proletariats that, as will see later, would prove as militant as their Fordist predecessors, and as committed to improving their very much lower living standards. Another was a vast expansion of transport activities, as container shipping, air-cargo, and road corridors sped commodities across the world, further intensifying carbon emissions. And a third was the political decomposition of mass worker of the global North: Falling or stagnant wages, layoffs, trades union defeats, and the collapse of great industrial regions into rustbelt shells, would eventually result in reactionary populist earthquakes with major consequences for climate politics. But before our narrative arrives at that point, we turn to the moment capital found it had an atmospheric carbon problem.

COGNITIVE LABOR

At the UN Climate Conference in Copenhagen in 2009, all the talk, both in the conference rooms where heads of state bargained, and in the streets where protestors marched, was of carbon emissions, tipping points, and pollution trading schemes; capital's climate crisis was on full display. So too was its mutation from the days of steam engines and assembly lines to those of computers and networks. Two years before the Intergovernmental Panel on Climate Change (IPPC) had unequivocally declared global warming to be both real and anthropogenic—on the basis of detailed computer climate modeling. The climate scientists had fought vigorous battles with fossil fuel-funded global warming deniers, who on the eve of the conference attempted to discredit their findings by releasing hacked and supposedly embarrassing email. At the Copenhagen meeting, a new type of social activist, climate protestors, organized with cell phones and multimedia news platforms, only to find their laptops and mobiles monitored and jammed by security forces. Global warming was being discovered, and responses to it bargained, and fought out within a new composition of technologies and labor that had been incubating since 1945, but only matured at the turn of the millennium.

In the mainstream narrative of the Anthropocene, global warming's slow industrial take off from the 1750s on, and the "great acceleration" from the 1950s, is followed at the start of the twenty-first century by a third phase in which scientific knowledge of its processes starts to inform public opinion begins. This was directly related to a change in the machinic and class composition of capital, for the scientists who for the first time reliably measure climate change and define its causes constitute an elite strata of techno-scientific labor working with the tools of "digital capitalism" (Schiller 1999). Computers and networks have their origins in the Second World War and Cold War, developed as cybernetic adjuncts of military planning, including that for nuclear war. As Bonneuil and Fressoz (2015, p. 59) explains, for the Pentagon, "Since the Cold War was global, the whole Earth became a strategic terrain to study"— outer space for satellites, atmosphere, and geomagnetism for ballistic missiles; oceanography for naval and logistics operations; polar ice for nuclear submarines; and so on.

It was from the digital systems built for these monitoring purposes that the data confirming the thesis of anthropogenic climate change was gathered. Paul Edward's (2010) describes the apparatus of weather stations, satellites, sensors, digitally archived records, and massive computer simulations on which comprehension of global warming rests. From within the institutions tending this "vast machine" (Edwards 2010) scientists such as Paul Crutzen, inventor of the term "Anthropocene," James Lovelock, creator of the "Gaia hypothesis," Roger Revelle, who tutored Al Gore on climate change and climatologist James Hansen notified capital it had a global heating problem, precipitating a series of international conclaves in Rio, Kyoto, Copenhagen, and Paris.

Digital, postindustrial capital also shaped the green movements of global North present at Copenhagen as both as lobbyists and protestors. Environmentalism has a long history, with roots in the wilderness preservation projects of capitalist elites, and a bad moment of adoption by fascist enthusiasts for blood and soil (Gottleib 2005). It was, however, fundamentally remade in the late twentieth century—Earth Day 1970 is the conventional landmark—as part of an explosion of new social forces. The deindustrialization of the global North was reshaping its workforce around techno-scientific or knowledge labor, attended to by a larger service sector, both increasingly mediated by technologies of the so-called "information revolution" as these diffused out from their military matrix into commercial use. This was the class composition of what *post-operaismo* theorists call "cognitive" or "immaterial" labor, engaged in intellectual, communicational, and affective, rather than industrial, tasks (Hardt and Negri 2000). Although "immaterial labor" is in many ways an unfortunate phrase, it does highlight how differently from the mass worker such a labor force might encounter a natural word, presented as a site of recreational and spiritual renewal, rather as an object that's more or less violent material transformation was a requirement of the wage.

The new environmentalism was initially part of the wave of social tumult in which countercultural revolts, and the anti-Vietnam war movement challenged military-industrial powers. Symptomatically, the West Coast of North America was a primary site of student and antiwar movements, information technologies, and environmental protest on issues from rainforest to whales, in whose cause green activists would carry out audacious protests, civil disobedience, and sabotage against corporate capital. In succeeding decades, however, the new forms of post-Fordist labor and production became increasingly assimilated into the normal operations of capital. The rebel experiments of early computer hackers became the industry of Silicon Valley whose "Californian ideology" (Barbrook and Cameron 1996) was fully compatible with the neoliberalism. In a parallel process, major sections of environmentalism were defanged by corporate donation and foundation funding, and adopted incremental, defensive, and market-based solutions to ecological devastation, alienated from the possibilities of alliance with industrial workers that had briefly blossomed in the 1970s by the "jobs versus the environment" opposition that intensified as neoliberal automation and offshoring took hold (Foster 1993).

It is a Marxist cliché to say that environmentalism is an urban-, middle- and upper-class movement. Nonetheless, by the time global warming was announced, large section of environmentalism *was* corporatized and professional-managerial in outlook (Klein 2014). At the same time, however, other, more radical strands were emerging. In the 1980s ecological justice movements in poor and racialized areas of United States and Europe, challenging the location of waste dumps, incinerators, and also the toxic dark side of Silicon Valley's postindustrial information technologies, sprang up. The eruption of an *alter-globalism* movement in the late 1990s connected environmental activism with larger issues of food sovereignty, urban planning, and sweated labor in the global South.

At Copenhagen, the divisions resulting from these changes were evident. While some ecological activists were on the streets under banners declaring "System Change Not Climate Change," mainstream green organizations were inside the conference, displaying a surprisingly passive acceptance of the official solutions to global warming (Hari 2010; Klein 2014). These solutions reflected the fact that climate change science was an offspring of the national security apparatus of the United States, which only a few years previously, had defeated its state socialism, and secured the global hegemony of neoliberal economics. It was therefore not surprising that the corrections proposed at Copenhagen were technocratic adjustments within the limits of the market system through schemes of emission trading and carbon offsets. Al Gore's 2006 film, *An Inconvenient Truth*, credited with inspiring wide public cognizance of global warming, but suggesting that it would be solved by enlightened corporate leadership and green consumerism, was a product of precisely this view. As the inadequacies of such an approach were measured by inexorably mounting emissions, both individual climate scientists and some green groupings would break from it, and fill the streets at official conferences with an angry protest. A yet greater sense of urgency would, however, come from very different zones of the planet factory.

ANTI-EXTRACTIVISTS

In October of 2010, ten months after of the failure of Copenhagen conference, thousands of social activists converged on the mountain city of Cochabamba in Bolivia for an event convened by that country's President, Evo Morales—the World People's Conference on Climate Change and the Rights of Mother Earth. Cochabamba had been the site of a historic victory against the privatization of water by a multinational company. It was also at the center of a region that had seen repeated conflicts between indigenous populations and fossil fuel corporations. Armed rebellion against oil drilling by the Huaorani, Amazonian forager hunters in Ecuador, had ended in forcible relocation from their traditional lands, while Peruvian eco-activists and indigenous farmers had fought the US-owned Yanacocha mine (Haller et al. 2007). The major resolution of the Cochabamba conference clearly situated climate change in the context of such struggles, denouncing the way "under capitalism, Mother Earth is converted into a source of raw materials."

As Michael Lowy writes, while many Marxist commentators criticize the mystical theology of "Mother Earth," "the powerful, radical anti-systemic social dynamic that... crystallized around these slogans" was undeniable (2015, p. 71). This dynamic is often dubbed "anti-extractivism," designating a wave of insurgencies against the operations of mining and oil corporations, part of the "environmentalism of the poor" (Martinez-Allier 2002) that defends natural habitats not for recreation or conservation but as a vital and sacred substrate of communal life. Anti-extractivism arises in what were once the peripheries of global capitalism, now ubiquitously penetrated and traversed by the infrastructures of energy corporations. While Latin America is a hub of such revolts, worldwide examples include the protracted, bloody low-level insurgency waged by communities of the Niger Delta against Shell BP, whose oilfields, and pipelines saturate rainforest wetlands with oil spills, acid rain, gas flares, and polluted air; Mongolian revolts against the environmental damage from China's coal mines; the armed resistance of indigenous Siberians against Russian oil extraction and guerrilla uprisings against oil companies in Papua, New Guinea (Haller 2007; McNeill and Engelke 2014). The protagonists are peasant, nomad, or indigenous populations in subsistence economies rapidly eroded by the expansion of the planet factory.

Anti-extractivist movements predate any scientific verification of global warming. Their immediate object is not prevention of climate change, but rather to halt the despoliation, toxification, and destabilization of the land by mining, pumping, or fracking, and its effect on water, forests, animals, and agriculture (though protesting communities may well be affected by drought, rising sea levels, or other of its climatic effects). However, since the "official" recognition of the carbon crisis, such protests have become supercharged with a new significance. Anti-extractivism movements threaten fossil fuel corporations with, at minimum, heightened costs of security, litigation and pacification, and, at maximum, abandonment of vast sunk capital projects, and even delegitimation of their entire industry. Thus although not *ab initio* against global warming, anti-extractivism has become articulated with it and now, in its ominous light, assume a new significance, and new adherents and allies.

Their characteristic forms of anti-extractivism action have therefore been hailed by climate change activists as a frontline defense against carbon emissions, a "blockadia" against global warming (Klein 2014). This connection has in turn aided the extension of anti-extractivism into the global North. Here it takes many forms, from protests against fracking in French vineyard and gold mining in northern Greece. The major manifestation is, however, "pipeline protests," such as those in Canada and the United States against the Keystone XL, Energy East, Northern Gateway, and Trans Mountain pipelines, and in Europe, against BP's Baku-Tbilisi-Ceyhan (BTC) project, which runs from Azerbaijan to Turkey's Mediterranean Coast (Szeman 2017). Participants include farmers and city dwellers fearing oil spills and green climate change activists. Still, in North America an especially important role continues to be played by indigenous peoples, for whom pipeline politics have become a focus of revived decolonization struggles.

Thus it was Sioux elders who in early 2016 established an encampment dedicated to cultural preservation and spiritual resistance to block the construction of Energy Transfer Partners' Dakota Access Pipeline on their traditional lands the northern United States. Thousands of people occupied the camp. In frequent clashes with police protestors were pepper sprayed and hosed down by water cannon in sub-zero temperatures; hundreds were arrested; at one point indigenous military veterans offered the camp armed support. While the primary objective of the protest was the protection of water from contamination by oil spills, Sioux spokesmen and indigenous commentators frequently linked the issue to climate change. President Obama gave the protest a brief victory when he denied a permit for construction of the pipeline but this was soon reversed by the incoming Trump administration, and the camp, weakened by winter conditions, was cleared. Indigenous communities across North America continued, however, to promise further resistance to the pipeline.

Anti-extractivism is a form of struggle very far indeed from the factory origins of *operaismo*, or indeed from the cyborg labor celebrated by *post-operaismo* theory. While it now involves a wide variety of participants, its indigenous and peasant origins are in what classic Marxism would regard as the residual traces of precapitalist social formations, defending land against capitalism's ever-ongoing processes of primitive accumulation and accumulation by dispossession. Such movements often divide the communities in which they arise, to whom energy corporations promise jobs and prosperity, and they are contentious even on the left. Anti-extractivist indigenous movements have come into conflict with regimes of Latin America's "pink tide" that regards revenues from oil, gas, and other extractive industries as necessary for development and poverty alleviation (Fuentes 2014). However, as the autonomists of Midnight Notes (2001) recognized at the time of the Zapatista revolt that galvanized alter-globalism movement (a revolt that, in its opposition to oil drilling in traditional territories was itself anti-extractivist), what is emerging in today's "fourth world war" of indigenous peoples colonial occupation should not be written down as some "primitivist" response to capitalism. It is rather a new synthesis bringing together elements of modernity and anti-modernity in rebellion against the logic of the planet factory, a rebellion that now plays a major role in radical politics against global warming.

MIGRANTS AND REFUGEES

In 2016, 5000 people drowned attempting to cross the Mediterranean from North Africa to Europe, an average of 14 a day; 3771 deaths had been recorded the year before. Over a third of a million successfully crossed, to reach, first, overcrowded refugee camps in Italy and Greece, and then attempt travel onwards into Europe, where their arrival caused political crisis. Many were fleeing a civil war in Syria in which about 400,000 had died since its outbreak in 2011. A scholarly study by Earth scientists at Columbia University published in 2015 pointed to evidence that a severe drought affecting Syria in 2007–2010, had resulted in an exodus from rural to urban areas, adding to the political instability that preceded the war. Those fleeing the drought went into "rapidly growing urban peripheries ... marked by illegal settlements, overcrowding, poor infrastructure, unemployment, and crime"; these areas "neglected by the Assad government" became "heart of the developing unrest" (Kelley et al. 2015, p. 3242). The likelihood of such a drought occurring was, the study estimated, doubled or tripled by global warming, and when it did occur exacerbated factors contributing to war, including "unemployment, corruption, and rampant inequality" (Kelley et al. 2015, p. 3241).

This suggestion of linkage between the Syrian war and global warming is controversial (Selby and Hulme 2015). Climate science can accurately ascertain the industrial activities that in varying quantities emit carbon emissions, and class composition theorists can then look at the social antagonisms driving and shaping these processes. When we turn from the class conflicts that send carbon emissions out into the atmosphere toward those ignited by the return to the earth of the heat trapped by those emissions, we face a more difficult task. It is impossible for a meteorologist to say if a given storm or drought is caused by global warming, though they may know that changing temperature of the atmosphere, land, and sea "weights the dice" toward more frequent extreme weather events. A similar problem, but of even greater magnitude, surrounds the political and social consequences of global warming, which are complex, diffuse and probabilistic.

No one has been more interested in this problem than the Pentagon. Since 2004 various branches of the US armed forces have issued a series of reports identifying the movements and unrests of populations displaced and stressed by the global warming as a danger to world order.

Though doubtless partly motivated by a need to find post-Cold budget justifications, senior US military officers have been remarkably persistent, even in the face of Congress's dominated by climate-change-denying Republicans, in warning that the loss of fresh supplies due to glacier retreat, hurricane devastation, droughts, and crop failure can "accelerant of instability." It is not so much the overt politics of social movements emerging in this context that alarms the Pentagon—though it does make a connection to terrorism—as chaotic disruption of social order that destabilize entire regions, block access to resources, impedes the free movement of labor, and raw materials, makes accumulation risky and unpredictable and acts as a general "threat multiplier" (Davenport 2014).

Such predictions have contributed to several accounts of imminent "climate wars" (Dyer 2008; Welzer 2012). The most sophisticated is that of Christian Parenti, whose *Tropics of Chaos* (2011) emphasizes that climate change cannot be considered an independent variable, but rather a vector that intersects with conditions already established by the uneven and combined development of global capitalism. By what is only partially a matter of meteorological mischance, Parenti points out, the region likely to be most severely affected by global warming is that of "undeveloped" countries lying between the Tropics of Cancer and Capricorn, an area where "the effects of climate change interacting with economic, social, and political problems will create a high-risk of violent conflict."

In his analysis, climate change amplifies previous crises rooted in the traumas of colonialism, Cold War militarism, and neoliberal privatization and deregulation of frail states. Here "current and impending dislocations of climate change intersect with the already-existing crises of poverty and violence... one expressing itself through the other." Water, Parenti argues, has long been a key driver of conflict, and with climate change bringing extreme weather with droughts and flooding, it will become an even greater issue. He then pursues this analysis through examination of ongoing mid-latitude crises in Afghanistan, India, Pakistan, Brazil, Mexico, and Africa, where large numbers of already impoverished and dispossessed peoples, their subsistence rendered precarious by agribusiness and industrialization, are now being struck by the meteorological consequences of global warming.

Since the publication of Parenti's study, events seem to verify his predictions. We have already alluded to the widely discussed suggestion that global warming helped set the scene for the Syrian civil war. In 2017 there is a very real possibility of four famines—in Somalia, South

Sudan, Nigeria, and Yemen—breaking out at once, endangering more than 20 million lives—driven by complex concatenations of drought and war. Where his analysis has been most strongly confirmed is around the issue of migrations fleeing the combined effects of climate change disasters, poverty, and war. He foresaw the United States and other developed countries becoming "armed lifeboats" with militarized borders and aggressive anti-immigration policies, where migrants are met with "the calumny, hatred, and ideological spittle of rightwing demagogues."

While anti-extractivism is a protest movement that has come be explicitly related to global warming, the tumults of migration and war are blind collateral damage from a heating planet, setting in motion movements and collisions of peoples driven by hunger, thirst, fear, and flood. To put matters in class composition terms, we can say that the survival struggles of migrants and refugees are those of peoples the planet factory has made into "surplus populations," rendered as parts of a reserve army of unemployed confined to global sacrifice zones where survival depends on "humanitarian aid, all kinds of illicit trade, agricultural survival, regulation by all various mafias and wars on a more or less restricted scale, but also by the revival of local and ethnic solidarities" (Theorie Communist 2011). These zones are now being rendered even less habitable by global warming and can only be escaped by illicit border-crossing journeys that stir the deepest anxieties of the planet factory's security guards.

New Proletarians

China's Pearl River Delta has always been subject to flooding. Now two things are changing. First, the floods are predicted to intensify, as global warming raises sea levels. Second, the Delta is full of factories. Waters that previously swamped rice fields now slosh through industrial plant that supplies the world with consumer goods. A one-meter rise in sea level would inundate 92,000 square kilometers of China's long, densely populated coastline, site of some of the most economically important cities in the world, displacing 67 million people (Kimmelman 2017). Many of these would be members of a new industrial proletariat, a proletariat whose super-exploited labors run factories directly involved in the generation of carbon emissions, who experience the consequences of atmospheric changes in the most dire and dramatic ways, and who may also be a major agent in any possible solution to the problem.

As we have seen, the decline of the industrial factory in the regions of advanced capital did not mark its disappearance from the planet, but, on the contrary, its diffusion around the globe. The main site of this migration was China. In 1978 the ruling Communist Party made a sharp turn from Maoist revolution to embrace the world market. By 2001 China's rulers had entered the World Trade Organization, abolished restrictions on foreign ownership, expanded its energy infrastructure by a huge expansion of coal plants, and were directing a vast movement of some 150–200 million people from rural areas into new manufacturing zones, from whence clothing, shoes, toys, furniture, appliances, light engineering goods, and electronics flowed out along supply chains of transnational corporations. At the start of the twenty-first century, these areas were not only the new "workshop of the world" but also its main "smokestack" (Malm 2016).

China's contribution to total global fossil fuel CO_2 nearly tripled from 1980 to 2010. In 2004 it became the largest national extractor of fossil fuels, and in 2006 it surpassed the United States as the largest emitter of CO_2; two-thirds of the global growth in carbon emissions between 2000 and 2007 happened in China with industry, rather than domestic heating, as "the most voracious sector" (Malm 2016, p. 341). All this has led some—especially right-wing American politicians—to declare China the main culprit of global warming. However, it still accounts for only 12% of the total greenhouse gases in the atmosphere, which "reflects the fact that others, such as the US and the EU, have been emitting far longer" (Pidcock 2016). Moreover, much of China's runaway carbon emissions were created producing goods made for consumption in the global North (Malm 2016, p. 331). The issue of China's responsibility to global warming is thus inextricable from that of global class divisions created by capital's imperial and colonial pattern of world development. Given China's leaders of repudiation of Mao, it is ironic that the climate crisis revives in a new ecological register his concept of global capital divided not just between capital and labor, but additionally between "First World" and "Third World," with the First World historically the major generator of global warming.

This international division is superimposed on a series of class conflicts internal to China. The primary attractor of foreign capital to China was low wages; in 2000 the wages of Chinese manufacturing workers were about 2% of their American equivalents. Many thought the migrant, poorly educated, largely female labor force drawn into Pearl River factories, protected only by state-compliant official trades unions, would remain a pool of cheap labor providing superprofits for foreign and domestic corporations. Successive generations of migrant workers have confounded this prediction with waves of social media coordinated wildcat strikes: Those of 2010, in which the protest suicides of Foxconn workers making Apple iPods drew international attention was only one moment, and a tipping point. Over a decade, Pearl River factory workers wages rose tenfold. With China an "emerging epicenter of world labor unrest," its ruling class found itself riding the tiger of the recently formed industrial working class, increasingly connected to the world, capable of massive militancy, and with an expectation of improving living standards (Silver and Zhang 2009).

This class force made any slowing of economic growth, had capital even considered it, socially risky, and might therefore be considered an obstacle to reduction to carbon emissions. But global warming also threatened serious disruptions to China's economic advance, with malign effects including not only the rising sea levels, but drought-driven reductions in agricultural production and flood carried epidemics of dysentery and typhus. Perhaps most serious is the interaction of global warming with China's terrifying air pollution problem. For the last several years vast fogs have shrouded megacities such as Beijing and Shanghai. Air pollution reportedly kills some 1.6 million Chinese people per year, over 4000 per day, are dying prematurely, and is linked with nearly one-third of all deaths in China (Fallows 2015).

The 2015 release of *Under the Dome*, a 103-minute-long documentary video on air pollution, hailed as China's version of the *Inconvenient Truth*, went viral on social media and gathered 150 million online views in days before being censored. Air pollution has its own class dimensions: The wealthy can afford expensive air filtration systems, drive to work, live in the less polluted sections of cities, flee to second homes on bad smog days and send their children to school abroad, while office workers and the urban poor cover their mouths with scarves and cough their way through low visibility streets. Beijing risks becoming a tale of two cities, "a place where the rich and poor don't even breathe the same air" (Berlinger et al. 2017). Some believe the problem has "reached a point where concern over air pollution throughout the country is threatening China's social stability" (Lu 2015).

Air pollution is not caused by CO_2, but by aerosol particles, some of which, such as black carbon soot, contribute to global warming; others, like light colored sulfates, actually have a cooling effect, radiating heat;

some climate scientists fear success in curbing China's air pollution might actually intensify warming. However, air pollutants are largely emitted by the same sources that give off carbon dioxide; when the Chinese state shuts down coal factories and closes cement factories it strikes at both problems. Conversely, it now appears climate change played a major role making "airpocalypses" more common; fast-melting Arctic ice and heavier Siberian snowfalls are changing winter weather patterns to make periods of stagnant air more common, trapping pollution, and building up of extreme levels of toxic air (Carrington 2017).

The double jeopardy of global warming and air pollution, and their combined effects on a population already massively restive in terms of strikes and environmental protests, are spurring the Chinese state to adopt dramatic clean energy measures, using what continue to be it significant socialist powers to direct economic development. It has placed a three year moratorium on new coal mines and started shutting down existing mines and pledged to end its use of coal by 2020. It has also become the manufacturing heart at the global solar energy boom, supported by tax credits and state subsidies. China has been the world's largest manufacturer of solar panels since 2008; six of the ten largest solar panel producing companies in the world are in China. In the first quarter of 2015, China installed solar capacity equal to France's entire output. This scale of installation has in turn built economies of scale in solar manufacturing and has contributed to a precipitous fall in the price of panels. Between 2008 and 2013, China's solar-electric industry dropped world prices by 80% (Fialka 2016).

And at the heart of this transformation is factory labor. Solar panels are made industrial plants that in China may employ up to 30,000 peoples, working with advanced robots to melt and slice blocks of silicon into wafers; etch, clean, furnace these cells; layer them with semiconductors; apply an antireflective coating; imprint electrical contacts; back them with aluminum, glass or plastic; test, sort, and connect them to form cell circuits and assemble them into solar panels now exported around the world, as basic component of a renewable energy programs (Campbell-Dollaghan 2014; Bwambale 2016). China's solar panel production is potentially toxic for both workers and communities around factories, and it is performed for wages which are still far lower than those for industrial work in the West. Nonetheless, one of the many paradoxes of China's industrialization is that its new proletariat may be the direct producers of the most hopeful developments on the horizon of the planet factory.

CONCLUSION: RECOMPOSITIONS OF CLASS AND CLIMATE

That hope is, however, hazy, not to say heavily overcast. The last days of 2016 gave the world the all but incredible sight of climate scientists frantically downloading their data on global warming from US government sites, lest it be destroyed or hidden by an incoming President. To understand this turn of events, we return to the force that at once instigates, presides over and is pursued and propelled by these conflicts in its planet factory—capital, the power for whom thousands of dead miners, laid off Detroit autoworkers, silenced climate scientists, dispossessed peasants, desperate migrants, and fog-choked proletarians, not to mention the fauna and flora of the entire terrestrial biosphere, are commonly just so many factors of production, albeit factors with an occasional unfortunate proclivity to rebel. Looking toward its ascendant heights, to the level of political elites and corporate leaders of a now global system, we can schematically identify four moments in climate change policy: (1) Sustained Neglect, (2) Failed Bargaining, (3) Late Greening, and (4) US Reactionary Reversal.

The first stage, sustained neglect, the longest chronologically, can be dealt with most briefly, because it encompasses the two and a half centuries during which, despite warnings, at first occasional and eccentric and then of intensifying authority, that fossil fuels could be having drastic effects on climate stability capital does nothing—or rather, led by the great energy and industrial corporations of the global North, does everything to magnify the profits to be gained from carbon-based accumulation, around which is shaped an entire civilization.[3]

In the second phase, running from 1992 up to about 2015, leading capitalist powers, confronted with conclusive scientific evidence, acknowledge the existence of a climate problem, but their attempts to deal with it in the framework of globalized market relations are a complete failure. The possibility of serious adverse effects on accumulation, voiced in documents such as the Stern Report, begins to arouse concern. Climate crisis becomes a topic for high-level policy conclaves. However, these negotiations are hamstrung by the competitive relations between capitalist regimes (there are by now no others), particularly the oil-driven US and coal-burning newly industrializing nations. No mandatory agreements on reducing carbon emissions are reached, and voluntary commitments often prove notional or are explicitly revoked. The start of the twenty-first century, when capital is ostensibly seeking a solution to its newly acknowledged problem, is in fact marked by a massive carbon emissions "explosion" (Malm 2016).

The third phase, overlapping with the second in the 2010s, sees the appearance of forms of green capital that detect profitable opportunities in climate crisis. The initial round is led by the financial sector and takes the form of emissions trading markets, whose convoluted cap and trade provisions foster fraudulent mitigation schemes, and have an almost null effect atmospheric carbon composition. However, as energy renewables become more commercially attractive, largely because of investments by China and other states fearful of climate-driven social tumults, the sector starts to expand in response to market signals. With the costs of solar and wind energy beginning to plummet there is "a struggle for supremacy between the dominant fossil capitalists and the emerging green capitalists" (*Aufheben* 2012, p. 52). It is not clear, however, that this belated greening comes in time to avert a disaster prepared over centuries; annual global carbon emissions stabilize from 2014 to 2016, but the accumulating atmospheric concentration of CO_2 continues to rise, reaching a new peak well beyond the limit for "safe" climate change.

The 2016 election of Donald Trump, an avowed climate change denier, as US President sees an abrupt setback to even these limited advancement. This fourth moment of "reactionary US reversal" demonstrates that climate change politics are inseparable from class politics. For the great failure of the "progressive" globalizing climate policies of the Obama administration proved to be its willingness to throw part of its industrial working class under the bus of the capitalist world-market, sacrificing the local factory worker to the construction of a larger planet factory. This gave the opportunity for the forces of the far right marshaled around Trump to constellate the depressions and despairs of the rust belt, racist fear of migrants and terrorists and climate refugees, anti-China job-loss animosity, fundamentalist antiscience sentiment and climate change denial around a "make America great again" slogan, under whose sign the new President could pose for photo ops with out-of-work Appalachian miners while nostalgically invoking the age of greatest US carbon emissions.

The Trump administration's early actions—the appointment of Exxon executive Rex Tillerson as Secretary of State; the removal by executive order of regulations to the coal, oil, and gas and automobile industries; cuts to the Environmental Protection Agency and hiring of climate-denier Scott Pruitt to run it; and, most significant, withdrawal on the first of June 2017 from the Paris Agreement on climate change—all seems species-level bad news. It is, however, impossible to foresee the consequences of a regime that every-day displays new scandals and may at any time collapse. Its industrial policy may prove largely irrelevant in the face the widespread adoption of renewable energy, not just by the rest of the world, but even by US states and municipalities. Perhaps the most likely outcome for the United States is a form of climate laissez-faire, by which commercially renewable energy use continues to expand as its price becomes attractive, but no attempt is made to suppress fossil capital, which continues to supply a portion of the planet factory's expanding energy appetite. The consequences for a planet where climate heating may be at or beyond critical tipping-points are highly uncertain.

Eco-socialists are skeptical capital can green itself (Tanuro 2014; Malm 2016). The scale of sunk investment in vast mining and drilling complexes; the problems of profitably enclosing potentially free sun and wind energy sources; the voracious energy appetites of a system reliant on growth to defuse inequalities all suggest leaving carbon fuel in the ground is beyond its logic. If so, hope of radical measures against global warming requires a political recomposition of global proletarians exploited or dispossessed in both the making and the consequences of carbon capital to demand "just transition" plans for workers whose livelihoods are at stake in the phasing out fossil fuel industries; millions of jobs in programs of clean energy industrial, transportation and infrastructural renewal; conversion of the "war on terror," with all its dread about chaotic surplus populations, into a war on climate catastrophe. Movements pressing for large-scale public works, distribution of free or cheap, renewable energy, demanding the reduction of carbon inequalities, and challenging the ideology of consumerism and limitless growth, could pose new revolutionary challenges to capital.

But even if international capitalism can make a renewable energy transition, questions of class and climate will continue to be deeply entangled. They will be integral to issues of work, profit and pay in new green industries; to the distributions of the costs of the energy transformation between capital and its subordinate classes; and to the questions of what populations are or are not protected or rescued from sea level rises and extreme weather disasters already in motion for decades to come. Global warming is as close to a universal crisis as can be imagined (topped only by an asteroid strike or thermonuclear war), but its outcome depends on contending class forces. As the journal *Aufheben* observes, "Even a crisis as serious as climate change does not create a unity of interests between capital and proletariat" (2012, p. 55). For the foreseeable future, the end of the world means further struggles in the planet factory.

NOTES

1. Malm (2016, pp. 288–292) adumbrates such a model, adapting Marx's "classic" circuit diagram to show the moments at which capital historically demands the expanding use of fossil fuels and emission of CO_2 in the production and consumption of commodities, but does not extend it deal with the heat generating consequences of these emissions.
2. This essay focuses fossil fuel emissions; class-composition analysis could also be applied to other forces affecting global warming, such as methane-generating factory farming or deforestation, but would considerably enlarge the scope of analysis.
3. From start of industrialization its critics pointed to the bad effects of factories on air and weather a hundred and fifty years ago that the possibility carbon emissions would overheat the earth was first scientifically advanced; as early as 1965 the President of the United States was briefed on this possibility.

Image 2.1   Revetment study (Mia Feuer, watercolor)

REFERENCES

Abramsky, K. (2010). *Sparking a Worldwide Energy Revolution: Social Struggles in the Transition to a Post-petrol World*. Oakland, CA: AK Press.

Angus, I. (2016). *Facing the Anthropocene: Fossil Capitalism and the Crisis of the Earth System*. New York: Monthly Review.

Aufheben. (2012). The Climate Crisis ...and the New Green Capitalism? *Aufheben* 21. http://ns210054.ovh.net/library/aufheben-21-2012.

Barbrook, R., & Cameron, A. (1996). The Californian Ideology. *Science as Culture, 26,* 44–72.

Berlinger, J., George, S., & Wang, S. (2017, January 16). Beijing's Smog: A Tale of Two Cities. *CNN*. http://www.cnn.com/2017/01/15/health/china-beijing-smog-tale-of-two-cities/.

Bonneuil, C., & Fressoz, J.-B. (2015). *The Shock of the Anthropocene: The Earth, History and Us*. London: Verso.

Bwambale, T. (2016, April 3). Inside One of the World's Largest Solar Panel Makers. *New Vision*. http://www.newvision.co.ug/new_vision/news/1421175/peek-inside-world-largest-solar-panel-makers#sthash.a4YrBiVM.VtLLZuQj.dpuf.

Caffentzis, G. (2013). *In Letters of Blood and Fire: Work, Machines, and the Crisis of Capitalism*. Oakland, CA: PM Press.

Campbell-Dollaghan, K. (2014, July 13). Go Inside the Factory That Makes the World's Solar Panels. *Gizmodo*. http://gizmodo.com/go-inside-the-factory-that-makes-the-worlds-solar-panel-1604361164.

Carrington, D. (2017, March 15). 'Airpocalypse' Smog Events in China Linked to Melting Ice Cap, Research Reveals. *The Guardian*. https://www.theguardian.com/environment/2017/mar/15/airpocalypse-smog-events-linked-to-global-warming-research-reveals.

Chancel, L., & Piketty, T. (2015). *Carbon and Inequality: From Kyoto to Paris Trends in the Global Inequality of Carbon Emissions (1998–2013) & Prospects for an Equitable Adaptation Fund*. Paris School of Economics. http://piketty.pse.ens.fr/files/ChancelPiketty2015.pdf.

Davenport, C. (2014, October 13). Pentagon Signals Security Risks of Climate Change. *New York Times*. https://www.nytimes.com/2014/10/14/us/pentagon-says-global-warming-presents-immediate-security-threat.html?_r=0.

Dyer, G. (2008). *Climate Wars: The Fight for Survival as the World Overheats*. Toronto: Random House.

Edwards, P. (2010). *A Vast Machine: Computer Models, Climate Data, and the Politics of Global Warming*. Cambridge, MA: MIT Press.

Fallows, J. (2015, August 14). China's Air Problem is Worse Than You Think. *The Atlantic*. https://www.theatlantic.com/international/archive/2015/08/china-air-pollution-harm-study/401315/.

Federici, S. (2012). *Revolution at Point Zero: Housework, Reproduction and Feminist Struggle*. Oakland, CA: AK Press.

Fialka, J. (2016, December 19). Why China Is Dominating the Solar Industry. *Scientific American*. https://www.scientificamerican.com/article/why-china-is-dominating-the-solar-industry/.

Foster, J. B. (1993). The Limits of Environmentalism Without Class: Lessons from the Ancient Forest Struggle of the Pacific Northwest. *Nature, Capital, Socialism, 4*(1), 11–41.

Foster, J. B., Clark, B., & York, R. (2010). *The Ecological Rift: Capitalism's War on the Earth*. New York: Monthly Review.

Fuentes, F. (2014, May 19). The Dangerous Myths of 'Anti-extractivism'. *Climate and Capitalism*. http://climateandcapitalism.com/2014/05/19/dangerous-myths-anti-extractivism/.

Gottleib, R. (2005). *Forcing the Spring: The Transformation of the American Environmental Movement*. Washington, DC: Island Press.

Haller, T., et al. (Eds.). (2007). *Fossil Fuels, Oil Companies and Indigenous People*. Berlin: Lit Verlag.

Hardt, M., & Negri, A. (2000). *Empire*. Cambridge, MA: Harvard University Press.

Hari, J. (2010, May 20). Polluted by Profit. *The Independent*. http://www.independent.co.uk/voices/commentators/johann-hari/polluted-by-profit-johann-hari-on-the-real-climategate-1978770.html.

Jorgenson, A., Schor, J., & Huang, X. (2017). Income Inequality and Carbon Emissions in the United States: A State-Level Analysis, 1997–2012. *Ecological Economics, 134,* 40–48.

Kelley, C., et al. (2015, March 2). Climate Change in the Fertile Crescent and Implications of the Recent Syrian Deought. *Proceedings of the National Academy of Sciences, cxii*(11), 3241–3246. http://www.pnas.org/content/112/11/3241.full.pdf.

Kimmelman, M. (2017, April 29). Rising Waters Threaten China. *The New York Times*, p. 1

Klein, N. (2014). *This Changes Everything: Capitalism vs. The Climate*. New York: Simon and Schuster.

Lowy, M. (2015). *Ecosocialism: A Radical Alternative to Capitalist Catastrophe*. Chicago: Haymarket.

Lu, R. (2015, March 6). China's Real Inconvenient Truth: Its Class Divide. *Foreign Policy*. http://foreignpolicy.com/2015/03/06/chinas-real-inconvenient-truth-its-class-divide/.

Malm, A. (2016). *Fossil Capital: The Rise of Steam Power and the Roots of Global Warming*. London: Verso.

Marinez-Allier, J. (2002). *The Environmentalism of the Poor*. Cheltenham: Edward Elgar.

McNeill, J. R., & Engelke, P. (2014). *The Great Acceleration: An Environmental History of the Anthropocene Since 1945*. Cambridge, MA: Belknap.

Midnight Notes. (1992). *Midnight Oil: Work, Energy, War, 1973–1992*. New York: Autonomedia.

Midnight Notes. (2001). *Auroras of the Zapatistas: Local and Global Struggles of the Fourth World War*. Oakland, CA: AK Press.

Mitchell, T. (2011). *Carbon Democracy: Political Power in the Age of Oil*. London: Verso.

Moore, J. W. (2015). *Capitalism in the Web of Life: Ecology and the Accumulation of Capital*. London: Verso.

Moore, J. (Ed.). (2016). *Anthropocene or Capitalocene? Nature, History and the Crisis of Capitalism*. Oakland: PM Press.

Nelson, S., & Braun, B. (2017). Autonomia in the Anthropocene: New Challenges to Radical Politics. *South Atlantic Quarterly, 116*(2), 223–235.

O'Connor, J. (1998). *Natural Causes: Essays in Ecological Marxism*. London: Guilford.

Oxfam. (2015, December 2). Extreme Carbon Inequality. *Oxfam Media Briefing*. https://www.oxfam.org/sites/www.oxfam.org/files/file_attachments/mb-extreme-carbon-inequality-021215-en.pdf.

Parenti, C. (2011). *Tropic of Chaos: Climate Change and the New Geography of Violence*. New York: Climate Change.

Pidcock, R. (2016, March 17). China Is Responsible for 10% of Human Influence on Climate Change, Study Says. *Carbon Brief*. https://www.carbonbrief.org/china-is-responsible-for-10-of-human-influence-on-climate-change-study-says.

Schiller, D. (1999). *Digital Capitalism: Networking the Global Market System*. Cambridge, MA: MIT Press.

Selby, J., & Hulme, M. (2015, November 29). Is Climate Change Really to Blame for Syria's Civil War? *The Guardian*. https://www.theguardian.com/commentisfree/2015/nov/29/climate-change-syria-civil-war-prince-charles.

Silver, B. (2003). *Forces of Labour: Workers' Movements and Globalization Since 1870*. Cambridge, UK: Cambridge University Press.

Silver, B., & Zhang, L. (2009). China as an Emerging Epicentre of World Labour Unrest. In H. -F. Hung (Ed.), *China and the Transformation of Global Capitalism* (pp. 174–187). Baltimore: John Hopkins Press.

Szeman, I. (2017). Introduction: Pipeline Politics. *South Atlantic Quarterly, 116*(2), 402–407.

Tanuro, D. (2014). *Green Capitalism: Why It Can't Work*. London: Merlin.

Theorie Communiste. (2011, May 15). The Present Moment. *Libcom.org*. https://libcom.org/library/present-moment-theorie-communiste.

United Nations. (2013). *National Accounts Main Aggregate Database*. http://unstats.un.org/unsd/snaama/dnlList.asp.

Welzer, H. (2012). *Climate Wars: How People Will Be Killed in the Twenty First Century*. Cambridge: Polity Press.

CHAPTER 3

Stuck in the Anthropocene: The Problem of History, Theory, and Practice in Jason W. Moore and John Bellamy Foster's Eco-Marxism

Alexander M. Stoner and Andony Melathopoulos

INTRODUCTION

Understanding history philosophically—as a theory—is not something that has occurred throughout time, but has recent origins, and is associated with a form of society in which change is anticipated and expected. Whereas intellectuals in pre-modern societies understood history in cyclical terms, Enlightenment thinkers helped solidify the conception of modernity as a radically new epoch, distinct from the past, in which history progresses forward into the future (Collinicos 1999, p. 13). The attempt to characterize the last 250 years as the Anthropocene must be recognized as part of this tradition.

A. M. Stoner (✉)
Department of Sociology & Anthropology,
Northern Michigan University, Marquette, USA

A. Melathopoulos
Department of Horticulture, Oregon State University, Corvallis, OH, USA

© The Author(s) 2018
j. jagodzinski (ed.), *Interrogating the Anthropocene*, Palgrave Studies in Educational Futures, https://doi.org/10.1007/978-3-319-78747-3_3

We must also recognize how differently history was regarded before the Anthropocene. Rather than an age of humans (i.e., the abrupt ascendency of humanity over biophysical nature), the conception of the world 250 years ago was the possibility of changing humanity itself. Captured in ideas such as Kant's "unsocial sociability," Rousseau's "perfectibility," and Adam Smith's idea that the pursuit of self-interest paradoxically results in higher levels of cooperation, these thinkers grasped the potential for humanity to be changed through "society." Indeed, the idea of history is bound up with the very concept of "society" (Adorno 1998 [1962]), which, in turn, was regarded as something other than simply a matter of technological advancement (e.g., steam engines), the reduction of material scarcity through material output (e.g., manufacture) and/or the discovery of new energy sources (e.g., coal). For these thinkers, the kind of change that was underway 250 years ago rendered, for example, "private vice" into something qualitatively different from the "war of all against all" (Hobbes 1994 [1668]), which characterized the past. The fact that "private vice" was now connected to "public virtue," was understood as being part of a deeper wholesale shift in what society was about. Consequently, for someone like Adam Smith, the invention of the steam engine was not an independent factor in history, but rather a product of the new ways people began relating to one another. In light of the emergence of a new form of sociality, humanity was obligated to take hold of society, and in turn, transform humanity itself.

As we will endeavor to demonstrate below, the issue of social action and historical transformation is the point that most befuddles Anthropocenarians. While the Anthropocene describes key changes in the relation between humans and their natural environment, this understanding does not lead convincingly beyond the present moment. Unlike the late eighteenth century, insight into history at the beginning of the twenty-first century appears entirely disconnected from human agency. Contemporary scientists have been able to pinpoint changes in earth system processes as a result of human behavior with an increasingly accurate degree of precision. However, although scientific research measures the severity of societally-induced environmental degradation, indicating the increased visibility of such degradation, increased visibility does not necessarily correspond with a greater ability to understand this degradation, let alone directly pursue action that might move toward ameliorating societally-induced environmental degradation. Indeed, throughout the latter half of the twentieth century societally-induced environmental degradation was compounded in relation to our awareness of these problems. In this chapter, we will refer to this paradox (wherein environmental degradation increases amid the growth of environmental attention and concern) as the *environment-society problematic* (see Stoner and Melathopoulos 2015, pp. 22–23).

The environment-society problematic, we contend, is not simply a product of fate (e.g., "how humans are"), but rather reflective of a growing gap between the broad awareness of environmental problems, the subjective dimension of societally-induced environmental degradation (Stoner 2014), and our ability to transform the objective dimension of the world in accordance with this awareness (Stoner and Melathopoulos 2015, p. 21). In this sense, the environment-society problematic indexes the ways in which our collective ability to self-consciously transform the socio-biophysical world have grown progressively dim with the advance of modern capitalist society.

We believe that the meaning of human agency in the Anthropocene has grown progressively dim because a key figure is neglected—Karl Marx. Below we will show that while Marx has been raised in an attempt to confront the problems with the Anthropocene (e.g., Foster 2016; Moore 2016), these prominent eco-Marxist critics fall well below the threshold of the Anthropocenarians they criticize. As we will explain, both the Anthropocenarians and eco-Marxist critics have omitted the problem of meaning in history and, in so doing, have elided the linkage of history, theory, and practice. As a result, both Anthropocenarians and eco-Marxist critics advance a profoundly *unhistorical* position (albeit in different ways), further naturalizing the society-environment problematic. We conclude by directing focus toward how the significance of history was understood by both Marx and Marxists who followed him.

CRUTZEN ET AL.

The push to recognize the last 250 years as the Anthropocene has two dimensions, both linked to a recognition of how the last 250 years are unlike the last 10,000 years—the Holocene. On the one hand, the Anthropocene is a matter of objective stratigraphic debate, centered on whether or not the sedimentary patterns of the last 250 years are distinct from the antecedent geological period. Of concern to this chapter is not the Anthropocene in its natural historic sense, but as a distinct period of human history.[1] The question of the meaning of the Anthropocene in light of human history has been a prominent feature of the work of two key Anthropocene advocates—Nobel Prize winner Paul Crutzen and fellow Earth Systems Scientist Will Steffen (see, e.g., Crutzen and Steffen 2003). Although much of their work describing the destructive scale of biophysical transformation currently underway is motivated by an implicit normative appeal for change in human society, their collaboration with prominent US environmental historian J.R. McNeill attempts to make explicit the historical meaning of our increasingly sophisticated knowledge of biophysical destruction (Steffen et al. 2007). For these three authors, the key transformation separating ancient from modern forms of social organization is the Industrial Revolution of the early nineteenth century (Graph 3.1). Although Steffen et al. give as much historical prominence to the Industrial Revolution as they do to the Neolithic revolution that ushered in civilization 10,000 years ago; they define the substance of both transformations in strictly quantitative terms. Consequently, for these authors the most striking difference between prehistoric and ancient society, and ancient and Industrial society, is their energetic efficiency: "industrial societies as a rule use four or five times as much energy as did agrarian ones, which in turn used three or four times as much as did hunting and gather societies" (616). How humanity found itself in the Anthropocene appears to these authors as a matter of contingency.

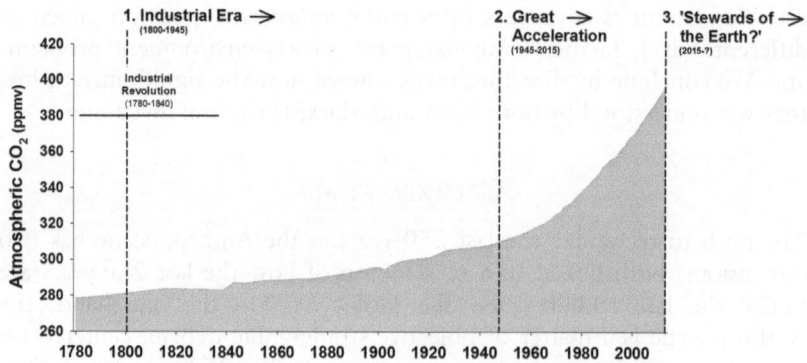

Graph 3.1   The level of destruction reflected in the rising levels of atmospheric CO_2

In spite of the accidental manner in which humanity found itself entering the Industrial Revolution, Steffen et al. ultimately advance a theory of history that centers on how we overcome the Anthropocene and become "Stewarts of the Earth" after two increasingly ecologically destructive phases: (1) the industrial era (French Revolution–WWII) and (2) Great Acceleration (WWII–present). Steffen et al. claim that the historical conditions for the transformation into the third phase ("Stewards of the Earth" (2015–?)) are not merely contingent, but instead bound up with key developments following WWII—namely, the emergence of environmentalism in the 1960s and the more recent phase of globalization in which they see a reduced "scope for the exercise of arbitrary state power and strengthening... role of civil society" (619). Not only is Steffen et al.'s description of the Anthropocene and its future unconvincing (civil society was arguably a stronger force in society at the beginning of the Anthropocene than it is now), as a theory of history, their account is incoherent. With what necessity would humans living in modern capitalist society self-consciously transition from capitalism into a future form of social organization, representing a historical transformation as profound as the rupture with history that came before it? According to Steffen et al. (2007, p. 619), "humanity is, in one way or another, becoming a self-conscious, active agent in the operation of its own life support system." However, this statement is more hope than theoretical insight.[2]

While Crutzen et al. mark a difference from the Industrial Revolution to the Great Acceleration, they characterize this change as an acceleration of what came before, as opposed to a fundamental transformation. However, Crutzen et al. do propose a fundamental, qualitative transformation into humanity's next phase "Stewards of the Earth." But how the current stage of history—the Anthropocene—could either accelerate or qualitatively change is seemingly disconnected from properties of society itself. Again, history appears not as a social creation, but as a matter of chance, as something that happens to society. History understood in this manner is the history most consistent with the society-environmental problematic, where growing awareness of environmental degradation does not lead society toward self-conscious transformation so as to rationally address this degradation, but rather the opposite—toward an incapacity to act in ways that are not predetermined from the outset. Crutzen et al. are aware of the gap between theory and practice, and in other fora they acknowledge the tremendous obstacles associated with getting beyond "politically mediated compromises that fall far short"

(Fischer et al. 2007, p. 623). In this sense, the account of history by Crutzen et al. might not be considered wrong, but rather a product of our current moment in which the inability to change society is reflected theoretically in our understanding of history.

The problem with characterizing the past 250 years as the Anthropocene is the extent to which it reflects our adaptation to the present, at the expense of being able to recognize what potential future the present might give rise to. In looking toward the past, Anthropocenarians recognize how thoroughly past and future human-environment relations are intertwined with one another. Anthropocenarians understand the present moment of the Great Acceleration in light of past human-environment dynamics that have run their course, and they struggle for a more sustainable future. In so doing, many Anthropocenarians—in spite of their attempt to understand the relationship between humanity and the natural environment historically—mistakenly presuppose the struggle for a more sustainable future because they fail to adequately confront the existing state of affairs. As a result, these scholars advance a profoundly *unhistorical* position.

ECO-MARXIST CRITICISM

Most critics of the Anthropocene evade its theory of history, preferring instead to attack the concept for being anthropocentric (Chernilo 2016), a function of the managerial quality of Earth System science (Luke 2015) and colonialism (Ahuja 2015; Schulz 2017). Where the critics fall short is in challenging how Crutzen et al.'s historical understanding of the past 250 years might yield theoretical insights into how a further wholesale transformation of humanity could be possible. Even eco-Marxist critics (e.g., Foster 2016; Moore 2016), who operate with an explicit theory of history, ignore the Anthropocene's central question concerning social action and historical transformation. By failing to object to the incoherence of the Anthropocene (as a theory of how change happens historically), eco-Marxist critics tend to either (1) neglect entirely the problem of change (denying that there is anything, in fact, unique about the last 250 years), or (2) confront the problem of change in an entirely superficial manner (e.g., "Marxists" who say the past 250 years was capitalism, and that only a revolution will end it—without being able to articulate, as Lenin did, how the basis of overcoming capitalism would have to be capitalism itself), so that the conditions of such change remain undisturbed.

This section discusses the work of two prominent eco-Marxists—Jason W. Moore and John Bellamy Foster. While both scholars understand the ecological crisis associated with the Great Acceleration as arising from contradictions internal to the capitalist system; they have vastly different ideas about how this system emerged and, relatedly, how we might plausibly move beyond it.

Moore

Jason W. Moore takes issue with the Anthropocene because it fails to account for deep transformations that have taken place in human society over the past five centuries. Moore attempts to locate the root of the problem deeper in history, in the Enlightenment, and with the conceptual separation of humans from nature in Cartesian Dualism. This separation, he asserts, led to the exploitation of both nature and humans in the service of developing capital. Moore takes from these insights the need to reconsider the Anthropocene as the Capitalocene. Below we review Moore's world-historical reconstruction of the Capitalocene, including his reinterpretation of the Great Acceleration. We will argue that Moore's shift in terminology (from the *Anthropocene* to the *Capitalocene*) fails to address the key limitation of the Crutzen et al.—namely, how one is to consider the potential for humanity to effect a qualitative change out of the Anthropocene/Capitolocene. Like Crutzen et al., Moore aspires to a philosophy of history, but can only describe what has happened.

The Capitalocene

According to Moore (2016, p. 84), the Anthropocene cannot adequately engage the big historical questions it poses because it remains conceptually bound to Cartesian dualism—above all, the duality of Nature/Society (according to which nature and humanity are ontologically discrete).[3] Not only is the Nature/Society dualism endemic to processes of violence and exploitation central to modern capitalism as a historical project, but according to Moore (2016, p. 79), such duality also obscures the history of the modern world.[4]

Moore has recently attempted to move beyond what he views as the impasse of Cartesian dualism by advancing a "world-ecology" framework.[5] Central to Moore's framework is the concept of *oikeios*, denoting "the relation through which humans act—and are acted upon by the whole of nature—in our environment-making" (Moore 2015, p. 4). Placing the *oikeois* at the center of his narrative strategy, Moore reconceptualizes historical change in terms of what he calls the *double-internality*, which he sums up with the catchphrase, "humanity inside nature, nature inside humanity" (Moore 2016, p. 79). Employing the double-internality to specify how capitalism works through nature (and vice versa), Moore advances a world-historical reconstruction not of the *Anthropocene* but of the *Capitalocene*:

> My central thesis is that capitalism is historically coherent—if 'vast but weak'—from the long sixteenth century; co-produced by human and extra-human natures in the web of life; and cohered by a 'law of value' that is a 'law' of Cheap Nature. At the core of this law is the ongoing, radically expansive, and relentlessly innovative quest to turn the work/energy of the biosphere into capital (value-in-motion). (Moore 2015, p. 14)

Moore (2015, p. 191) stresses that capital's law of value is not solely economic, for its "accumulation (as abstract labor) is historically materialized though the development of scientific and symbolic regimes necessary to identify, quantify, survey, and otherwise enable not only the advance of commodity production but also the ever-more expansive appropriation of cheap natures." Accordingly, Moore (2015, p. 191) deemphasizes the historical prominence Crutzen et al. give to the eighteenth century Industrial Revolution, directing focus instead toward antecedent transformations (of land, labor, science, and knowledge) during the so-called "long sixteenth century" that facilitated the "Age of Coal." In this regard, Moore (2016, pp. 85–86) highlights three interrelated historical processes (ca. 1450–1640) central to the rise of industrialization: (1) primitive accumulation; (2) new forms of territorial power and imperialism; and (3) new ways of knowing and making the world. Moore stresses the *dialectical relation between these three historical processes* as science, the economy, and the state emerge to serve capital accumulation (Moore 2016, p. 86). The key shift proceeding the Industrial Revolution—namely, the inversion of the labor-land relation and the ascendance of labor productivity as a metric of wealth (Moore 2015, p. 189)—cannot be explained *solely* in regards to technical advancements following 1450, but must instead be regarded as unfolding on the basis of "Cheap Natures" made possible by the new technics of global appropriation.[6]

Moore explains that, beginning with the Cartesian revolution and the "age of exploration" in the Atlantic world, knowledge production and environment-making both produce and are produced by processes of primitive accumulation and imperialism. Historically, relations of appropriation enabled the expanded accumulation of abstract social labor. Revolutions in cartography during the fifteenth century, and more important, the emergence of abstract time throughout the latter half of the fourteenth century are prime examples (cf. Postone 1993). Hence, Moore's (2016, p. 90) contention that "great 'economic revolutions,' propelling labor productivity within the commodity system, are always accompanied by 'new' imperialisms, 'new' sciences, [and] 'new' forms of state power." Taken together, Moore sees these processes as indicative of capitalism's law of value as a peculiar way of organizing nature—that is, of "putting the whole of nature to work for capital" (Moore 2016, p. 86). For Moore, cheap labor and cheap nature play a similar role in accounting for historical dynamics.

For Moore, at the heart of capitalism's law of value is the contradictory dialectic of *exploitation* and *appropriation*. Gleaning insight from Marx, Moore is concerned with how the capacity to do work (by human *and* extra-human natures) is transformed into value, measured in terms of socially necessary labor time.[7] However, such relations of exploitation operate properly insofar as their reproduction costs can be held in check (Moore 2015, p. 16). According to Moore (2015, pp. 16–17), labor exploitation is contingent upon "the historical-geographical connections between wage-work and its necessary conditions of expanded reproduction. These conditions depend on massive contributions of unpaid work, outside the commodity system but necessary to its generalization."

Hence, for Moore, the contradiction at the heart of capitalism's law of value (between exploitation and appropriation) revolves around the disproportionality between "'paid work,' reproduced through the cash nexus, and 'unpaid work,' reproduced outside the circuit of capital but indispensable to its expanded reproduction" (Moore 2016, p. 92). Given capitalism's tendency toward rising labor productivity (which implies accelerating biophysical throughput) the system "must appropriate ever-larger spheres of uncapitalized nature" (Moore 2016, p. 92). Thus, Moore (2015, p. 96) writes, "Great advances in labor productivity, expressing the rising material throughput of an average hour of work, have been possible through great expansions of the ecological surplus." However, according to Moore (2016, p. 92), "the whole system works (…) because capital pays for only one set of costs, and works strenuously to keep all other costs off the books. Centrally, these are

the costs of reproducing labor-power, food, energy, and raw materials."
While such costs can be externalized in the short-term, Moore (2015,
pp. 95–96) notes that "ultimately new sources of work/energy must be
found, and appropriated. Thus, every long accumulation cycle unfolds
through new commodity frontiers." For Moore, understanding capitalism as a civilization that, for the first time, "mobilized a metric of wealth
premised on labor rather than land productivity" (2016, p. 110) explains
much of the early modern landscape transformation—from the appropriation and exhaustion of forests and soil in Brazil, Scandinavia, and
Poland in the long seventeenth century; and from the sugar and slave
frontiers in the New World, to the expansion of the fossil fuel frontier
after the eighteenth century (Moore 2016, p. 110).

The End of the Capitalocene?

According to Moore (2016, p. 93), the Great Acceleration is the result
of the law of value, which is "premised on advancing labor productivity within a very narrow zone: paid work." For Moore, the slowdown of
labor productivity growth since the 1970s and the subsequent growth
of global neoliberalism mark the signal crisis of the Capitalocene; that is,
the end of the era of Cheap Nature. Moreover, Moore (2016, p. 110)
contends that capitalism today has exhausted its Cheap Nature strategy,
as the "progressive enclosure of capitalism's Cheap Nature frontiers" has
made it increasingly impossible to appropriate Cheap Nature in the same
way.

The central question for Moore is not the beginning of the
Anthropocene, but rather the end of the Capitalocene. While for the
end of Cheap Nature spells the end of capitalism, according to Moore
(2016, p. 114), this does not automatically lead to liberation. Rather,
for Moore, adequately addressing our current ecological predicament
requires putting the concept of "work" at the center of analysis. Here
Moore argues the need to critique the specific ways human and non-human natures are put to work for capitalism. As he explains, "Popular
strategies for liberation will succeed or fail on our capacity to forge a
different ontology of nature, humanity, and justice—one that asks not
merely how to redistribute wealth, but how to remake our place in
nature in a way that promises emancipation for all life" (Moore 2016,
p. 114). We agree. However, we do not think Moore's world-ecological
reconstruction of the Capitalocene moves us in this direction.

Given Moore's concern with capitalism's law of value, it makes sense that he would be drawn to Marx. But while Moore draws from Marx, he does so in a way that raises issues. Moore is certainly correct to point toward the inversion of the labor-land relation, including the ascendance of labor productivity as a metric of wealth, as central to the emergence of value. However, by asserting that the distinction between subject and object (of labor) is a mere cover for expropriation, Moore fails to specify what is qualitatively distinct between traditional society and the Capitalocene. Within Moore's framework, this transition amounts to nothing other than a more "rational" domination of human and non-human natures and, therefore, a more efficient extraction of the surplus (i.e., an accelerated form of traditional society).

Indeed, the essence of Moore's argument that cheap labor and cheap nature play a similar role in accounting for historical dynamics is not altogether different from Proudhon (1840 [1890]), who explains history—from ancient to feudal, and then bourgeois society—in terms of the "theft" of "property relations" (according to which property is stolen from small-holding peasants and upwardly redistributed). Here it is worth bearing in mind that Marx devoted significant passages, throughout his various writings, toward a critique of Proudhon. For example, Marx (1865) noted Proudhon's failure to understand scientific dialectics, and he criticized Proudhon for garbling the categories of political economy "into pre-existing *eternal ideas*," as opposed to "*the theoretical expression of historical relations of production, corresponding to a particular stage of development in material production.*" For Marx, Proudhon's critique of property relations was one-sided, considering only their "legal aspect as relations of volition" while omitting their essence as relations of production (ibid.). In much the same way, Moore asserts that the distinction between subject and object is a mere cover for "rational" domination, including a more efficient extraction of the surplus. Is the modern distinction between subject and object entirely negative, as Moore argues, leading one to believe that there is nothing in modernity that is worth redeeming? Here Moore appears to adopt what we take to be an incoherent postmodern position (despite his claims to the contrary). Instead of working through the dialectic of subject and object, Moore attempts to transcend this dialectic, thereby stepping outside of history, in order to create a post-Cartesian ontology.

Moore's inability to adequately distinguish traditional society and the Capitalocene also means that he is unable to fully grasp the modern bourgeois articulation of labor. Although Moore identifies capitalism's law of value as a systemic process, he does not fully appreciate the fact that value works through human and non-human natures in different ways. Specifically, Moore fails to recognize abstract social domination, structured by a particular form of labor, as that which distinguishes capital historically. Given the importance of Marx's category of socially necessary labor time to Moore's analysis, this is a crucial omission.

Moore views the capitalist work regime (beginning with the emergence of abstract time and the development of socially necessary labor time as the magnitude of value) as a process of violent abstraction—something that is simply done *to* humans (and non-human natures). Again, according to Moore, the subject-object relation of traditional epistemology is a mere cover for capitalism's violence and exploitation. Marx's critical theory, on the other hand, aims to specify exactly how, through concrete forms of social practice, both subject and object are produced. Marx's theory of practice breaks with the subject-object dualism of traditional epistemology to conceptualize objectivity and subjectivity as mediated through social practice (Postone 1993, p. 217). For Marx, labor constitutes forms of social objectivity and social subjectivity. Praxis, as such, can then be analyzed in terms of structures of social mediation (Postone 1993, pp. 218, 220). Marx was not concerned with labor in general but rather its particular *alienated* form, which under modern capitalism, structures social relations (including the relation between people and nature) (op. cit.). Crucially, alienated labor is a process of *self-generated domination*, constituted by the capitalist production of value, in which individuals do not consciously control but rather are controlled by that which they produce. Within this contradiction, Marx was able to specify immanently and critically the possibility of capital's supersession, giving rise to the possibility of freedom and human agency. Marx's critical theory specifies the twofold task of critique and transformation via immanent critique, which begins by accounting for its immersion in history. But critical theory is not merely descriptive; it also seeks to specify the possibility of qualitative social transformation (Melathopoulos and Stoner 2015). Indeed, the dynamic and contradictory nature of modern capitalist society is what normatively compels and analytically enables Marx to develop tools capable of elucidating critical recognition of the problematic features of modern capitalist society and the related consequences that result from how our lives are created. Moore's approach, on the other hand, dismisses entirely the duality of subject and object. Consequently, he is unable to confront the issue of domination and its legitimacy.

The categories of Marx's analysis, such as value, labor, and commodity, are reflective of a historically-specific form of social life characterized by internal contradictions (e.g., the opposition between abstract and concrete, general and particular) and, as such denote alienated modes of being comprehended by thought. In contrast, Moore emphasizes the contradiction between "paid work" and "unpaid work" as that which points beyond the Capitalocene. Yet, because Moore does not adequately distinguish traditional society, the Enlightenment, and the Capitolocene, he is unable to fully recognize that, while the opposition between "paid work" and "unpaid work" develops within capitalism, it does not point beyond capitalism. Gleaning insight from Marx, critique must instead be rooted in the double-sided nature of capital's dual social forms, and it must be directed toward the dynamic totality of social relations that he called capital. Unfortunately, Moore's rejection of subject-object dualism forecloses the possibility of such immanent critique. While Moore's Capitalocene alternative offers much historical nuance, like the Anthropocenarians he criticizes, this insight is confined to descriptive history.

Foster

J.B. Foster is one of the most well-known eco-Marxists. Foster (2016, p. 394) explains that the world today is in a period of transition of immense consequence, "represented by the advent of the Anthropocene, coupled with the emergence of what could be called the Age of Ecological Enlightenment." For Foster (2016), the Anthropocene is describing something real, which he likens to a "'second Copernican Revolution', fundamentally altering the way in which human beings perceive their relation to the earth" (393). Foster (2016, p. 394) understands the emergence of Anthropocene discourse in relation to the concomitant rise of the global environmental movement. Foster understands the global environmental movement as a positive, progressive force fighting against "bad" capitalists who gain short-term benefit while their actions drive the Great Acceleration (ibid.).

Below we elaborate Foster's position in regards to the Anthropocene, starting with his theory of metabolic rift. We then discuss how Foster understands the Great Acceleration. And finally, we turn to discuss Foster's solution to the problem of freedom embodied in the Anthropocene argument.

The Anthropocene and Metabolic Rift

Perhaps more than any other Marxist thinker in the past two decades, Foster has popularized "Marx's ecology" and, in so doing, has made significant headway in dispelling the conventional view of Marx as an enemy of nature (see Foster 1999). Foster is most well-known for his concept of metabolic rift (Foster 1999, 2000), which is an attempt to conceptualize nature-society interaction within capitalist society and is rooted in Foster's interpretation of Marx as a social theorist concerned with the fundamental metabolism between humans and nature. Foster explains the theoretical premise of his approach as follows:

> It was in *Capital* that Marx's materialist conception of nature became fully integrated with his materialism conception of history. In his developed political economy, as presented in *Captial*, Marx employed the concept of 'metabolism' (*Stoffwechsel*) to define the labor process as 'a process between man and nature, a process by which man, through his own actions, mediates, regulates and controls the metabolism between himself and nature.' Yet an 'irreparable rift' had emerged in this metabolism as a result of capitalist relations of production and the antagonistic separation of town and country. Hence under the society of associated producers it would be necessary to 'govern the human metabolism with nature in a rational way,' completely beyond the capabilities of bourgeois society. (Foster 2000, p. 141)

Taking a lead from Marx, Foster places human labor at the center of his analysis of environment-society relations under modern capitalism. Marx understood labor and production in general as the metabolic relation between humanity and nature (see, e.g., Marx 1976 [1867], p. 283). For Marx, production is a social process bound to the "universal metabolism of nature" in which material use values are "appropriated from the 'natural world' and transformed by production into social use values to fit 'human needs'" (quoted in Foster 2013, p. 5). However, with the emergence of capitalism, beginning with the dissolution of landed property, labor becomes alienated from its "prior communal relation to the Earth," which, under the capitalist mode of production is "'historically dissolved' in its entirety" (Marx, quoted in Foster et al. 2010, p. 283; see also Marx 1973 [1857/58], p. 497). According to Foster (1999, p. 384), the metabolic rift became most acute during the so-called "second agricultural revolution" (ca. 1830–1880), finding expression in both the local (the "antagonistic division between town and country") and the global as "whole colonies saw their land, resources, and soil robbed to support the industrialization of the colonizing countries" (cf. Marx 1973 [1857/58], p. 485).

Foster attempts to specify the tension between capitalist production and the natural environment by emphasizing the contradiction between use-value and exchange value.[8] Interpreting Marx in this way, Foster (2013, p. 2) enumerates the growth imperative of capitalist production:

> It is not *use value*, fulfilling concrete, qualitative needs, that constitutes the aim of capitalist production, but rather *exchange value*, generating profit for the capitalist. The abstract, purely quantitative nature of this process, moreover, means that there is no end to the incentive of seeking more money or surplus value, since M′ leads in the next circuit of production to a drive to obtain M″, followed by the drive to obtain M‴ in the circuit after that, in an unending sequence of accumulation and expansion. (Foster 2013, p. 2, original emphases)

In other words, the tendency toward rising productivity and ever-greater expansion is contingent upon the increasing exploitation of land and labor, which in capitalism are regarded as "free" gifts of nature. The concept of rift thus illustrates how labor exploitation and environmental degradation are co-determined by the same historical process; namely, capital's unending quest to accumulate and expand.

The Great Acceleration, Monopoly Capitalism, and Contemporary Environmentalism

More recently, Foster has attempted to update his theory of metabolic rift for the twenty-first century, in order to confront today's "epochal crisis"—a term Foster uses to name "the convergence of economic and ecological contradictions in such a way that the material conditions of society as a whole are undermined, posing the question of a historical transition to a new mode of production" (2013, pp. 1–2). Specifically, Foster attempts to account for the transition from liberal/competitive to monopoly capitalism, which reaches its full maturity after WWII and, in terms of its ecological impact, is analogous to the Great Acceleration.

Drawing on Baran and Sweezy's (1966, pp. 131–139) analysis of the "interpenetration effect," Foster emphasizes the intermixing of sales costs with production costs, which, under monopoly, become virtually indistinguishable. Foster explains that with monopoly capitalism, the contradiction at the heart of the metabolic rift (between use-value and exchange value) is warped even further "through the displacement of natural-material use value by *specifically capitalist use-value*—the only real 'use' of which is to enhance exchange value for the capitalist" (2013, p. 2). In competitive capitalism, only the minimal costs of production, along with the minimal costs of packaging, transportation, and distribution could be recognized as "socially necessary costs of purveying a product to its buyer" (Baran and Sweezy 1966, p. 131). Under these conditions, the product itself "could be legitimately considered an object of utility satisfying a genuine human need" (op. cit., p. 132). Under monopoly capitalism, however, the intermixing of the sales effort with production costs had grown so remarkably that "the use-value structure of the economy could no longer be viewed as a rational expression of production costs" (Foster 2013, p. 6). Gleaning insight from Baran and Sweezy, Foster amends Marx's M-C-M' formula to account for the transition from competitive to monopolistic capitalism. With the displacement of natural-material use value by capitalist use value, Foster (2013, p. 2) contends "Marx's general formula for capital, as it pertains to production itself, has metamorphized into $M\text{-}C^{K}\text{-}M'''$" (where C^{K}=capitalist use values, i.e., wasteful and destructive commodities such as military hardware, massive sales efforts, excessive packaging, as well as the appreciation of financial assets). In other words, as a result of the interpenetration effect, "an ever-larger proportion of what were considered costs of production [are] in fact forms of waste imposed by the system, i.e., specifically capitalist use values (C^{K})" (Foster 2013, p. 6).

Monopoly capitalism has developed in a highly uneven fashion, particularly with finance capital's rise to prominence and the advent of global neoliberalism during the last quarter of the twentieth century. Seeking the lowest unit labor costs worldwide, multinational corporations have shifted the majority of industrial production to export zones in the Global South (Foster 2013, p. 4). Labor and raw materials appropriated from the periphery tend to realize their value in the consumption-based centers of wealthier nations, who in turn export polluting technologies and hazardous waste back to the Global South in a process known as *ecologically unequal exchange* (Frey 2015).[9]

For Foster (2013, p. 8), the epochal crisis is "traceable to the growing distortion, displacement, and degradation of natural-material use values." While it is impossible to calculate the economic output imputable to that which is exploited and/or wasted, Foster (2013, p. 7) claims that "the sheer magnitude of such unproductive expenditures" paves the way "for the potential of a more rational, more sustainable society to satisfy real human needs."[10] Moreover, Foster (2013, p. 9) contends that "the objective forces today are progressively erasing previous distinctions between workplace exploitation and environmental degradation—as capitalism universally undermines all real-material conditions of production."

Ecological Revolution?

For Foster, the socio-ecological devastation brought about by the Great Acceleration under conditions of monopoly capitalism gives rise to a growing environmental awareness (beginning in the late 1960s and cumulating in the Anthropocene discourse), resulting in the global environmental movement. Foster (2016, p. 394) views the contemporary environmental struggle as an opposition between environmentalists (who pay, or otherwise direct attention toward, the "costs" of production) and capitalists (who gain short-term benefit as their actions drive the Great Acceleration). Unlike Crutzen et al., who respond to the challenge of the Anthropocene by presupposing the existence of the "Stewards of the Earth," Foster (2016, p. 394) understands this possibility, including what this transition might entail, in relation to capitalism's converging economic and ecological crisis.

According to Foster (2015), the most recent phase of global neoliberal capitalism has exacerbated the social and ecological tensions which underlying such unequal exchange, giving rise to "a nascent 'environmental proletariat—a broad mass of working-class humanity who recognize, as a result of the crisis of their own existence, the indissoluble bond between economic and ecological conditions" (Foster 2015). Because the poor are disproportionately impacted by the unpaid costs of capitalist production, Foster insists that a socialist revolution, or what he terms "ecological revolution," is more likely to take place in the Global South.

According to Foster, the growth of the global environmental movement (e.g., the climate justice movement) is premised upon an emergent "environmental working class" whose members recognize the socially irrational character of capital-induced environmental degradation. Hence, for Foster (2016, p. 394), it is not surprising that the Great Acceleration parallels the rise of the global environmental movement, as the socially irrational nature of capitalist production was recognized by a new generation of Left environmentalists during the 1960s and 1970s.

However, Foster's assumption regarding the deepening of capitalism's internal contradictions, which gives rise to a nascent environmental working class, does not withstand scrutiny.[11]

Indeed, Foster's position is deeply flawed. Historically, the environmental discontents expressed by contemporary environmentalism failed to engender changes in social structure conducive to moving beyond the societally-induced environmental degradation that characterizes this period. In fact, the exact opposite occurred, as the growth of environmentalism throughout the 1970s and 1980s coincided with the advent of neoliberal global capitalism whose penetration continues to define our current moment. Unfortunately, Foster offers no plausible explanation for the environment-society problematic. Perhaps because of this, Foster continues to predict an impending ecological revolution, but like Crutzen et al., he fails to provide insight into how such a wholescale future transformation might be possible.

In asserting the reality of an emerging environmental working class, Foster ignores the cultural mediations between his ideal-typical model of class attitudes toward the environment and the real consciousness of classes (cf. Feenberg 1979). As Andrew Feenberg (1979) pointed out more than three decades ago in his critical commentary on Barry Commoner, workers' "objective position" in relation to environmental harms does not necessarily lead to their militant anti-capitalist opposition to environmental harms. Moreover, working-class support for Donald J. Trump in the United States draws into relief how weak the connection between environmentalism and labor has grown since Commoner's moment. Today, action on climate change is more liable to be mistaken with a defense of the status quo (and the Democratic Party) than any plausible strategy toward the independent political articulation of workers. While Foster should be cognizant of his own social-historical embeddedness, his insistence on an impending "ecological revolution," premised on an emerging environmental working class, suggests otherwise.

Indeed, how Foster understands the solution to the crisis of the Anthropocene is tightly coupled with how he understands the source of the crisis. As noted above, Foster's ecological critique of capitalism is rooted in the contradiction between exchange value (profit) and use-value (land and labor). As the fungability of the latter is accelerated by capital's endless quest to accumulate, rising labor exploitation and increasing environmental degradation occur apace. Like any good

Marxist, Foster understands the task of critique—that it is not enough to name and confront the current stage of capitalist development; one must also specify the possibilities of moving beyond the present crisis. Unfortunately, Foster's use of Marx's categories forecloses insight into a future beyond capitalism.

Contrary to Foster's theory of metabolic rift, Marx does not root his critique of alienation in the opposition between "first nature" (i.e., the original identity of humanity and nature) and "second nature" (the social metabolic order of capital). Although the conceptualization of an original identity of humanity and nature only applies to pre-bourgeois forms, this conceptualization only becomes possible with the development of bourgeois relations of production, where the original identity of humanity and nature turns into its equally abstract opposite; that is, "the radical divorce of labor from its objective natural conditions" (Schmidt 2014 [1962], p. 82). Marx's position neither confirms the social metabolic order of capital as transhistorical nor denies the existence of a material substratum independent of labor. Rather, Marx treats both conceptualizations of nature as the expression of alienated social relations, which is why the opposition of "first nature" and "second nature" is not an opposition of noncapitalist and capitalist moments. The fact that the original identity of humanity and nature can be, and has been projected backward onto all of human history is itself reflective of capital's specific historical logic—namely, its abstract generality (cf. Postone 1993, pp. 17–18).

Foster's approach is rooted in an uncritical and ahistorical affirmation of "labor" (as the universal metabolic relation between society and nature), which then becomes disrupted in the social metabolic order of capital, giving rise to an irreparable rift. However, in capitalism cause and effect [of alienation] are inverted. We must therefore recognize critically the non-identity of these subjective-objective moments because, although contradictory and even opposed to one another, the subject and the object of labor *appear* unified in social practice (cf. Stoner 2014, p. 633).[12] Marx registers this problem regarding the failure to advance beyond the terms of bourgeois society in his critique of the famous French socialist, Pierre-Joseph Proudhon. For although capital appears to be the result of alienated labor, Marx (1988 [1844], p. 81) contends that alienated labor is the real cause of capital and private property. This is why Marx (81–82), in his polemic against Proudhon, argues: "A *forcing-up of wages* (...) would therefore be nothing but *better payment for the slave.*" If, logically speaking, alienation is more fundamental than private property, then it is conceivably possible to abolish private property but not alienation, so that society therefore becomes "an abstract capitalist" (82). For Marx, the reason to take up Communism is not its one-sided and obvious features, but because it really points beyond the aspects that are obvious to the participants. Foster, by contrast, wants to pit workers vs. capitalists as use-value vs. exchange value, or environmentalists vs. affluent society, or the Global South vs. monopoly. Each time, however, Foster misses Marx's point of being able to advance through surface phenomena critically *and* imminently.

CONCLUSION

Following the emergence of what earlier thinkers termed market (or civil or bourgeois) society, Marx further specified the problem of social transformation under the conditions of the Industrial Revolution during the early nineteenth century. Following Kant, Rousseau and Adam Smith, Marx viewed the significance of capitalism before the Industrial Revolution in terms of the freedom of humanity to transform itself: "the absolute elaboration of [humanity's] creative dispositions, without any preconditions other than antecedent historical evolution which the totality of this evolution—i.e., the evolution of all human powers as such, unmeasured by any previously established yardstick—an end in itself" (Marx 1973 [1857/58], p. 488). But with the Industrial Revolution, Marx registers another shift in which "this complete elaboration of what lies within man, appears as the total alienation, and the destruction of all fixed, one-sided purposes as the sacrifice of the end in itself to a wholly external compulsion" (ibid.).

Marx recognized that the conceptualization of a radical rupture with the past is only possible within the context of bourgeois society, in which the productivity of humanity had fully moved from the rural forms of peasant production to production in cities—and the rise of Third Estate—the class of "commoners" who were to be judged not on the basis of tradition or divine orders but on their capacity to "work". As Marx (1973 [1857/58], p. 109) writes, "World history has not always existed; history as world history is a result." And, in *The German Ideology* Marx and Engels (1845) state: "We know only a single science, the science of history," which, as Marx understood far too well, is a product of bourgeois society. Later, in a letter to JB Schweizer (in which Marx attacks Proudhon for his superficial knowledge of political economy), Marx (1865) explains that this "new" science can only be derived "from a critical knowledge of the historical movement, a movement which itself produces the *material conditions of emancipation.*"[13]

In developed capitalism, "individuals *seem* independent (...) free to collide with one another and to engage in exchange within this freedom" (Marx 1973 [1857/58], pp. 163–164). But for Marx, such personal independence is illusory, for this relation takes place in a context of objective dependence, in which a significant portion of the population does not consume what they produce, but rather sells their labor as a commodity in order to receive a wage to buy commodities from others.[14] Hence, freedom cannot be actualized within the narrow bourgeois form. While capital appears to serve humanity's end; in actuality, capital is the end to which humanity serves. This contradiction becomes particularly acute after the Industrial Revolution. Following Hegel, Marx understood the categories of political economy, such as commodity, labor, and value as categories of bourgeois society, and he illuminated these categories as self-contradictory. For example, capital renders labor obsolete (thereby generating the conditions of its own supersession) *and* necessary (i.e., a structural precondition for the capitalist production of value) at the same time. It is this dialectic, related to the internal contradictions of the value form of wealth, which Marx grasped, and which remains pertinent today (see Stoner and Melathopoulos 2015).

Marx's emphasis on modes of production allowed him to identify and explain historical transformations. History becoming world history is a historically-specific reflection of bourgeois society. Marx recognized the crisis of bourgeois society in capital in which the tremendous "development of all human powers" is determined by the need to produce surplus value (measured in socially necessary labor time). But capital also points beyond itself insofar as it furnishes the conditions under which the development of man's potentials might become an end in itself. According to Louis Menand (2003), "Marxism was founded on an appeal for social justice, but there were many forms that such an appeal might have taken. Its deeper attraction was the discovery of meaning, a meaning in which human beings might participate, in history itself." Marx is able to advance a praxis adequately attuned to the meaning of capital in history because he was able to reflect upon the present from the perspective of capital pointing beyond itself.

Meaning in history is precisely what the Anthropocene fails to grasp. The Anthropocene says something about history—the history of the present—to be sure, but it does so "one-sidedly" (change comes from factors exogenous to the system). Crutzen et al. (2011), for example, identify "runaway" economic growth as a prominent driver of the Great Acceleration. In doing so, Anthropocene scholars implicitly recognize human activity as increasingly constrained by structures we ourselves create—a conceptualization, which, descriptively, is not too dissimilar from what Marx understood as capital's abstract domination. Whereas Marx was able to trace such domination, including its possible overcoming, back to existing social relations, the Anthropocene is unable to do so, thereby naturalizing what is historically-specific about the relationship between modern capitalist society and the natural environment. As a result, Anthropocenarians do not recognize that the system's abstract generality *is its historical specificity*. Beyond measuring biophysical indicators in ice cores, the Anthropocene needs to be recognized as theory on what history is—not as descriptive history, but as a theory of how humanity might fundamentally be transformed.

Without a critical knowledge of history (Marx's "new" science), it is impossible to understand the environment-society problematic (increasing environmental degradation amid growing environmental attention and concern). The environment-society problematic is not the result of Cartesian dualism (Moore 2016) or capitalist "interests" (Foster 2016). Rather, as environmental degradation is becoming increasingly visible and less deniable, the paradoxical process at work remains largely concealed. Recognition of global ecological problems is not actual (critical) recognition if it is confined to the form of appearance of ecological concern within existing social conditions. Indeed, we should not mistake growing environmental awareness as being other or more than what it is, namely, a product of alienation and reification (see Stoner and Melathopolous 2015). For it is only on the basis of such critical recognition that politics is possible.

In this chapter, we addressed why a philosophy of history has become so difficult and how, in light of this difficulty, practical activity to change things (such as environmentalism) fails to act on history, but rather ends up rationalizing how powerless such activity actually is. This is how activity that might have seemed radical in the 1960s New Left has found itself opposed to change in the present. Activity that does not attend to the growing gulf between theory of change in history and action, paradoxically, becomes a form of historical inertia.

Notes

1. Both Anthropocenarians as well as their eco-Marxist critics aspire to a philosophy of history. While these scholars may not recognize their endeavors as such, both implicitly attempt to circumscribe the meaning of this history. Below we confront this dimension of the Anthropocene—as a theory of history.

2. In *Freedom in the Anthropocene: Twentieth-Century Helplessness in the Face of Climate Change* (2015), we advance a critical reading of this history through three theorists who span the Great Acceleration. Georg Lukacs and his critique of reification in the wake of the Russian Revolution in 1923, Theodore W. Adorno and his critique of identity thinking in his *Negative Dialectics* after the collapse of revolutionary politics by the 1930s and evaporation of the political tasks of Marxist theory, and lastly Moishe Postone's critique of traditional Marxism in the wake of the return to Marx by the 1960 New Left (1993).

3. Specifically, Moore is referring to Descartes' separation of mind and body and, more generally, the era of scientific revolution mostly associated with Descartes. Moore notes three aspects central to the Cartesian dualistic worldview that emerged during this time: (1) the imposition of "an ontological status on entities (substances), as opposed to relationships (that is to say, energy, matter, people, ideas and so on become things)"; (2) the imposition of an either/or logic (rather than both/and); and (3) "the 'idea of a purposive control over nature through applied science'" (Glacken, quoted in Moore 2015, pp. 19–20). Although the distinction between humanity and the rest of nature predates capitalism, according to Moore, this duality takes on a new meaning in history with the emergence of capitalism. As he notes, "Never before (…) had a civilization organized around a praxis of external nature: a world-praxis in which representations, rationality, and empirical investigation found common cause with capital accumulation in creating Nature as external" (Moore 2015, p. 17).

4. "The story of Humanity and Nature conceals a dirty secret of modern world history. That secret is how capitalism was built on excluding most humans from Humanity—indigenous peoples, enslaves Africans, nearly all women, and even many white-skinned men (Slavs, Jews, the Irish). From the perspective of imperial administrators, merchants, planters, and conquistadores, these humans were not Human at all. They were regarded as part of Nature, along with trees and soils and rivers—and treated accordingly (…) The symbolic, material, and bodily violence of this audacious separation—Humanity and Nature—performed a special kind of 'work' for the modern world. Backed by imperial power and capitalist rationality,

it mobilized unpaid work and energy of humans—especially women, especially the enslaved—in service of transforming landscapes with a singular purpose: the endless accumulation of capital" (Moore 2016, p. 79).

5. Our discussion of Moore's "world-ecology" is limited to aspects of this approach which we deem relevant to the purposes of the present chapter. Readers are encouraged to see Moore (2015) for a detailed elaboration.

6. As Moore (2015, p. 17) explains, "So important is the appropriation of unpaid work that the rising rate of exploitation depends upon the fruits of appropriation derived from Cheap Natures, understood primarily as the 'Four Cheaps' of labor-power, food, energy, and raw materials."

7. To be more precise, value is the objectification of abstract social labor (socially necessary labor time). In Marx, however, there is a distinction between value and its measure, which is necessarily expressed through exchange. Historically, the key transition where the emergence of abstract social labor as a general social phenomenon, is when time expenditure (measured in "independent" abstract temporal units) "is transformed from a result *of* activity into a normative measure *for* activity" (Postone 1993, pp. 214–215).

8. Marx employed the formula M-C-M' (Money-Commodity-Money)—where the difference between M and M' is necessarily only quantitative; that is, "buying in order to sell dearer" (1976 [1867], p. 256)—to capture "the general formula for capital, in the form in which it *appears* directly in the sphere of circulation" (1976 [1867], p. 257, emphasis added).

9. Here it is worth bearing in mind that the recent explosion of "green" technologies, including the "greening" of many cities in the Global North, is made possible by outsourcing dirty industry elsewhere (Parr 2013, p. 14).

10. According to Foster (2013, pp. 8–9), "This potential is manifested in the exploitation, waste, idle capacity, displacement of use values, and rapacious destruction of real wealth that characterizes the present system. The gigantic misuse of human and natural resources that constitutes the modern capitalist economy means that we already have the potential several times over to redirect production and consumption to meet human needs and to practice conservation on a global level, creating a society of ecological sustainability and substantive equality."

11. The notion that a socialist revolution in advanced capitalism is something that can be entertained as a possibility is highly unlikely in even the most crisis-prone advanced capitalist societies, for the very simple reason that the majority of individuals have identities that would resist the prospect of socialist revolution with all means available. It is implausible that the kind of revolution that would be required, at the necessary scale, could

practically occur, not just at present, but either for a long time to come, or at all. There are many reasons for this (e.g., people's identities, pre-conceived notions, the power of ideology, the lack of vision and imaginary, the warped mode of the social, the lack of solidarity except in small groups, the prevalence of alienation, the absence of a notion of reconciliation, and the resistance to all constructive efforts by those opposed to them).

12. As Postone (1993, p. 222) explains: "In capitalism, both moments of people's relation to nature are a function of labor: the transformation of nature by concrete social labor can, therefore, seem to condition the notions people have of reality, as though the source of meaning is the labor-mediated interaction with nature alone. Consequently, the undifferentiated notion of 'labor' can be taken to be the principle of constitution, and knowledge of natural reality can be presumed to develop as a direct function of the degree to which humans dominate nature."

13. Significantly, Marx continues in this letter to explain that, while Proudhon recognizes that "every economic relation has a good and a bad side (...) he sees the good side expounded by the economists; the bad side denounced by the socialists. He borrows from the economists the necessity of eternal relations; he borrows from the socialists the illusion of seeing in poverty nothing but poverty (instead of seeing in it the *revolutionary and destructive aspect which will overthrow the old society*)" [emphases added].

14. "These objective dependency relations also appear, in antithesis to those of *personal* dependence (the objective dependency relation is nothing more than social relations which have become independent and now enter into opposition to the seemingly independent individuals; i.e. the reciprocal relations of production separated from and autonomous of individuals) in such a way that individuals are now ruled by *abstractions*, whereas earlier they depended on one another" (Marx 1973 [1857/58], p. 164).

Image 3.1   Arctic tall ship study (Mia Feuer, watercolor)

REFERENCES

Adorno, T. W. (1998 [1962]). Progress. In H. W. Pickford (Ed.), *Critical Models: Interventions and Catchwords* (pp. 143–160). New York: Columbia University Press.

Ahuja, N. (2015). Intimate Atmospheres: Queer Theory in a Time of Extinctions. *GLQ: A Journal of Lesbian and Gay Studies, 21*(2–3), 365–385.

Baran, P. A., & Sweezy, P. M. (1966). *Monopoly Capitalism: An Essay on the American Social and Economic Order.* New York: Monthly Review Press.

Chernilo, D. (2016). The Question of the Human in the Anthropocene Debate. *European Journal of Social Theory.* Online First: June-02-2016. https://doi.org/10.1177/1368431016651874.

Collinicos, A. (1999). *Social Theory: A Historical Introduction*. New York: New York University Press.

Crutzen, P. J., & Steffen, W. (2003). How Long Have We Been in the Anthropocene Era? *Climate Change, 61*(3), 251–257.

Feenberg, A. (1979). Beyond the Politics of Survival. *Theory and Society, 7*(3), 319–361.

Fischer, J., et al. (2007). Mind the Sustainability Gap. *Trends in Ecology & Evolution, 22*(12), 621–624.

Foster, J. B. (1999). Marx's Theory of Metabolic Rift: Classical Foundations for Environmental Sociology. *American Journal of Sociology, 105*(2), 366–405.

Foster, J. B. (2000). *Marx's Ecology: Materialism and Nature*. New York, NY: Monthly Review Press.

Foster, J. B. (2013). The Epochal Crisis. *Monthly Review, 65*(5), 1–12. Available at: http://monthlyreview.org/2013/10/01/mr-065-05-2013-09_0/. Accessed January 19, 2017.

Foster, J. B. (2015). The Great Capitalist Climacteric: Marxism and "System Change Not Climate Change". *Monthly Review, 67*(6). Available at: https://monthlyreview.org/2015/11/01/the-great-capitalist-climacteric/. Accessed April 2, 2018.

Foster, J. B. (2016). Marxism in the Anthropocene: Dialectical Rifts on the Left. *International Critical Thought, 6*(3), 393–421.

Foster, J. B., Clark, B., & York, R. (2010). *The Ecological Rift: Capitalism's War on the Earth*. New York: Monthly Review Press.

Frey, R. S. (2015). Breaking Ships in the World-System: An Analysis of Two Ship Breaking Capitals, Alang-Sosiya, India and Chittagong, Bangladesh. *Journal of World-Systems Research, 21*(1), 25–49.

Hobbes, T. (1994 [1668]). *Leviathan*. Indianapolis and Cambridge: Hackett.

Luke, T. W. (2015). On the Politics of the Anthropocene. *Telos, 172*, 139–162.

Marx, K. (1865). On Proudhon (Letter to JB Schweizer). Available at: https://www.marxists.org/archive/marx/works/1865/letters/65_01_24.htm. Accessed March 23, 2017.

Marx, K. (1973 [1857/58]). *Grundrisse*. New York: Penguin Books.

Marx, K. (1976 [1867]). *Capital, Volume 1*. New York: Penguin Books.

Marx, K. (1988 [1844]). *Economic and Philosophic Manuscripts of 1844*. New York: Prometheus Books.

Marx, K., & Engels, F. (1845). *The German Ideology*. Available at: https://www.marxists.org/archive/marx/works/1845/german-ideology/ch01a.htm. Accessed March 27, 2017.

Melathopoulos, A., & Stoner, A. M. (2015). Critique and Transformation: On the Hypothetical Nature of Ecosystem Service Value and Its Neo-Marxist, Liberal and Pragmatist Criticisms. *Ecological Economics, 117*(C), 173–181.

Menand, L. (2003). Foreword. In E. Wilson (Ed.), *To the Finland Station* (pp. vii–xx). New York: New York Review of Books.

Moore, J. W. (2015). *Capitalism in the Web of Life: Ecology and the Accumulation of Capital.* New York: Verso.

Moore, J. W. (2016). The Rise of Cheap Nature. In J. W. Moore (Ed.), *Anthropocene or Capitalocene? Nature, History, and the Crisis of Capital* (pp. 78–115). Oakland, CA: Kairos Books.

Parr, A. (2013). *The Wrath of Capital: Neoliberalism and Climate Change Politics.* New York: Columbia University Press.

Postone, M. (1993). *Time, Labor, and Social Domination: A Reinterpretation of Marx's Critical Theory.* New York: Cambridge University Press.

Proudhon, P. J. (1840 [1890]). *What Is Property? An Inquiry into the Principle of Right and of Government.* New York: Humboldt Publishing Company. Available at: https://www.marxists.org/reference/subject/economics/proudhon/property/index.htm. Accessed March 23, 2017.

Schmidt, A. (2014 [1962]). *The Concept of Nature in Marx.* New York: Verso.

Schulz, K. A. (2017). Decolonizing Political Ecology: Ontology, Technology and 'Critical' Enchantment. *Journal of Political Ecology, 24,* 125–143.

Steffen, W., et al. (2007). The Anthropocene: Are Humans Now Overwhelming the Great Forces of Nature. *Ambio: A Journal of the Human Environment, 36*(8), 614–621.

Steffen, W., Grinevald, J., Crutzen, P., & McNeill, J. (2011). The Anthropocene: Conceptual and Historical Perspectives. *Philosophical Transactions of the Royal Society A: Mathematical, Physical and Engineering Sciences, 369*(1938), 842–867.

Stoner, A. M. (2014). Sociobiophysicality and the Necessity of Critical Theory: Moving Beyond Prevailing Conceptions of Environmental Sociology in the USA. *Critical Sociology, 40*(4), 621–642.

Stoner, A. M., & Melathopoulos, A. (2015). *Freedom in the Anthropocene: Twentieth-Century Helplessness in the Face of Climate Change.* New York: Palgrave.

CHAPTER 4

Making Our Way in a World of Our Making: The Anthropocene, Debt-Money, and the Pre-emptive Production of Our Future

Matthew Tiessen

ANTHROPOCENIC ARTIFICE AND THE NEW NATURES OF CONTROL

The Anthropocene is a geological period defined by recent and ongoing human activities that invite decisive responses by humanity in order to grapple with contemporary and future ecological, biological, and social change. In this essay, I describe a variety of current "natural" and cultural scenarios and draw on some authors not typically associated with weather disturbances and atmospheric change (Deleuze, Shaviro, Marx) to argue that the concept of the Anthropocene is also one that has the potential to be mobilized in service of unanticipated and emerging forms of social, spatial, political, financial, and ecological control. Indeed, we can perhaps imagine a not too distant future wherein the apparent urgency of the Anthropocene era's challenges may require global responses that bypass democratic deliberation and necessitate—or are said to necessitate—new expressions of global, regional, and local power. At the same time I want to emphasize that emerging modalities of control in the Anthropocene will

M. Tiessen (✉)
School of Professional Communication, Ryerson University,
Toronto, Canada

© The Author(s) 2018
j. jagodzinski (ed.), *Interrogating the Anthropocene*, Palgrave Studies in Educational Futures, https://doi.org/10.1007/978-3-319-78747-3_4

likely be sold to the public as being in their interest (and in many instances that might in fact be the case). It is the potential proliferation of these emerging expressions of power in the name of the Anthropocene that, I want to suggest, we must approach with caution as we seek solutions to tomorrow's social, ecological, and spatial challenges. Or as Simon Dalby puts it: "How the Anthropocene is interpreted, and who gets to invoke which framing of the new human age, matters greatly both for the planet and for particular parts of humanity" (2016, p. 33). Put differently, perhaps the question worth attending to as we hurtle into the future might not be whether humanity will experience a "good Anthropocene" (Ellis 2011) or a "bad Anthropocene," but rather: What will the Anthropocene mean (prior to deciding whether it's good or bad)? In other words, how will the Anthropocene be defined (not merely as a geological epoch but as a socio-cultural phenomenon or vector of control) and, who will do the defining (and for whom)? So, although when faced with the problems posed by the Anthropocene, the temptation will be to find solutions and to solve problems, perhaps we'd be better served by asking questions, by reflecting on what agendas are afoot and on whose interests are being served.

Indeed, we could go so far as to suggest that emerging collective enunciations and mobilizations in the name of the Anthropocene have the potential to bring with them their own forms of what Deleuze would describe as "new fascisms." For Deleuze, the old-style fascisms are no longer the threat they once were. Their tactics are too blunt, too obvious, too predictable. Like everything else, he observes, fascisms change and evolve with the times, and as we speak "new fascisms are being born ... and prepared for us" (2006, pp. 137–138). These new fascisms are not only about the "politics and economy of war" but masquerade as seemingly progressive "global agreements on security, on the maintenance of a 'peace' just as terrifying as war" (p. 138). Deleuze warns us that "all our petty fears will be organized in concert, all our petty anxieties will be harnessed to make micro-fascists of us; we will be called upon to stifle every little thing, every suspicious face, every dissonant voice in our streets [and] our neighborhoods" (p. 138). It is the potential proliferation and emergence of these micro-fascisms—or some variant thereof—in the name of a phenomenon like the Anthropocene that, I want to suggest, we must be wary of.

OPEN FOR BUSINESS AND THE PERPETUATION OF THE SAME IN THE ANTHROPOCENE

Bloomberg recently announced to its plutocrat and wanna-be-plutocrat readers that "The World has Discovered a $1 Trillion Ocean!" This purported "discovery" is, of course, the "new" ocean that was once the ice-filled Arctic. The Arctic, Bloomberg writes, "is open for business, and everyone wants a piece" (Roston 2016). This is what the Anthropocene looks like through the eyes of the status quo, of those eager to extend our neo-fascistic present well into the future (with the support of our democratically elected and financially supported political elite). As Bloomberg explains:

> there's no doubting the melting of the Arctic ice cap, and the unveiling of resources below, presents mind-boggling opportunities for energy, shipping, fishing, science, and military exploitation. [...] The financial measure of opportunities available [...] is difficult to estimate, but $1 trillion may be a solid first-pass. [...] The Arctic is warming faster than any other part of the globe, [and World Economic Forum member Jan-Gunnar Withner notes that:] "These changes are like nothing we have seen. We don't have anything to compare with it in history." (Roston 2016)

I suppose this is a good thing, isn't it? A new market primed for profit thanks to the Anthropocene! It certainly fulfills the need for more of the status quo: infinite extractive growth on our finite and fragile planet (Brown et al. 2011; Jackson 2011; Mueller and Passadakis 2009; Pretty 2013). We can all speculate, of course, about who will benefit from this Anthropocene-induced windfall (and that the beneficiaries will certainly not be donating funds to fight climate change). But where, we might ask, is all this money—a new 1 trillion—going to come from? The answer is a strange one indeed, one I think is important when grappling with the Anthropocene, and one that even mainstream economists are only beginning to fully grasp: The 1 trillion dollars the melting arctic promises will simply be brought into existence out of nothing—*ex nihilo*—by private banks in the form of loans made out to private corporations and investors (with interest owing, of course).

See, in today's financial ecologies (Haldane and May 2011), 97% of the money in existence, namely digitally generated credit (rather than the comparatively small amount of paper money and coins that are created by governments) is created out of nothing by the banks when, for example, we sign on the dotted line for a mortgage, a student loan, automotive financing, credit card purchases, or government spending (Aglietta 1979; Benes and Kumhof 2012; Moore 1989, 1979; Nichols 1992; Rochon 1999; Seccareccia 2012; Terzi 1986). In other words, there is no "borrowing" going on at all, there is only money creation, and—most importantly—the decision to create money/credit—for whom? for what cause? in support of what agenda?—is decided by a global banking oligarchy in accordance with its very specific and all too often short-sighted profit-driven agendas and priorities. And because all of this credit comes into existence with interest owing, there is always more debt needing to be paid back than there is money in circulation; and because of this the creation of money—that is, the perpetual borrowing of new money into existence—must continue ad infinitum so that debts will be able to be paid, inflation will be able to continue, and the current financial state of affairs will remain afloat. This is why, for example, interest rates have recently been so historically low—to encourage not "borrowing" for overpriced houses for you to flip or rent out on AirBnB, but to encourage the manufacturing of the money needed to keep the system of consumption, 2% inflation, and infinite growth going. Or as John Smithin explains in a recently edited volume by monetary reformers Louis-Philippe Rochon and Mario Seccareccia on the role of money in our global economies:

> To "keep the show on the road" in the future [...] it is clear that there is always going to have to be more borrowing activity [to keep inflating the economy], from one source or another, essentially *ad infinitum*. Given the structure of this financial system, and the specific nature of its money, the fact is that someone, somewhere, must always be willing to go into debt in order to generate profits for others. This statement is not an attempt to indicate that capitalist monetary production is nothing more than a "Ponzi scheme." That is not the point at all. It is merely a logical statement about the arithmetic of this particular social technology. It is the way things are, given the assumed set of social relations, or social ontology. (Smithin 2013, p. 47)

Keeping the "show on the road" is what the status quo maintaining Bloomberg headline—which purports to be about new "discoveries"—is really about. So in the age of the Anthropocene let's begin to take seriously, as Canadian economist Mario Seccareccia does, the fact that economic production itself is "a process of debt formation" (1988, p. 51) and that in today's world the *ex nihilo* manufacturing of bank credit is the dominant mode of financing production and of ecological destruction (Seccareccia 1996). "The flow of investment spending in a monetary economy is primarily a process of debt formation," he observes. "That is to say, in order for capital accumulation to proceed, firms must first become indebted to the financial sector, the purveyor of credit advances" (Seccareccia 1988, p. 51). And what better medium to encourage debt/credit creation than an all-new geological era—the Anthropocene!—with all its costs, expenses, emergent risks, security needs, and disciplinary imperatives?

I have written elsewhere about the significance and power of privately owned banks who have been granted the ability to create money out of nothing in potentially infinite quantities (Tiessen 2013, 2014a, b, 2015) and touch on this topic here because questions related to who gets to create money, and for what purpose, are critical if we are to pursue an ecological future—let alone an Anthropocene—that isn't simply another, greener, or greenwashed, version of contemporary, extractive form of capitalism (Common and Perrings 1992; Spash 2012). In other words, when we begin to imagine what might be possible in, or what forces might shape, the Anthropocene, we must not lose sight of the fact that whoever is in control of bringing (i.e. loaning) money into existence with a few keystrokes on a computer gets to decide what that money is being created for (Nesvetailova 2014). Consider that all wars, all coal plants, all oil rigs, all nuclear weapons programmes, and all tar sands projects begin with a signature of a borrower willing to take on debt and risk and the deposit of newly created money into someone's bank account by a bank in service of profit; consider too, that all the world's solar panels, wind farms, hospitals, universities, organic farms, etc. get created the same way. Who is doing the choosing and for what purpose? The purpose of creating loans is, primarily, the pursuit of profit, preferably short-term profit. So, who is making today's ecocidal loans? Who will make them tomorrow? What are their names? Who will profit most from the Anthropocene as a concept, rallying cry, public policy issue? Which banks? Loans for how much? What are the terms? When will the loans be created? Why do we allow such decisions to be made by purely financial interests in search of pecuniary gains and shareholder value? How is the public served when profit is the dominant paradigm? How does the environment benefit or lose? In sum, it's worth pondering the fact that most of today's institutionalized barbarism and eco-destruction—a vast majority of life's most heinous crimes and catastrophes—were facilitated through bankers' use of computers, privately and digitally created money, credit checks, and crediting international bank accounts. Even Marx understood the potential for the credit-creators to preemptively impose upon us the future's imperatives with their money-conjuring power. As he explains:

> the banking system, by its organization and centralization, is the most artificial and elaborate product brought into existence by the capitalist mode of production […] Banking and credit, however, thereby also become the most powerful means for driving capitalist production beyond its own barriers and one of the most effective vehicles for crises and swindling. (Marx 1981, p. 742)

Contemporary financial orthodoxy—and power—is worthy of our attention because of the fact that when the control of credit creation is in private and for-profit hands (Eisen 2016; Schnurr 2016), our collective ability to design *our* Anthropocene—or any other geological era of our own making that we deem name-worthy—democratically or strategically is short-circuited and taken from us. Indeed, under the current global financial apparatus the public interest—if recent financial history is any indication—may not even exist. Indeed, the public's role, it often seems, is simply to pay interest!

Perhaps most disturbingly, tomorrow's Anthropocene may have, in a very real way, already been foreclosed in so far as the banks will be the ones who decide what it will look like by making financial decisions they are able to benefit from. The economist Richard Werner has done a lot of recent work explaining to mainstream economists—many of whom ignore the role of credit money—about the role of money in an economy, the role of debt, the ways money is actually created, and the relationship of these processes to power. Crazily, most mainstream economists do not actually consider the way money's created or who gets to create it as being significant enough to be included in their financial models. This is a problem since, as Werner explains, "Recognition of the banks' true role [in the economy] is the precondition for solving many of the world's problems, including: the problem of recurring banking crises, unemployment, business cycles, underdevelopment, and depletion of finite resources" (Werner 2014, 2015). Werner explains (at length) the current global money situation as follows:

> The empirical facts are only consistent with the credit creation theory of banking. According to this theory, banks can individually create credit and

money out of nothing, and they do this when they extend credit. When a loan is granted by a bank, it purchases the loan contract (legally considered a promissory note issued by the borrower), which is reflected by an increase in its assets by the amount of the loan. The borrower "receives" the "money" when the bank credits the borrower's account at the bank with the amount of the loan. The balance sheet lengthens. Through the process of credit creation 97% of the money supply is created in the [world] today.... Not surprisingly, the use to which bank credit is put to determines its effect, namely whether bank credit is extended for productive, consumptive, or speculative purposes. One reason for the neglect of the institutional and operational details of banks in the research literature in the past decades is likely the fact that no law, statute or bank regulation explicitly grants banks the right (usually considered a sovereign prerogative) to create and allocate the money supply. As a result, many economists, finance researchers, lawyers, accountants, even bankers, let alone the general public, have not been aware of the role of banks as creators and allocators of the money supply. (Werner 2014, p. 71)

Remember too, that governments are also beholden to the whims of the financial gatekeepers in so far as the agendas the political classes put into motion require the borrowing of money which, in turn, requires those agendas being vetted—in one way or another—by those who have been given the right to create money for us (and their friends the credit-rating agencies). Perhaps a new era of public banking or government-issued money (as articulated, for example, in the "Chicago Plan" following the Great Depression) could allow for greater democratic governance and "people power" to be expressed through financial means (Benes and Kumhof 2012; Dittmer 2015)? Regardless, private domination of money creation needs to change by incorporating public financing in the public interest for public purposes (such as responding to Anthropocene issues).

Criticisms of monetary orthodoxy are rarely outright or overarching criticisms of money and banking in and of themselves, but they do critique the often invisible mechanisms of power and control that preemptively allow those in control of credit (not to mention those capable of shaping rhetoric related to speculative futures, e.g. the Anthropocene) to pull strings behind the scenes. Based on their decisions regarding investment capital, interest rates, profitability, return on investment, etc., the financial sector will decide what our world looks like tomorrow (and—unless real change occurs—the powerful mechanism they use will remain for the most part unknown to the public). The crucial question here is how do we live sustainably in a world of scarcity when money can be produced by for-profit entities infinitely (restricted only by banks' ability to contain or off-load, risk). Indeed, it is money's ability to be "automagically" created that is the real ecological risk. Moreover, the potentially destructive capacities of money and the infinite elasticity of credit creation are compounded by the fact that as resources get more scarce they also become more profitable (Baveye et al. 2013; Clark 1973).

My cautions about the potential uses and abuses of the concept of the Anthropocene are in light of the potential profits to be made off of multiple aspects of this ecological adventure. After all, the banks benefit by creating money for both oil exploration and oil spill cleanup. In response to this conundrum, we need a financial system that reflects these new realities. Moreover, as economist Mathew Forstater observes, the problem is not that today's money is limited or scarce (since it's created out of nothing), it's that it's not usefully distributed and that those who control and allocate credit control the future (Forstater 2003; see also, Wray 2015, p. 147).

Money is, among other things, simply an enabler of human desire, and the Anthropocene can be thought of as the era when it becomes clear that our desires, our ambitions, our capacities for consumption exceed the capacities of our planet, our home. How, then, we might ask, will beneficial ecological change occur if private banks continue struggling to recognize that ecofriendly change can be economically profitable on a massive scale? And how will any eco-initiative that threatens the status quo be able to survive the financial power and policy decision-making capacity of the banks, particularly if they continue pursuing profit at all (ecological) costs? Deleuze understood the power of money to be used by its creators to shape the future. He wrote: "[b]eyond the state it's money that rules [and] money that communicates" (1995, p. 152). For Deleuze and co-author Guattari, the money form—whether commodity money, credit-money (which we use today), central bank money, digital money, etc.—determines the ways ideologies, politics, and culture get expressed and shapes the dispositions and desires of those who are beholden to money's built-in demands (demands like profit, repayment, etc.). For them, money itself is a tool for initiating, modulating, and multiplying desire. They observe that "the productive essence of capitalism can itself function only in this necessarily monetary or commodity form that controls it." They posit that money's flows map onto and determine flows of desire—both collective and individual: "It is at the level of flows, the monetary flows included, and not at the level of ideology, that the integration of desire is achieved" (Deleuze and Guattari 1983, p. 239). In other words, money's effects *precede* and even *determine* ideology (what we often call neoliberalism, e.g., then becomes an *effect* of, or a response to, monetary phenomena since privatization of public infrastructure, e.g., is necessary because it isn't usually profitable, and ballooning government debts are owing). That is, it is the bankers who are the enablers and instigators of the proponents of neoliberalism, and it is the bankers who are the gatekeepers that separate us from our desires and their fulfillment—whether we're living in the Anthropocene or not. Moreover, it is we and our governments who give them permission—whether wittingly or unwittingly—to do what, thanks to us, they are capable of doing to the best of their ability! Deleuze and Guattari note too that "in a sense" it's the banking system that "controls the whole system and the investment of desire" (1983, p. 230), and highlight that a primary contribution of Keynes was his having reintroduced desire "into the problem of money" (ibid.). Indeed, given the financial sector's current credit generating capacities we might imagine that in a world of ecological collapse and resource scarcity, despite periodic debt-driven and speculative catastrophes, the last thing standing will be a banker ready to offer a loan in order to extract interest from that final purchase at the end of the line for the predatory Ponzi scheme (Minsky 2015; Nesvetailova 2008; Tiessen 2014a).

For Deleuze and Guattari, engaging the question of desire through money, and considering money through desire is what "must be subjected to the requirements of Marxist analysis" since, as they posit, Marxist economists unfortunately:

> too often dwell on considerations concerning the mode of production, and on the theory of money as the general equivalent as found in the first section of *Capital*, without attaching enough importance to banking practice, to financial operations, and to the specific circulation of credit money which would be the meaning of a return to Marx, to the Marxist theory of money. (1983, p. 230)

The most salient passages by Marx on money to which Deleuze and Guattari may be referring speak powerfully to our current macro-economic moment insofar as they examine the libidinal and affective dimension of what Marx called "vulgar economics"—a type of monetarily based economics wherein capital—as privately created credit—generates a debt that becomes "an independent source of wealth"; this form of usurious capital is, in Marx's view, "a godsend" for the money "lenders" and capital owners since its profits are "no longer recognizable" and since the capitalist production process obtains via the affordances of the money form "an autonomous existence" (Marx 1981, p. 517). In such a money system, capital—as virtual production capacity conjured from nothing as credit with debt owing—"becomes a commodity" with a "self-valorizing quality" and a "fixed price"—namely "the prevailing rate of interest" (ibid.).

Marx observes that as interest-bearing capital (credit-)money gains a sort of nonhuman urgency (if not agency) all its own, appearing magically "as a mysterious and self-creating source of interest, of its own increase" (1981, p. 516). Money as autopoietic flow of credit and debt parasitically preys on human desire, guilt, and greed. As Marx explains:

> In interest-bearing capital, therefore, this automatic fetish is elaborated into its pure form, self-valourizing value, money breeding money, and in this form it no longer bears any marks of its origin. The social relation is consummated in the relationship of a thing, money, to itself. Instead of the actual transformation of money into capital, we have here only the form of this devoid of content. *As in the case of labour-power, here the use-value of money is that of creating value, a greater value than is contained in itself.* Money as such is already potentially self-valorizing value, and it is as such that it is lent, this being the form of sale for this particular commodity. *Thus it becomes as completely the property of money to create value, to yield interest, as it is the property of a pear tree to bear pears. And it is as this interest-bearing thing that the money-lender sells his money.* (Marx 1981, p. 516, my italics)

This passage by Marx echoes Deleuze and Guattari's thinking about the role of money as an autopoietic medium of power and control and anticipates the ecocidal economic logic of credit money in the Anthropocene. Of course, perpetual growth cannot keep pace with the exponentially compounding growth of debt. For Marx in the *Third Critique*, money's demands (for debt repayment) infect all that it puts a price on and all desires that it promises to fulfill. Money's surplus value, Marx observes, accrues to it "as such": "Like the growth of trees, so the generation of money seems a property of capital in this form of money capital" (Marx 1981, p. 517). Marx's biologically based monetary metaphors are especially striking and provide fertile ground for describing money's capacity for growth and capacity to subsume reality according to its logic; he explains that "interest is the specific fruit of capital, the original thing," while profit—from actually selling goods and services with some necessary use value—becomes "a mere accessory and trimming added in [capital's] reproduction process. The fetish character of capital and the representation of this capital fetish is now complete" (1981, p. 516).

Worth emphasizing, however, is that Marx perhaps misinterprets "vulgar economics" as being an "irrational" form of capital, when in fact, the cold rationality of this logic—perpetrated by private interests on public borrowers and debtors—is the height of sound and profitable thinking (at least if power, accumulation, and control are your objectives). This rationality—"this ingrown existence of interest in money capital as a thing" (1981, p. 518) or capital as that which "reproduces itself and increases in reproduction, by virtue of its innate property as ever persisting and growing value" (1981, p. 519)—is one that depends on mythologies, illusions, hopes, and dreams and, although the true motives are concealed, manipulates desire with—literally—mathematical precision: "the ability of money as a commodity to valorize its own value independent of reproduction—the capital mystification in the most flagrant form" (1981, p. 516). Credit-money, as Marx observes, is not unlike a parasite for centralizing power and wealth by expropriating the wealth of those to whom it is extended. Crucially, this is not a mistake, nor is it irrational! Credit—according to this view—is extended like the fishing net of a deep-sea trawler, capturing everything in its path and laying waste to the fertile sea floor (or ground) from which wealth was generated in the first place. As Marx explains:

> Where the means of production are fragmented, usury centralizes monetary wealth. It does not change the mode of production, but clings on to it like a parasite and impoverishes it. It sucks it dry, emasculates it and forces reproduction to proceed under ever more pitiable conditions. Hence the popular hatred of usury, at its peak in the ancient world, where the producer's ownership of his conditions of production was at the same time the basis for political relations, for the independence of the citizen. (1981, p. 731)

Again, Marx's analysis could not be more clear:

> All wealth that can ever be produced belongs to capital in its capacity as
> interest-bearing capital, and everything that it has received up till now is
> only a first installment for its all-engrossing appetite. By its own inher-
> ent laws, all surplus labour that the human race can supply belongs to it.
> Moloch. (1981, p. 521)

Sometimes theorists predict or at least pine for the end of today's
rapacious form of capitalism, expecting it to collapse in on itself or
self-destruct. But, again the banks win since self-destruction and collapse
are the very lifeblood of money and credit creation since it clears the way
for new, more intensive rounds of money production (i.e. borrowing).
Steven Shaviro, in a recent blog post, anticipates the banker-built
Anthropocene of the near future as follows:

> Among all its other accomplishments, neoliberal capitalism has also robbed
> us of the future. It turns everything into an eternal present. The highest
> values are supposedly novelty, innovation, and creativity, and yet these
> always turn out to be more of the same. The future exists only in order to
> be colonized and made into an investment opportunity. In other words,
> we cannot hope to negate capitalism, because capitalism itself mobilizes
> a far greater negativity than anything we could hope to mount against
> it. The dirty little secret of capitalism is that it produces abundance, but
> also continually transforms this abundance into scarcity. It has to do so,
> because it cannot endure its own abundance. And yet, none of these con-
> tradictions have caused the system to collapse, or even remotely menaced
> its expanded reproduction. Instead, capitalism perpetuates itself through
> a continual series of readjustments. Nearly all of us, as individuals, have
> suffered from these blockages and degradations; but Capital itself has not.
> (Shaviro 2013)

Of course, the readjustments identified by Shaviro are not readjustments
at all but merely the perpetuation of the same—that is, the issuing of
more credit by finance capital following the inevitable saturation of the
debt carrying capacity of the financial Ponzi scheme. After all, capitalism
is a complex adaptive system and credit-driven capital survives by perpet-
uating, reconstituting, and manufacturing new desires, not satisfying or
satiating them. Or as Naomi Klein puts it in her book *This Changes
Everything*: "We are stuck because the actions that would give us the best

chance of averting catastrophe—and would benefit the vast majority—
are extremely threatening to an elite minority that has a stranglehold
over our economy, our political process, and most of our major media
outlets" (2014, p. 18).

NATURE'S BIOLOGICAL BENEFITS AND THE CULTIVATION OF FUTURE ECO-DESIRES

Recognition of the role of finance is critical if we are to move towards
a more ecologically balanced Anthropocene. Dalby concurs, observing that the direction "we collectively push the planet in is shaped by
human actions and decisions made mostly by the rich and powerful of
our spacies" (2016, p. 35). But how might we begin, as Jedediah Purdy
says, to become "different people" (in Andersen 2015) in a way that
puts us on course for a more ecologically friendly future? Is it possible
that the proverbial 99% could also play a part in steering or pushing our
planet in an ecologically sustainable direction? Indeed, assuming a non-
linear and unpredictable future, today's elite may not even be in position to "push" the planet of tomorrow. Regardless, one thing that will
be important if we are to move towards a "good Anthropocene" is for
human beings to *desire* this future, whatever it may look like. These days,
individuals in the West often gravitate towards material and physical
experiences they believe to be pleasurable, holistic, healthy. Consider the
explosion of consumer interest in such things as organic agriculture and
animal products (Batte et al. 2007; Moser 2015), higher quality direct
or fair trade coffees, and yoga (Askegaard and Eckhardt 2012). People
are drawn to these things because they value what they perceive to be
the mental and physical benefits they receive from paying a bit more for
something a bit better and a bit more in tune with the environment that
surrounds them. Sometimes people *demand* disruption, if the benefits
are tangible enough and can be easily experienced. Consider, for example, such demanded or desired disruptions as the admittedly for-profit
business models of Airbnb or UBER (both of which perpetuate contract
labor and precarity). But how do we create parallel, desirable realities and
societies appropriate to the Anthropocene? This pursuit of quality, senso-
rially beneficial, and authentic experiences finds expression in emerging
research on the physical benefits of being in natural environments, even

urban natural environments. In Japan, for instance, doctors are prescribing "forest bathing" and researchers are increasingly finding that immersion in urban and/or rural greenspaces is biologically and psychologically beneficial to peoples' health and wellbeing (Pietilä et al. 2015; Pröbstl-Haider 2015; Romagosa et al. 2015). Similarly, walking among trees has been shown to lower cortisol levels and feelings of stress (Thompson et al. 2012). Research in this area might inform our reflections upon the Anthropocene by providing us with scientific permission to engage with nature in ways we don't normally do in our busy urban lives.

Again, it's one thing for things to be good for us (like exercise), it's another for things to be enjoyable or desirable. Perhaps in the future we will begin to address both our need of exercise and our "nature deficit disorder" (Louv 2008, 2012) by getting outside and finding that we actually like it and that it feels good, and perhaps this engagement with natural environments and greenspaces could contribute to a desire for more conservation and less ecological destruction. If the public's appetite for cage-free chickens can convince McDonalds to require that their chicken suppliers go 100% cage free within five years, perhaps similar consciousness raising in other areas can result from the intersection of scientific research and people's desire for more pleasure and less stress in their harried lives.

It strikes me also that a positive way forward is not so much to try to empathize with nonhuman agents and give voice to animal and plant agencies (as important as these may be), but rather to expand or enlarge our sense of what it could possibly mean to be a bit *more* anthropocentric: that is, to think of taking care of the earth as synonymous with taking care of ourselves. As the technophilic authors of, for example, the Ecomodernist Manifesto declare: "To say that the Earth is a human planet becomes truer every day" (Asafu-Adjaye et al. 2015, p. 6). This could mean trying to imagine a future for humanity that is, for example, not simply in service to capital. Or this could mean developing an increased recognition that we're only harming ourselves when we destroy the ecosystems and environments that support us and give us life. Perhaps to think *more* anthropocentrically would mean developing a new appreciation for our integration with and interdependence on the world around us, to begin to imagine ourselves as biological beings composed of permeable membranes and flows, not to mention the nonhuman beings in our gut microbiome and beyond. How, in other words, can we as human beings better attune ourselves to earth's flows? How do we create relationships with nonhuman others of mutual beneficence—the relationships that, as Spinoza might say, will begin to show us what we're capable of doing. Or as Buckminster Fuller once famously observed: "You never change things by fighting the existing reality. To change something, build a new model that makes the existing model obsolete."

As Petra Hroch argues, one assumption about the Anthropocene, "as an account of the present moment," is that it "forces us to confront [...] the fact that further [conventionally] Anthropocentric thinking will not create a viable response to today's eco-crises" (2014, p. 4). But is anthropocentrism capable of changing, becoming more expansive and integrative, becoming more open to the outside? Surely, if the great acceleration of ecological destruction we're currently living through only began during the last century or two (speeding up post World War II), then our relationship to the natural world and to ourselves is one that can change drastically across times and spaces. This change will, of course, continue. As engineering and ethics professor Brad Allenby recently observed in the *New York Times*, the term Anthropocene "tends to reify humans as they are now" (2011). This, he observes:

> may well require adjustment in the future, since the suite of emerging technologies – nanotechnology, biotechnology, robotics, information and communication technology, and applied cognitive science – is rapidly making the human a design space. [...] And as humans increasingly integrate with the technology around them, and as the evolution of that technology continues to accelerate, it is questionable that what we will have in a couple of decades is still "Anthro" [as we currently understand it]. It is not just Earth systems, but the human itself, that are in the midst of radical and unpredictable change, and it is probably premature to evaluate what sort of system will come out of the process. (2011)

If anything, these unpredictable processes of change will result in the modification of ourselves, the modification of the environments that surround us, and the modification of our relationships to these environments and to our emerging identities. Whitehead once remarked that "Successful organisms modify their environment"; moreover, he added, "Those organisms are successful which modify their environments so as to assist each other." This law, he observes, "is exemplified in nature on a vast scale" (2011, p. 256). We can only hope that these modifications of ourselves and the worlds around us will be as mutually beneficial as possible by the time the next geological era—human created or not—makes the Anthropocene just a bad—or maybe good?—memory.

Image 4.1   Oil rig (Mia Feuer, watercolor)

REFERENCES

Aglietta, M. (1979). *A Theory of Capitalist Regulation: The US Experience.* London: New Left Books.

Allenby, B. (2011). What's Next, The Cognocene (The Age of Anthropocene: Should We Worry?). *The New York Times.* Retrieved from http://www.nytimes.com/roomfordebate/2011/05/19/the-age-of-anthropocene-should-we-worry/whats-next-the-cognocene.

Andersen, R. (2015). Nature Has Lost Its Meaning. *The Atlantic*. Retrieved from http://www.theatlantic.com/science/archive/2015/11/nature-has-lost-its-meaning/417918/.

Asafu-Adjaye, J., Blomquist, L., Brand, S., et al. (2015). *An Ecomodernist Manifesto*. Retrieved from http://www.ecomodernism.org/manifesto-english.

Askegaard, S., & Eckhardt, G. M. (2012). Glocal Yoga: Re-appropriation in the Indian Consumptionscape. *Marketing Theory, 12*(1), 45–60.

Batte, M. T., Hooker, N. H., Haab, T. C., & Beaverson, J. (2007). Putting Their Money Where Their Mouths Are: Consumer Willingness to Pay for Multi-Ingredient, Processed Organic Food Products. *Food Policy, 32*(2), 145–159.

Baveye, P. C., Baveye, J., & Gowdy, J. (2013). Monetary Valuation of Ecosystem Services: It Matters to Get the Timeline Right. *Ecological Economics, 95*, 231–235.

Benes, J., & Kumhof, M. (2012). *IMF Working Paper: The Chicago Plan Revisited*. International Monetary Fund. http://www.imf.org/external/pubs/ft/wp/2012/wp12202.pdf.

Brown, J. H., Burnside, W. R., Davidson, A. D., DeLong, J. P., Dunn, W. C., & Hamilton, M. J. (2011). Energetic Limits to Economic Growth. *BioScience, 61*(1), 19–26.

Clark, C. W. (1973). Profit Maximization and the Extinction of Animal Species. *Journal of Political Economy, 81*(4), 950–961.

Common, M., & Perrings, C. (1992). Towards an Ecological Economics of Sustainability. *Ecological Economics, 6*(1), 7–34.

Dalby, S. (2016). Framing the Anthropocene: The Good, the Bad and the Ugly. *The Anthropocene Review, 3*(1), 33–51.

Deleuze, G. (1995). *Negotiations, 1972–1990*. New York, NY: Columbia.

Deleuze, G. (2006). *Two Regimes of Madness: Texts and Interviews, 1975–1995*. Los Angeles, CA and Cambridge, MA: Semiotexte (MIT Press).

Deleuze, G., & Guattari, F. (1983). *Anti-Oedipus: Capitalism and Schizophrenia*. Minneapolis: University of Minnesota Press.

Dittmer, K. (2015). 100 Percent Reserve Banking: A Critical Review of Green Perspectives. *Ecological Economics, 109*, 9–16.

Eisen, B. (2016, February 17). A New Worry for Bank Investors: Bail-In Risk. *Wall Street Journal*. Retrieved from http://www.wsj.com/articles/a-new-worry-for-bank-investors-bail-in-risk-1455705000.

Ellis, E. C. (2011). The Planet of No Return: Human Resilience on an Artificial Earth. *Breakthrough Journal, 2*, 39–44.

Forstater, M. (2003). Public Employment and Environmental Sustainability. *Journal of Post Keynesian Economics, 25*(3), 385–406.

Fuller, B. (n.d.). *The Buckminster Fuller Institute*. Retrieved from https://www.bfi.org/ideaindex/projects/2015/greenwave.

Haldane, A., & May, R. (2011). Systemic Risk in Banking Ecosystems. *Nature,* 469, 351–355.

Hroch, P. (2014). *Sustaining Intensities: Materialism, Feminism and Posthumanism Meet Sustainable Design.* Edmonton: University of Alberta. Retrieved from https://era.library.ualberta.ca/files/z603qx859.

Jackson, T. (2011). *Prosperity Without Growth: Economics for a Finite Planet.* New York, NY: Routledge.

Klein, N. (2014). *This Changes Everything: Capitalism vs. the Climate.* New York, NY: Simon and Schuster.

Louv, R. (2008). *Last Child in the Woods: Saving Our Children from Nature-Deficit Disorder.* Chapel Hill, NC: Algonquin Books.

Louv, R. (2012). *The Nature Principle: Reconnecting with Life in a Virtual Age.* Chapel Hill, NC: Algonquin Books.

Marx, K. (1981). *Capital: A Critique of Political Economy* (Vol. 3). London: Penguin.

Minsky, H. P. (2015). *Can "It" Happen Again?: Essays on Instability and Finance.* New York, NY and London: Routledge.

Moore, B. J. (1979). The Endogenous Money Stock. *Journal of Post Keynesian Economics,* 2(1), 49–70.

Moore, B. J. (1989). On the Endogeneity of Money Once More. *Journal of Post Keynesian Economics,* 11(3), 479–487.

Moser, A. K. (2015). Thinking Green, Buying Green? Drivers of Pro-environmental Purchasing Behavior. *Journal of Consumer Marketing,* 32(3), 167–175.

Mueller, T., & Passadakis, A. (2009). Green Capitalism and the Climate: It's Economic Growth, Stupid. *Critical Currents,* 6, 54–61.

Nesvetailova, A. (2008). *Ponzi Finance and Global Liquidity Meltdown: Lessons from Minsky* (Working Papers on Transnational Politics). London: City University of London. Retrieved from http://www.city.ac.uk/__data/assets/pdf_file/0005/83993/CUWPTP002.pdf.

Nesvetailova, A. (2014). Innovations, Fragility and Complexity: Understanding the Power of Finance. *Government and Opposition,* 49(3), 542–568.

Nichols, D. M. (1992). *Modern Money Mechanics.* Chicago: Federal Reserve Bank of Chicago.

Pietilä, M., Neuvonen, M., Borodulin, K., Korpela, K., Sievänen, T., & Tyrväinen, L. (2015). Relationships Between Exposure to Urban Green Spaces, Physical Activity and Self-Rated Health. *Journal of Outdoor Recreation and Tourism,* 10, 44–54.

Pretty, J. (2013). The Consumption of a Finite Planet: Well-Being, Convergence, Divergence and the Nascent Green Economy. *Environmental & Resource Economics,* 55(4), 475–499.

Pröbstl-Haider, U. (2015). Cultural Ecosystem Services and Their Effects on Human Health and Well-Being: A Cross-Disciplinary Methodological Review. *Journal of Outdoor Recreation and Tourism,* 10, 1–13.

Rochon, L. P. (1999). The Creation and Circulation of Endogenous Money: A Circuit Dynamique Approach. *Journal of Economic Issues, 33*(1), 1–21.

Romagosa, F., Eagles, P. F. J., & Lemieux, C. J. (2015). From the Inside Out to the Outside In: Exploring the Role of Parks and Protected Areas as Providers of Human Health and Well-Being. *Journal of Outdoor Recreation and Tourism, 10,* 70–77.

Roston, E. (2016). The World Has Discovered a $1 Trillion Ocean. *Bloomberg. com.* Retrieved from http://www.bloomberg.com/news/articles/2016-01-21/the-world-has-discovered-a-1-trillion-ocean.

Schnurr, L. (2016, March 22). Canada to Introduce "Bail-In" Bank Recapitalization Legislation. *Reuters.* Retrieved from http://www.reuters.com/article/us-canada-budget-banks-idUSKCN0WO2Y5.

Seccareccia, M. (1988). Systematic Viability and Credit Crunches: An Examination of Recent Canadian Cyclical Fluctuations. *Journal of Economic Issues, 22*(1), 49–77.

Seccareccia, M. (1996). Post-Keynesian Fundism and Monetary Circulation. In E. Nell & G. Deleplace (Eds.), *Money in Motion.* London: Macmillan.

Seccareccia, M. (2012). Financialization and the Transformation of Commercial Banking: Understanding the Recent Canadian Experience Before and During the International Financial Crisis. *Journal of Post Keynesian Economics, 35*(2), 277–300.

Shaviro, S. (2013, November 17). More on Accelerationism. *The Pinocchio Theory.* Retrieved from http://www.shaviro.com/Blog/?p=1174.

Smithin, J. (2013). Credit Creation, the Monetary Circuit and the Formal Validity of Money. In L.-P. Rochon & M. Seccareccia (Eds.), *Monetary Economies of Production: Banking and Financial Circuits and the Role of the State* (pp. 41–53). Northampton, MA: Edward Elgar.

Spash, C. L. (2012). New Foundations for Ecological Economics. *Ecological Economics, 77,* 36–47.

Terzi, A. (1986). The Independence of Finance from Saving: A Flow-of-Funds Interpretation. *Journal of Post Keynesian Economics, 9*(2), 188–197.

Thompson, C. W., Roe, J., Aspinall, P., Mitchell, R., Clow, A., & Miller, D. (2012). More Green Space Is Linked to Less Stress in Deprived Communities: Evidence from Salivary Cortisol Patterns. *Landscape and Urban Planning, 105*(3), 221–229.

Tiessen, M. (2013). Monetary Mediations and the Invisible Overcoding of Potential: Nietzsche, Deleuze & Guattari and How the Affective Diagrammatics of Debt Have Gone Global. *MediaTropes: An Interdisciplinary eJournal Devoted to the Study of Media and Mediation, 4*(1), 47–64.

Tiessen, M. (2014a). Coding the (Digital) Flows: Debt-by-Design and the Econo-Blogospheres' Transparency-Driven Infowar. *Cultural Studies<=> Critical Methodologies, 14*(1), 50–61.

Tiessen, M. (2014b). Giving Credit Where Credit's Due: Making Visible the Ex Nihilo Dimensions of Money's "Agency." *TOPIA: Canadian Journal of Cultural Studies* (Special issue on "The Financialized Imagination and Beyond"), 30–31 (Fall 2013/Spring 2014), 290–300.

Tiessen, M. (2015). The Appetites of App-Based Finance. *Cultural Studies, 29*(5–6), 869–886.

Werner, R. A. (2014). How Do Banks Create Money, and Why Can Other Firms Not Do the Same? An Explanation for the Coexistence of Lending and Deposit-Taking. *International Review of Financial Analysis, 36,* 71–77.

Werner, R. (2015). Richard Werner Speaking in Moscow on the Central Bank Issue. *YouTube.* Retrieved from https://www.youtube.com/watch?v=9Um-9wR46Ir4&feature=youtu.be&t=51m35s.

Whitehead, A. N. (2011). *Science and the Modern World.* Cambridge, UK: Cambridge University Press.

Wray, L. R. (2015). *Modern Money Theory: A Primer on Macroeconomics for Sovereign Monetary Systems* (2nd ed.). London: Springer.

PART II

Planetary Projections

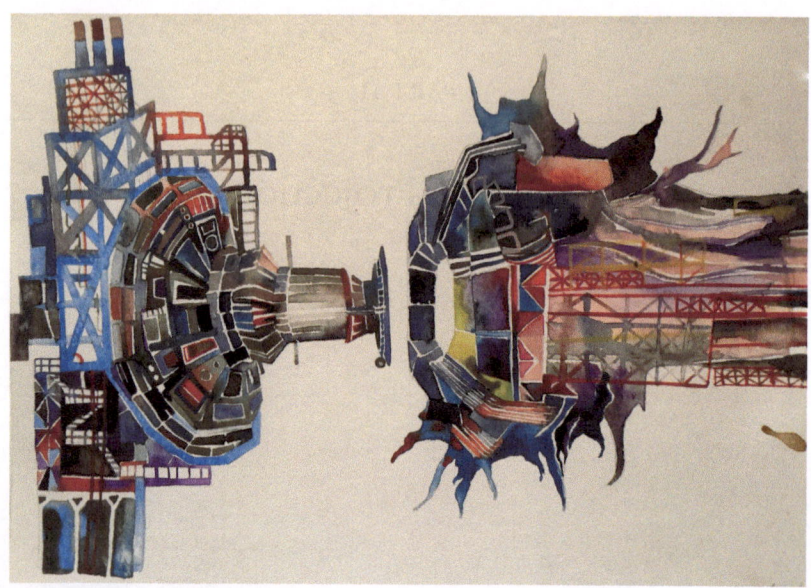

Image II.1   Turbo (Mia Feuer, watercolor)

CHAPTER 5

Catch 'Em All and Let Man Sort 'Em Out: Animals and Extinction in the World of Pokémon GO

Jason J. Wallin

NO ROOM LEFT FOR NATURE

We find ourselves increasingly in a world *without* animals. Disappearing at a rate not evidenced since the mass extinction of dinosaurs some 65 million years ago, it has been projected that up to 50% of *all* animal species will face extinction by mid-century. Already, speculations suggest that the loss of animal life is accelerating at 1000–10,000 times that of typical "background" extinction rates (De Vos et al. 2014). As increasing consensus across the research demonstrates, this ecocatastrophic scenario is distinctly human in design. Perpetrated by unlimited human development, myopic and short-term economic decision-making, and unrestrained consumerism, we now face the sixth mass extinction of life on this planet (David Suzuki Foundation 2012).

The extinction of animal[1] life in the Anthropocene is precipitated in part by the disconnection of the nature from human activity. A 2012 report by the UK-based National Trust, for instance, articulates the ways in which nature has *already* disappeared from the lives of the young.

J. J. Wallin (✉)
University of Alberta, Alberta, Canada

© The Author(s) 2018
j. jagodzinski (ed.), *Interrogating the Anthropocene*, Palgrave Studies in Educational Futures, https://doi.org/10.1007/978-3-319-78747-3_5

Over the course of a single generation, the number of British children who play in "wild places" has dropped from 50% to less than 10% (Moss 2012). An approximation of this separation appears in US statistics, where the number of children with outdoor hobbies fell by half over the six-year period from 1997 to 2003 (Moss 2012). A similar survey of approximately 700 Canadian youth found that *half* perceived themselves to be too busy with school work and chores to participate in outdoor activities, with 70% spending less than one hour per day outside (David Suzuki Foundation 2012). These figures point to a stark realization of life in the Anthropocene: *that culture no longer has room for nature.* Outside of the breeding facilities and slaughterhouses most never see, the animals left to us are largely selected for their affection and significance to their Masters (Baudrillard 1994).

The chasm between Man and animal is as much an index of the Anthropocene as are the specific ways in which the animal has today been liquidated and replaced. While it might be said that cultural life has receded from a natural world it no longer has time for, it is equally apparent that animals proliferate in the cultural imaginary (Kalof and Fitzgerald 2007). Such proliferation, however, cloaks the perpetration of a war on animals symptomatic of the contemporary moment. As Baudrillard (1994) develops, the symbolic significance of the animal has undergone radical transformation from earlier periods where, for example, the world of animals functioned as a primary order of reference Baudrillard claims that mythology recounts the divine and sacred character of the animal in premodern society, and further, the "reversible enchainment" of human and animal life (p. 134). In short, Baudrillard contends, "one cannot simply do whatever one wants with nature" (p. 131). Prior to the "reign of the human" born from the advance of humanism then, Baudrillard argues that there existed no division of species or distinct categorical opposition between man and animal (p. 133). In reference to Amerindian thought, Viveiros de Castro (2014) similarly argues that before the distinction of animal and Man, each were considered *human* in the most expansive and morphogenic sense of the term.

THE LIQUIDATION OF ANIMALS

Once considered sacred and wed to humans in mutual fate, the animal is today made to occupy an inferior racial world (Baudrillard 1994). The degraded universe of the modern animal is intimate to the production of the rational humanist subject, whose existence relies upon the characterization of the animal as irrational, lower, and as Descartes (1985) infamously conjectured, driven by rudimentary mechanical response. The animal becomes everything rational man *is not* (Derrida 2008). The abyss produced herein between human and animal lies at the heart of humanism, through which rational Man exists in oppositional affiliation to the degraded and excluded world of the animal (Wolfe 2009). The exclusion of the animal in modern society hence marks the disappearance of its enchainment to the fate of humans—a disappearance accelerated in the Anthropocene, in which we witness not only the annexation of the animal, but also the liquidation of its difference. That is, in their relegation to an inferior racial world, the animal becomes domesticated *by* and *for* Man. This is to suggest that today, the animal is *always-already* conceived from the vantage of a culture it is made to serve, if in distinction to the presumed supremacy of Man or through those modes of death and extinction to which the animal is continually submitted.

The animal is liquidated not only through the hell of industrial breeding and extermination, but in its having become *sentimentalized* and liberated into the ambit of human significance (Baudrillard 1994). In dialogue with Claire Parnet, Deleuze (2008) remarks on such sentimentality as it is dramatized in the orthodoxies of psychoanalysis. What has become most revulsive in our relationship with animals, Deleuze argues, is the way the animal has been rallied to all-too-human images of psychical life. Freud's clinical analysis, for instance, presumes the animal's automatic correspondence to the familial scene of Oedipus wherein the monstrosity and fascination of wolves, rats, and horses evoked by Freud's clinical patients are hermeneutically reworked into the figure of *the father, the mother or child* (Buchanan 2014). And so it goes with such Disney blockbusters as *Finding Nemo* and *Finding Dory*, where the animal is ultimately "found" in its rehabilitation within the familial, domestic order. For Deleuze, what is repellant in this scenario is that our relationship to animals has become distinctly human, or rather, conceived from the perspective of human desire and the empire of human meaning (Deleuze 2008; Baudrillard 1994, p. 137). This process of animal anthropomorphization is rejoined in Stivale's (2011) response to Haraway (2007), who condemns Deleuze's ostensible disinterest in real animals. Where Haraway critiques Deleuze for maligning domesticated animals, Stivale articulates how the focus of Deleuze's critique is not animals per se, but rather, the ways in which humans come to relate to animals. Hence, when Deleuze remarks that to like cats or dogs is foolish, it is not the animal that Deleuze finds distasteful, but rather, the human's relationship with animals. That is, the problem of animals evoked by Deleuze in *L'Abecedaire* does not amount to a stock rejection or contempt of the animal, but rather, of its individuation into those familial and political structures to which they are forcibly rallied.

THE ALL-TOO-HUMAN ANIMAL

Beyond the clinical hermeneutics of Freud, the psychical domestication of animals suffuses the cultural imaginary of the West. While non-domesticated animals disappear at an unprecedented rate, animals of the Anthropocene proliferate within the psychical models of liberalism and humanism (Baudrillard 1994). Herein, we no longer discover the monstrosity or alienness of the animal but through their correlation to a normative model of the human unconscious against which they are held to account for their misdeeds and transgressions recall that one of the earliest moving images of the animal in Western culture is the Edison Company's 1903 electrocution of Topsy the Elephant, who having been anthropomorphically attributed a derailed criminal unconscious in the media, is sentenced to summary execution (Gregersdotter et al. 2015). Today, the filmic genre of animal horror cinema has expanded this vector of anthropomorphization in such films as *Jaws*, where the animal is attributed a derailed unconscious ultimately resolved through human punishment and negation. This process finds its corollary in the moral panic surrounding Topsy's execution and the presumption that the animal unconscious is capable of producing "ugly" desires that, like those of humans, must be managed and regulated.

The anthropomorphic attribution of a human unconscious to the animal extends into the trope of revenge featured throughout the animal horror film genre, where the animal is summoned to challenge the world of Man and save us from our culture (Gregersdotter et al. 2015). For its horrifying dramatization of animal resistance to the ambit of human desire, the trope of animal revenge remains *all-too-human* for both its ascription of human motivation to the animal and nostalgic scene of animal resistance when the endgame of animal extinction in the Anthropocene does not change (Baudrillard 1994). The trope of animal revenge is today rallied to human sentiment in that the idea of revenge already exemplifies the animal's domestication within human thought. While research in the field of animal horror studies convincingly interprets animal revenge in terms of its commentary on and confirmation of the degraded and distant world Man has produced for animals, it occurs also that the very premise of revenge is vulgar insofar as its nostalgia for reversibility now appears remote given the sequestered status of the animal to breeding facilities, experimental laboratories, and now, the psychological fashions of humanism where, like the films of *Disney*, animals have become annexed by the emotional life of humans in which they are continually made to speak.

THE CHANGING FACE OF MONSTROSITY

The meaning of monstrosity has changed in the era of the Anthropocene. Once intimately enchained to a world rich in metamorphic and ambivalent potential, animals have today been remade in spectacular fashion and given over to the world of Man. *King Kong's* (1933) prescient commentary on the broken pact between Man and animal in the Anthropocene is illustrated through *Kong's* forcible removal from the jungle and enslavement to spectacular stardom in the music-hall (Baudrillard 1994, p. 135). Here, Baudrillard argues, *King Kong* articulates the liquidation of monstrosity in modern culture, or rather, the disappearance of a kind of bestial kinship and exchange between animals and humans. As an index of many animal horror films produced since the 1930s, *Kong* figures as a figure of liberation from a culture in which monstrosity has been purged, or rather, where all of inhumanity has been rallied to the side of Man. As is now well established in monster studies literature, the monstrous animal has come to figure in the anxieties of a culture that no longer has time for nature, or rather, of a culture that has ostensibly overcome in its rational and humanist aims the "primitive belief" in the uncanny reversibility of nature (Levina and Bui 2013). Under the ambit of humanism and rationality, nature's monstrous reversibility has long been relegated to captivity and domesticity.[2] Of course, animal horror film intervenes in this assumption by disrupting the very order of reality presupposed by Man. Returning us to monstrous scene of exchange between animals and humans from which culture is born, animal horror film often functions to unsettle the representation of animals as a commodity of the butchery, of the laboratory, and of liberal psychology by restoring it as a problematic to the image of life advanced by rational and humanist Man (Singer 1987). By example, we might think here of such "natural horror films" as the Canadian produced *Backcountry* (2014) insofar as it not only reverses the settler colonial presumption of mastery over the land, but also returns to the forefront of human-animal relations the prospect of animality remote to the ambit of human desire and anthropocentric bias.

In the era of the Anthropocene, the monstrosity of animals has assumed a degraded status, or rather, a status that no longer seems to evoke the sacrificial tableaux or relations of bestial transversality intimate to premodern societies. As an indexical example, we might look at the popularity of the mobile game Pokémon GO and its powers of fascination. The popularity of Pokémon GO is, of course, intimate to the now near global recognition of the Pokémon brand. Pokémon has radically eclipsed its circulation solely within children's markets, now expanding across a diversity of markets and media, including television, video games, mobile games, cosplay, and *otaku* cultures. Following Kubo Masakazu's characterization of Pokémon, Allison (2003) writes that the underpinning allure of Pokémon is predicated on the "cuteness" (*kawaii*) with which it is imbued. Against the broader socioeconomic context of hypercompetition and pressures intimate to the contemporary moment, Allison argues, Pokémon function as a soothing antidote and vehicle to another, gentler world. This shift in desire is evident in Japanese cultural production more broadly, insofar as the popularity of Pokémon illustrates a turn away from the ideals of *moretsu* (hard work) to *byūchifuru* (a synonym of "cute"). Further still, Pokémon have come to figure in all the characteristics of a mythical childhood. This is to say that Pokémon are not *simply* cute, but figure in the polymorphous, metamorphic, and animist qualities of childhood fantasy that like the world of animals, is now annexed by the world of adults.

The kind of *monster* from which the etymology of Pokémon is drawn marks the transformed status of monstrosity in the modern era. Purged of threat, Pokémon are relegated to a captive form of bestiality as domesticated *pocket monsters* that like their *Tamagotchi* predecessor, are characterized by their dependence and general nonthreatening companionship (Allison 2003). This is not to say that Pokémon do not retain aspects of the premodern relationship between humans and animals. As Allison (2006) argues, the allure of Pokémon stems from their evocation of a world enchanted with metamorphic possibility and the rehabilitation of nature imbued with difference and multiplicity.[3]

Here, we might imagine, Pokémon functions as a magical transport from the horrors of the contemporary moment in which both biodiversity and difference have been vastly eroded under the motors of progress and the ecocidal machinations perpetuated in its name. In distinction to the anthropomorphization of animals that has become commonplace in the contemporary cultural imaginary, Pokémon retain their "natural" status as creatures both distinct and distant, yet enchained to the lives of their human companions. In albeit bastardized fashion, it might be ventured that Pokémon ostensibly rehabituates a mythical relationship between humans and animals predicated on both care and the intimate relationship of training and sacrifice. That is, Pokémon not only features an emphasis on loving domestication, but insofar as Pokémon are trained to battle on behalf of and as proxies of their human companions, the mythical tableaux of animal sacrifice that Baudrillard (1994) contends founds the birth of culture.

CUTEPANION SPECIES IN THE ANTHROPOCENE

For their widespread cultural allure and involvement in human life, Pokémon constitute a kind of contemporary companion species. For akin to the qualities of animal companions, Pokémon are not simply *pets*, or at least, do not *begin* as such. Rather, the relationship between Pokémon trainers and their "animal" companions might more adequately be understood in terms of the reciprocal relationship of training and the intimacies of communication into which Pokémon and trainer enter. Pokémon trainers must not only learn the capabilities of their "animal" companions, but to do so, must learn to converse and pay proper attention to their powers to act and be acted upon. In this way, the relationship between the Pokémon trainer and Pokémon is not *necessarily* a reenactment of dominion over animals in which, like medieval and Renaissance iconography, often emplaces the domesticated animal at their owner's feet (Berger 1980). Rather, the relationship between Pokémon and its human is seemingly predicated on the cultivation of attentive interrelation by which trainer and Pokémon mutually enhance their powers through metamorphic and symbolic becoming. Simply, Pokémon do not rest at their trainer's feet in domestic resignation, but act *alongside* them. In this scene of human and animal interrelation, it is not only the animal that is changed. Within the world of Pokémon, both trainers and their animal companions are transformed through an inter-species pact, which, while commencing with the Pokémon's capture, seems ultimately oriented to the realization of cooperative and symbolic metamorphoses. In this regard, it seems that Pokémon might be an exemplary, albeit strange instantiation of natureculture in the Anthropocene.

The world of Pokémon is not simply born from the inexorable imbrication of nature and culture where, for example, the quasi-AI world of Pokémon is rallied to the culture of training and domestication. Rather, the enchanted world of Pokémon figures in the rehabilitation of nature *qua* culture. Simply, and as Allison (2006) suggests, Pokémon restore an image of nature *as it is for-us*. While the world of Pokémon figures in the reenchantment of culture via the return of human-animal intimacy and sacrifice, this metamorphic tableau seems also to repeat in the aspirations of rational humanism and its strategies of capture and mastery. It is in this way that Pokémon are not *merely* a form of companion species, but might be understood to function more broadly as a way of thinking both animals and "nature" in the era of the Anthropocene (Fudge 2008). Remarkable here is Baudrillard's (1994) conjecture that the disappearance of the animal into breeding facilities, experimental laboratories, and the "psy" language of liberalism is concomitant with the rise of a cultural nostalgia for animals and the symbolic dilation of the given world that animals once induced.[4] Put differently, the disappearance of animals through annexation and extermination today runs corollary to their sentimental degradation via which their dependence upon humans, human affection and charity is continually communicated (Baudrillard 1994). The global phenomenon of Pokémon might be situated alongside such sentimentalization insofar as the world of Pokémon recapitulates animality purged of monstrosity, or rather, of its distance from and reversal of humanism as it is commensurate with the ideal of anthropocentrism and mastery over reality.

In an era conspicuous for its war on animals, Pokémon speculates on bestial intimacy and human-animal alliance predicated on care and inter-relation. Rehabilitating a mythical pact of care and alliance with animals, Pokémon speculates on the dilation of our referential universe and metamorphic enchainment with animals. For these compelling speculative fantasies, it occurs that Pokémon inheres the very forms of violence against which its ethos of care and alliance ostensibly reject. As Baudrillard (1994) conjectures, the contemporary sentiment with which animals are regarded is itself a gesture of violence. Insofar as sentimentality is marked by the liquidation of violence, Baudrillard contends, the animal's powers of reversibility—from revenging the ambit of human desire to dilating *all-too-human* referential universes for thought—are degraded. The relegation of animals to this degraded world from which they are recuperated through human sentiment is for Baudrillard indicative of our profound hatred of them. The scene of human-animal alliance figured in Pokémon ultimately transpires upon such a scene of hatred. Despite its speculation on human-animal alliance through training and mutual alliance, Pokémon seems ultimately contracted with the fantasy of an animal *agon* in which Pokémon are compelled into inter-animal relations founded in opposition and negation. PETA's (People for the Ethical Treatment of Animals) admonishment of the mobile game Pokémon GO is situated upon this scene of captive bestiality and its fantasy conscription of the animal into war against other animals (Horvath 2016).

In vulgar resemblance of the very animal extinction we are today living alongside, Pokémon GO dramatizes the rehabilitation of a "nature" but to reenact its destruction and relegation to the hell of zoos and laboratories. PETA's claim argues that the animal's circulation as an object of leisure and consumption in Pokémon GO is a direct corollary to attitudes informing upon "actual" instances of animal exploitation and abuse, if not more generally an attitude of indifference to their fate. The fantasy of controlling nature and submitting the animal, we are asked to imagine, transpires into everyday beliefs about animals. Despite the largely disproven correlation between the performance of violence in video gaming and actual behavior (see, e.g., Jones 2002), it remains that the conditions of the Anthropocene are born from a mixture of belief and behavior, ideation and actuality. To take its most fervent cultural critics seriously, Pokémon GO might be thought as a redramatization of the war on animals by other means.

GOTTA CATCH 'EM ALL

To venture a summation at this juncture, it might be adequate to suggest that the adoration of the animal and its imagined alliance with the human in Pokémon is imbricated with hatred. The two poles conjoined, Pokémon and Pokémon GO transpire the complexity of an Anthropocene imaginary that on one hand revels in the reenchantment of culture through diversity and inhuman metamorphosis, and on the other, its annexation and submission to the hell of war and extermination. What was undoubtedly a significant aspect of Pokémon GO's initial novelty was its capacity to metamorphose reality through digital augmentation as to imagine a world reenchanted with diversity and difference. The augmented reality of Pokémon GO produced new conditions of excitement, chance, and diversity from under the ordered habits of social space, leveraging the temporary re-exploration of the given world. As mainstream news was quick to celebrate, Pokémon GO had ostensibly mobilized the sedentary bodies of predominantly youth gamers by rethinking all of reality as an undiscovered land. Millions of players flooded into this new world to first discover, then resolve its difference (Bliss 2017).

Pokémon GO's modes of desiring-production, or rather, the image of desire it presupposes are intimate to the imagination of the Anthropocene. Already, Pokémon GO's technological metamorphoses of reality attest to the simulated status of nature in an era when "nature" has become irrevocably invested within the culture. More adequately, the Anthropocene era might be characterized for both the constructed status of nature through which Man (*anthropos*) presumes dominion as well as the complex entanglements of nature and culture today signified by the imbricated term *naturecultures*. The constructed status of nature is, in part, evident in the very gestures of Pokémon GO, which figure in the impulses of settler colonialism and its strategies of capture, annexation, and domination over the Other. Redoubling the franchise catchphrase "*Gotta Catch 'Em All*," the prospect of commanding such rare monsters as *Porygon* and *Scyther* constitutes a driver for player desire. Yet, it is not sufficient that players simply capture Pokémon, for akin to the practices of "colonial humanism," the captive creatures must be transformed from distinct "natives" into productive actors from the vantage of their captors (Wilder 2007). That is, Pokémon GO imagines the transformation of the "native animal" from its static, territorial state to its fluid conscription under the ambit of human desire and production.

The strategies of colonial humanism are elsewhere enacted in the game via its orientation to capturing and striating territories through opposition and coding. Players not only annex the territory of the Pokémon through capture, but also control the territory by occupying and producing hierarchies of dominion in the *agon* of the Pokémon gym. As if to repeat a particular aspiration of Renaissance humanism, the enchanted world of Pokémon is ultimately made to conform to the world *as it is for-us* (Thacker 2011). In what might be considered a leitmotif of the Anthropocene, Pokémon GO anthropically subverts its rehabilitation of an inhuman world by conditioning it according to the very gestures that have brought us to the brink of extinguishing animals and other "inhumans" born in distinction to rational Man. Despite the technological innovation of Pokémon GO, the performative gestures of the game are hardly new, dating back to the imperial impulses of Middle Age and Early Modern humanism and its pretext for the domination of *less-than-human* life.

CHASING THE CHTHULUCENE

While the performative gestures of Pokémon GO redramatize the anthropocentric subversion of life under the ambit of Man, it concomitantly articulates a new set of conditions intimate to the Anthropocene. To borrow loosely from Haraway (2015), Pokémon GO figures in the annexation of a more-than-human "chthulucene" via the accelerative and large-scale reterritorialization capital accumulation of "matter" for and profit (Haraway 2015). Herein, Pokémon GO is not simply an index of anthropocentrism and its implications upon planetary life, but of the boundary event between the Anthropocene and the exploitative cheapening of life intimate to what Moore (2017b) has dubbed the *Capitalocene*. Beyond the investment of matter within the image of Man intimate to the anthropocentric pretext of the Anthropocene, Moore argues that the Capitalocene marks the investment of matter in capital, or rather, the realization that the history of capital is always-already implicated in the history and future of climatology and environmentalism (Moore 2017b). Put differently, Moore avers that the current ecological and environmental crisis that characterizes the Anthropocene is not the effect of an abstract humanity (*Anthropos*), but rather, the crowning achievement of *capitalogenesis* or rather, the annexation of reality by capital.

Pokémon GO functions to recapitulate the boundary event of the Anthropocene/Capitalocene in significant ways, foremost amongst which is the investment of its "natural" world and rehabilitation of animality within the ambit of capital. To date, net revenue earnings for Pokémon GO have reportedly eclipsed the 1 billion-dollar (USD) mark (Takahashi 2016), with an estimated daily revenue of 3 million dollars (USD) from in-app purchases. Herein, we might extrapolate that the rehabilitation of nature fantasized in Pokémon GO is always-already born from an investment in capital, or as Moore (2017b) avers of the imbricated status of capitalism and environmentalism, of the investment of *nature-in-capital*. Beyond the obvious collapse of Pokémon Go's quasi-AI image of "nature" with the forces of capital is the contemporary reterritorialization of the human/animal relationship within the Capitalocene, where the animal is no longer anthropomorphically invested in the image of the human, but within the flows of capital itself. Herein, the animal pocket monsters featured in Pokémon GO are not merely reflections of human desire, but are made to reflect in the monstrous forces of capital to which they are ultimately conscripted in an aim for maximum profitability.

Beyond the obvious connection of Pokémon GO and market economy, the game constitutes an incisive commentary on the exploitation of labor-power under capital. Pokémon are not simply made to work *for-us*, but further, we find in the presumable misery of their captive bestiality our leisure and entertainment (Fisher 2014). Herein, Pokémon GO dramatizes the degraded libidinal relationship of human and animal. For where animals once figured in the metamorphic potential and reversibility of the world *for-us*, we today find their radicality and ambivalence disappeared into the sentimental kitsch-pop simulacrum of human pleasure and spectacle. What the augmented reality of Pokémon GO ultimately revealed was less a vector of escape from the *given* world than the dystopic realization that life—and virtual, larval life in particular—is ultimately "the meat puppet of Capital" (Fisher 2014). As a parallel to the disavowal inherent in James Cameron's *Avatar*, Pokémon GO's re-enchantment of the world occurs only by virtue of the augmented technologies that themselves presuppose the disappearance and destruction of the "natural world" (Fisher 2014). That is, the reenchanted world of Pokémon GO is not only born from the latent fantasies of imperial conquest and domination of reality, but the material exploitation of the planet through the extraction of tech-dependent Coltan and the internet's annual 300 million ton CO_2 carbon footprint (Clark and Berners-Lee 2010; Scranton 2015). Alike the capitalist fantasy of interminable exchange and growth that has catalyzed the ecocidal vectors of the Anthropocene, Pokémon GO habilitates an image of the world liberated from the problem of extinction—a disavowal intimate to the production of the capitalist subject. As an index of the event boundary of the Anthropocene/Capitalocene, Pokémon GO proliferates an image of animality without end, but for the pleasure of its interminable exploitation and "happy" consumption.

Pokémon GO covers over the fact that nature has already disappeared, if not within the ambit of humanist desire, then via those Global Positioning Systems that obliterate the territory they precede (Baudrillard 1994). The "real" world[5] with which the game is grafted is of such tangential interest to many players that reports of vehicular deaths, near-deaths from falling (Hernandez 2016) and hypothermia (Broomhead 2016), and incidences of actual animal attacks against players multiply (Scott 2016).[6] In this manner, Pokémon GO exemplifies Baudrillard's theorization of the hyperreal in that the game occludes the fact that culture no longer has room for nature. That is, Pokémon GO is not simply a simulation, but a simulation that covers over the brutal realization that nature is always-already simulated. That we no longer have room for nature is dramatized not only through the war on animals perpetuated in the game, but more significantly, in the act of archiving pocket monsters within the game's encyclopedic *Pokedex*.[7] While this facet of the game has been rightly linked with the humanist impulse of collecting and archiving the world[8] if but to reveal and master it, it is too indicative of a cultural response to extinction in the Anthropocene. That is, in the wake of planetary animal and inhuman extinction, Pokémon GO figures as a performative act of preservation through the animal's archival permanence in digital cryostasis. Yet, the war on animals figured in their forced combat and preservation in encyclopedic archives are much the same, where one is condemned to the hell of the arena and the other, the hell of the zoo.

The Precarious Future of Animals

Pokémon GO is a way of rethinking nature in an era marked by the geophysical impacts of humans upon the planet, yet additionally constitutes a commentary on the investment of human activity within the capitalocene, or rather, within relations shaped by the privileged status accorded to capital accumulation and the liquidation of *things* to their market value (Moore 2017a). What is apparent about the world of Pokémon GO is not only its implicit affirmation of accumulation and consumption but further, the non-consequence of such activities upon its augmented game-world. In Pokémon GO, the labor of the inhuman can be inexhaustibly mined. While admittedly courting hyperbole, Pokémon GO conjoins a particular form of disavowal intimate to late capitalism whereby the prospect of extinction must be continually allayed and deferred if only to perpetuate the presumption that the present order will repeat indefinitely (Culp 2015). That extinction *actually* extends from the pulsional motors of the capitalocene, as it does in such cases as black market poaching, habitat loss for industrial development and biodiversity loss through animal breeding and engineering, demonstrates that there is no room left for nature that cannot be remitted to its reordering under capital, or as Fisher (2014) contends in consideration of the "nativist" *Na'vi* in James Cameron's *Avatar*, imagined in its *first instance* through the technologies of capital. As it is in Pokémon GO, nature is *always-already* a special effect of planetary surveillance technologies and the capitalist entertainment complex. The orientation to the inhuman, as it is thought in Pokémon GO, commences with its deterritorialization and subsequent "cheapened" reinscription in value, or as Moore (2017a) terms it, within the metrics of locating value-in-nature.

While Pokémon GO performs a particular relationship with the world coextensive of the anthropocene/capitalocene, it might concomitantly be thought as an index of human precarity. Within the algorithmic unconscious of Pokémon GO, for instance, inheres the very process of capture that characterizes a key focus of the game. That is, Pokémon GO functions as an assemblage for capturing the affective labor of the player. Just as the labor of the inhuman is overtly exploited within the augmented gamespace of Pokémon GO, so too the perambulatory labor of the player is fed into to *Niantic's* billion-dollar digital marketplace. As a parallel to the machinations of the capitalocene, the affective labor of the player is transformed into a surplus of free labor. Following Deleuze's (1992) prescient speculations on the rise of "open" surveillance technologies in late capitalism, Pokémon GO dramatizes the inscription of affective labor within widely distributed networks of social control. That is, within the augmented world of Pokémon GO, players willingly or perhaps ambivalently invest themselves within the control assemblage of GPS tracking, data mining, and surveillance by private corporations. As Deleuze characterized the synopticon, the power of Pokémon GO is not merely its capacity to track players at a planetary level, but rather, to algorithmically anticipate and motivate their behaviors, in effect freeing desire but on behalf of circuiting it to the new conditions of immaterial labor and the valorization of movement, exchange, and genetic restlessness with which it is commensurate (Larsen 2010). Like the AI-animals[9] that populate the augmented world of Pokémon GO, we too are deterritorialized that the energy of our flight might be recaptured for profit in short, what happens to the Pokémon within *Niantic's* worldwide arcade has *already* happened to us.

Pokémon GO does neither simply constitute a pretext for the extinction of animal life, nor is it a precursor of nature's disappearance. What makes Pokémon GO significant is its revelation that animality and nature have already been surgically remade. As an index of the Anthropocene, Pokémon GO is significant in that it redoubles the aspiration of humanism and its aim to remake the material world in the image of its will. This process of anthropic subversion by which world is rallied to the empire of meaning is intimate to the domestication of the inhuman and the evacuation of its monstrosity and difference (Thacker 2011). Yet, for such domestication, the correlation of the world to the ambit of human desire intimates a profound narrowing of referential universes. That is, where the animal once figured as a fulcrum for the metamorphoses of human cosmology and as a figure capable of liberating us from culture, they are today remade as the sentimental kitsch-pop of Capital (Baudrillard 1994). Such cheapening is intimate to our own fate in that the evacuation of the negative intimate to Pokémon GO parallels a more general disavowal of reversibility and monstrosity attendant to the Anthropocene. That is, where Pokémon GO speculates on the interminable labor of the inhuman within the ambit of human desire, it excludes those forces of negativity and reversal by which we might contend with both the accelerative force of capitalism and the extinction of inhuman and human life its pulsional motors perpetuate. Here, Pokémon GO is both a eulogy on the disappearance of the wild and an affirmation that the wild might live on as a sign in the Anthropocene—one inexorably tied to its cultural production by human beings and moreover, the impulses of Capital by which it is exhausted of its radicality.

Notes

1. Although I consider that I too am an animal, I use the term animal throughout this essay to designate "non-human animals."
2. Such captivity, of course, does not preclude the event of reversal. An indexical example here is the outbreak of "mad cow disease" in the later 1980s and throughout the 1990s and its direct result of having surgically remade cows into genetically programmed and hormonally manipulated simulacra (Baudrillard 2003).
3. As players might know, Pokémon differ from themselves through the process of evolution, where for example, a creature like Squirtle evolves into Wartortle and then Blastoise at increasing levels of experience.
4. The past tense is intended to suggest the liquidation of animals in the West and is not intended to imply that for particular peoples (i.e., the

Indigenous people of North America) animals remain central to their universe of reference.

5. The contact zone that Pokémon GO's augmented technology establishes with the "real" world is a course not devoid of its accursed share (Bataille 1998). The game's smooth space has produced new vectors of social movement. Beyond the much-lauded movement of sedentary gamers into the outdoors, players have trespassed on private property and sparked social anxiety and paranoia for their modes of movement and aggregation. Numerous reports document how players' movement has been seized upon by criminals, who have used in-game "lures" to ambush and extort players. In another instance, three Pokémon GO players in San Diego stumbled upon a dead body. Clearly, the game and its gestures remain inflected by the Real.

6. Incidences of play deaths and injuries are archived on the *Pokémon Go Death Tracker* at http://www.Pokemongodeathtracker.com.

7. As a corollary to the scientific naming of animals and arrangement of animals through taxonomic emplacement, The *Pokedex* functions as an electronic encyclopedia devoted to the articulation of Pokémon types, their evolutionary powers, and behaviors.

8. Such an orientation to the natural world inheres those practices of nature photography and bird watching that obliquely antedate Pokémon GO.

9. Of course, the fauna of Pokémon GO were *never* animals, per se. They were *always-already* the animal thought by and *for* humans. The genetic algorithms of the Pokémon select for their passivity. They neither attack or flee. They are predestined to obey the will of their trainers. While the dramatization of *capture* and *control* intimate to Pokémon GO perpetuates an anthropocentric conceit that *might* figure in actual instances of animal abuse, it is in the very design of Pokémon GO that we might also apprehend the extent of the animal's disappearance in the contemporary moment.

Image 5.1   Crane lifting up (Mia Feuer, watercolor)

References

Allison, A. (2003). Portable Monsters and Commodity Cuteness: Pokémon as Japan's New Global Power. *Postcolonial Studies, 6*(3), 381–395.

Allison, A. (2006). *Millennial Monsters: Japanese Toys and the Global Imagination*. Oakland: University of California Press.

Baker, B. (Producer), Baker, M. (Producer), & MacDonald, A. (2014). Backcountry (Motion Picture). Canada: D Films.

Bataille, G. (1998). *Visions of Excess: Selected Writings, 1927–1939*. Minneapolis: University of Minnesota Press.

Baudrillard, J. (1994). *Simulacra and Simulation* (S. F. Glaser, Trans.). Ann Arbor: University of Michigan Press.

Baudrillard, J. (2003). *Screened Out* (C. Turner, Trans.). New York: Verso.

Berger, J. (1980). *About Looking*. New York: Pantheon Books.

Bliss, L. (2017). *Pokémon GO Has Created a New Kind of Flaneur*. Retrieved February 13, 2017, from http://www.citylab.com/navigator/2016/07/Pokémon-go-flaneur-baudelaire/490796/.

Broomhead, M. (2016). *Young Woman Suffers Multiple Injuries During Pokémon Hunt in Chesterfield*. Retrieved February 4, 2017, from http://www.derbyshiretimes.co.uk/news/young-woman-suffers-multiple-injuries-during-Pokémon-hunt-in-chesterfield-1-8279580.

Buchanan, I. (2014). Schizoanalysis and the Pedagogy of the Oppressed. In M. Carlin & J. Wallin (Eds.), *Deleuze, Guattari, Politics and Education* (pp. 1–14). New York: Bloomsbury.

Clark, D., & Berners-Lee, M. (2010). *What Is the Carbon Footprint of ... the Internet?* Retrieved January 25, 2017, from https://www.theguardian.com/environment/2010/aug/12/carbon-footprint-internet.

Culp, A. (2015). *Dark Deleuze*. Minneapolis: The University of Minnesota Press.

David Suzuki Foundation. (2012). *Youth Engagement with Nature and the Outdoors: A Summery of Survey Findings*. Retrieved January 4, 2017, from http://www.davidsuzuki.org/publications/downloads/2012/youth%20survey%20findings%20summary.pdf.

Deleuze, G. (1992). Postscript on the Societies of Control. *October, 59*(Winter), 3–7.

Deleuze, G., & Parnet, C. (2008). *Gilles Deleuze's ABC Primer* (P. Boutang, Director). Retrieved December 13, 2010, from http://www.langlab.wayne.edu/Cstivale/D-G/ABC1.html.

Derrida, J. (2008). *The Animal That Therefore I Am*. New York: Fordham University Press.

Descartes, R. (1985). The Philosophical Writings of Rene Descartes. In J. Cottingham, R. Stoothoff, & D. Murdoch (Eds.). Cambridge: Cambridge University Press.

De Vos, J. M., Joppa, L. N., Gittleman, J. L., Stephens, P. R., & Pimm, S. L. (2014). Estimating the Normal Background Rate of Species Extinction. *Conservation Biology, 29*(2), 452–462.

Fisher, M. (2014). Terminator vs. Avatar. In R. Mackay & A. Avanessian (Eds.), *#Accelerate#: The Accelerationist Reader* (pp. 335–346). Falmouth, UK: Urbanomic Media.

Fudge, E. (2008). *Pets*. Stocksfield, UK: Acumen.

Gregersdotter, K., Hållén, N., & Höglund, J. (2015). *Animal Horror Cinema: Genre, History, and Criticism*. New York: Palgrave.

Haraway, D. (2007). *When Species Meet*. Minneapolis: University of Minnesota Press.

Haraway, D. (2015). Anthropocene, Capitalocene, Plantitionocene, Chthulucene: Making Kin. *Environmental Humanities, 6*, 159–165.

Hernandez, D. (2016). *Pokémon Go Players Fall off 90-Foot Ocean Bluff.* Retrieved February 4, 2017, from http://www.sandiegouniontribune.com/sdut-Pokémon-go-encinitas-cliff-fall-2016jul13-story.html.

Horvath, S. (2016). *Peta's L.A. Office Is Pokémon 'Safe Zone' Following Pokémon Go Release.* Retrieved February 5, 2017, from http://www.peta.org/blog/peta-Pokémon-safe-zone-following-Pokémon-go-release/.

Jones, G. (2002). *Killing Monsters: Why Children Need Fantasy, Super Heroes, and Make-Believe Violence.* Oakland: University of California Press.

Kalof, L., & Fitzgerald, A. (2007). *The Animals Reader: The Essential Classic and Contemporary Writings.* New York: Berg.

Larsen, L. B. (2010). *Zombies of Immaterial Labor: The Modern Monster and the Death of Death.* Retrieved April 17, 2010, from http://www.e-flux.com/journal/zombies-of-immaterial-labor-the-modern-monster-and-the-death-of-death/.

Levina, M., & Bui, D. T. (2013). *Monster Culture in the 21st Century: A Reader.* New York: Bloomsbury.

Moore, J. W. (2017a). *World Ecological Imaginations: Power and Production in the Web of Life.* Retrieved February 3, 2017, from https://jasonwmoore.wordpress.com.

Moore, J. W. (2017b). The Capitalocene, Part I: On the Nature and Origins of Our Ecological Crisis. *The Journal of Peasant Studies, 44*(3), 594–630. https://doi.org/10.1080/03066150.2016.1235036.

Moss, S. (2012). *Natural Childhood.* Retrieved February 5, 2017, from https://www.nationaltrust.org.uk/children-and-nature.

Scott, B. (2016). *Pokémon GO Player Bit by Real Snake in North Texas Park.* Retrieved January 25, 2017, from http://www.nbcdfw.com/news/weird/Pokémon-GO-Player-Bit-by-Real-Snake-in-Park-386849431.html.

Scranton, R. (2015). *Learning to Die in the Anthropocene.* San Francisco, CA: City Lights Books.

Singer, P. (1987, January). Animal Liberation or Animal Rights? *The Monist, 70*(1), 3–14.

Stivale, C. (2011). *When Philosophers Meet (Sort of): Animals, Deleuze, and Haraway.* Colloquium on Twentieth/Twenty-First Century French Studies, San Francisco University (Organizers), San Francisco, CA, March 31–April 2.

Takahashi, D. (2016). *Pokémon GO is the Fastest Mobile Game to Reach $600 Million in Revenues.* Retrieved February 2, 2017, from http://venturebeat.com/2016/10/20/Pokémon-go-is-the-fastest-mobile-game-to-hit-600-million-in-revenues/.

Thacker, E. (2011). *In the Dust of This Planet: Horror of Philosophy* (Vol. 1). Washington, DC: Zero Books.

Viveiros de Castro, E. (2014). *Cannibal Metaphysics* (P. Skafish, Ed., Trans.). Minneapolis: Univocal Publishing.

Wilder, G. (2007). Colonial Ethnology and Political Rationality in French West Africa. In H. L. Tilley & R. J Gordon (Eds.), *Ordering Africa* (pp. 336–375). Manchester, UK: Manchester University Press.

Wolfe, C. (2009). *What Is Posthumanism?* Minneapolis: University of Minneapolis Press.

CHAPTER 6

Intervals of Resistance: Being True to the Earth in the Light of the Anthropocene

Janae Sholtz

Deleuze and Guattari have made many creative interventions in the arenas of the social and political, and scholars have often theorized the importance of their work for providing a new framework from which to launch questions of the political.[1] Yet the political significance of their work lies not just in their historical political engagements or in their assessment of any particular politics, but in the way that their philosophy as a whole imparts new life to questions about the political as such. Deleuze and Guattari's rhizomatic ontology resists the metaphysical priority of essence so entrenched in past political formations, while accounting for the multiple networks and forces that underlie these illusory projections of wholeness. My claim is that Deleuze's realignment of ontology upon a spatio-temporal axis points us toward a new political future, the possibility for which is predicated upon a certain cultivation of awareness, which is to say, a new pedagogical project to rethink our relationship to the earth through the lens of the present.

Deleuze and Guattari (1994) call this coupling of a specific ontological vision with the concrete social formations geophilosophy. The preface of philosophy with "geo" indicating that this analysis is first and foremost

J. Sholtz (✉)
Alvernia University, Reading, PA, USA

j. jagodzinski (ed.), *Interrogating the Anthropocene*, Palgrave Studies in Educational Futures, https://doi.org/10.1007/978-3-319-78747-3_6

a deterritorialization of philosophy itself, away from its traditional anthropocentric center (Flaxman 2012, p. 88). Likewise, the shift to becoming over being indicates a new model of the communality, which is constituted through the perpetual assessment of temporal and material situatedness of the human within the immanent whole.[2] In order to maintain the prerogative of temporal singularity, rather than ahistorical universality belonging to the totalizing systems of the past, one would need to extend Deleuze and Guattari's geophilosophical analysis of the modern "cosmic" age as post-industrialist, information-driven, and rife with virtualized economic capacities to our present situation. This time, geologically speaking, has been progressively hailed as the Anthropocene and, economically speaking, represents a certain acceleration of production/destruction and perpetual expansion of control, the radicalized, hyperbolized state of 'late' capitalism, which Deleuze anticipates.

In this chapter, I develop an account of the Anthropocene, which addresses the ways that human beings have become alienated from the earth and mired in pessimistic resignation with regard to the possibility of making significant transformations in our politico-economic situation. I suggest that this malaise is the correlate of a general suspension of imagination, in other words, an inability to imagine a radically different future, and is linked to the issue of pedagogy in light of the Anthropocene. Next, I will argue that the task for thinking and learning, i.e. the pedagogical imperative, is to initiate an ontological shift in awareness. I argue that such a shift in ontological awareness is a necessary pedagogical tool in our attempts to navigate the epoch of the Anthropocene. This would involve attuning our selves, both conceptually and materially, to the level of the imperceptible forces, intensities, and affects that populate the earth—the cosmic level of being—which is to say, radical immanence. Finally, I suggest a two-pronged account of what this attunement would require (1) developing an awareness of the level of force and intensity by which the cosmic arises and operates—what I am going to call a sensitivity to affect and immanence. One of the paradoxes that such a project encounters is that such a sensitivity to our thorough embeddedness in immanent conditions demands that we attend to that which remains below our normal thresholds of attention. This sensitivity to immanence would then operate at the level of affect, rather than cognitive perception. In order to address this, I will draw upon Deleuze's robust conception of the autonomous nature of affect and considers it in light of the work of those involved in affect studies who have theorized the communal and contagious power of affect. We have to consider how we can cultivate modes of attentiveness and openness to this affective dimension. Therefore, the second requirement (2) is the invention of practices and ways of being that allow for or precipitate this development, which I am going to explore through the creative potential of art to infuse philosophy with intervals and slowness that help us to cultivate modes of attentiveness and openness to this affective dimension.

THE GEOLOGICAL TIME OF THE ANTHROPOCENE

Presently, it seems that we are mired in disillusionment and apathy concerning the future: it seems that our belief in the political process is broken and our very humanity is becoming redundant in the face of the globalized, corporatized market and that we have no new ideas or vision and any possibility of thinking that things can be otherwise. We are thus besieged by a kind of fatalistic realism. When there are forces that relentlessly oppose changing the system from which they are fed, even while barreling toward impending catastrophe, *how can anything change?* With regard to issues of corporatization, environmental exhaustion, and political disempowerment, the refrains of our age arise: 'it's too late, there is nothing that can be done—irreversible!' To my mind, this malaise is the correlate of a general suspension of imagination, an inability to imagine a radically different future, and the solution is linked to the issue of pedagogy, of how to think, in light of the Anthropocene. The question, 'what can be thought—*differently?*' may embolden us to think 'what must be done—*differently.*' These questions should weigh upon us as the most profound, the most urgent, and yet the most unforseeable. It is perhaps in this space of ambiguity, between the horrifying specter of foreclosure and our refusal to accept it, that we can begin to imagine differently.

As a means of foreshadowing of my response to these questions, I refer to Benjamin's auspicious quote from 1929: "They alone shall possess the earth who live from the powers of the cosmos" (Benjamin 2008, p. 58). Benjamin is speaking of a particular ecstatic and communal experience of the cosmos that has been lost to modernity. He argues that ignoring this rapturous, affectively-charged contact has been the error of modern man, and, while our influence has expanded to planetary scales, it is with the spirit of technological mastery rather than awe-inspired respect. This attitude of possessive domination occludes any genuine experience of the cosmos, such that our relationship with the earth needs to be re-invented. I believe such a shift in ontological awareness is exactly what is needed for a pedagogy of the future, for our attempts to navigate the epoch of the Anthropocene, and by invoking Deleuze's ontology, as that which can help us better understand capitalist processes of deterritorialization characteristic of the Anthropocene and to resist its inertia, I distinguish my position from what might seem to be well-trodden ground by thinkers such as Heidegger or Benjamin.[3]

First, it is important to clarify what we take to be the indices of the Anthropocene. Generally, the Anthropocene indicates a new epoch in which humans are no longer just biological, but geological agents—in other words, that we, as human beings, have changed geology, not just our history, or our culture. There is ample evidence for our geological agency: the making of a new mineral epoch—through the artificial separating out of metals (500 million tons of aluminum for instance); changing the geological strata—through addition of 6 billion tons of plastics; 500 billion tons of concrete; a trillion bricks a year; not to mention atmospheric alteration, as we've doubled the amount of nitrogen at the earth's surface (Vitousek et al. 1997; Zalasiewisc and Schwagerl 2015). For many, acknowledging this has precipitated the realization that the age of the human risks destroying the earth. From this point of view, philosophy must think about the end, a new version of the age-old philosophical imperative, to learn how to die. However, as some have been pointed out (Zalasiewisc and Schwagerl 2015; Saldanha 2015, p. 211), the Anthropocene is not necessarily or merely anthropocentric, and to commit this erroneous assumption is to ignore our interrelatedness with the biosphere, rather than to "take seriously the earth and the human as two branches of the same abstract machine" (Dukes 2016, p. 516). Even recognizing that there is something *like* a new epoch can allow us to think, imagine and act differently. So, from a more optimistic perspective, it means thinking the conditions of the anthropocene, which is to say, *beyond* the conditions of the human, in order to think of a different future and new ways of inhabiting this planet.

Here is where we must think very precisely, about the kind of pedagogy that this entails. The Anthropocene invites the recognition that since our "activities [have] transform[ed] the earth... [we] must therefore take responsibility for the future of the planet" (Stengers 2015, p. 9). But, we must also ask ourselves, is this merely one side of the same Promethean coin? Can we solve our problems by operating from a paradigm of human agency that has created them? It really depends on what we mean by 'take responsibility'—green capitalism, bio-genetic technologies, more 'growth and development', or should we think of responsibility as the necessity of resisting solutions that operate within the same framework. Stengers, for instance, advocates the latter approach, arguing that, rather than eco-conservation inserted into the system (of course, here we mean advanced Capitalism), we must entirely "reinvent modes of production and of cooperation that escape from the evidences of economic growth and competition" (Stengers 2015, p. 24), becoming conscientious objectors to the slavish worship of growth that pits humans against environment, and humans against humans, in an ever-increasing exploitative spiral. I suggest that this same choice exists philosophically, and that we must enact a paradigm shift rather than merely critique the same conceptual plane. Given that the economic and the philosophic are ineluctably bound to each other, we must also ask what are the conditions that Capitalism entails to which philosophy must respond?

We find just such realizations in Deleuze and Guattari's *A Thousand Plateaus*, where they develop the concept of the cosmic to address the illusionary wholeness and substantiality that has undergirded our concept of the earth, as that which speaks to our modern era of capitalist deterritorialization, and in *What is Philosophy?*, in which they develop geophilosophy as a mode of thinking that engages an inhuman temporality in order to liberate our philosophical and political imaginations. In both cases, what is called for is the de-centering of our selves in order *to be true to the earth*. This is the potential that we want to explore.

GEOPHILOSOPHY AND BEING TRUE TO THE EARTH

The following passage from *A Thousand Plateaus* will serve as the launching point for navigating the ambiguities of this strange situation in which human and earth have so intertwined themselves:

> Finally, it is clear that the relation to the earth and the people has changed, and is no longer of the romantic type. The earth is now at its most deterritorialized: not only a point in a galaxy, but one galaxy among others. The people is now at its most molecularized: a molecular population, a people

of oscillators as so many forces of interaction... *The question then became whether molecular or atomic 'populations' of all natures (mass media, monitoring procedures, computers, space weapons) would continue to bombard the existing people in order to train it or control it or annihilate it – or if other molecular populations were possible, could slip into the first and give rise to a people yet to come* … . (Deleuze and Guattari 1987, pp. 345–346, my emphasis)

Here it is clear that any uptake of geo-philosophy has to account for the intersection of a new ontological vision with the concrete social formations indicative of the modern 'cosmic age'—post-industrialist, information-driven, virtualized economic capacities. As this quote illustrates, we are already caught up in these processes, and rather than understanding them, we have become captured by them. Presently, it seems that we have become unequal to the forces that our activities unleashed, and we are swept along at a blistering pace for which we have yet to develop a language, or conceptual framework. Though we can say that we have become geological agents rather than merely biological ones, our productions have overtaken our bodies and our minds. It is these virtual intensive passages of information, the fluidity of modes of production that dominate our world, that have, in effect, changed our relation to the earth and world. But has there been an equal shift in the conception of the human and its place within this scheme—in our ability to, as Deleuze would say, become worthy of the events that happen to us?

Another key passage, this time from *What is Philosophy?*, presents us with a rejoinder to this earlier provocation:

We lack creation. We lack resistance to the present. The creation of concepts in itself calls for a future form, for a new earth and people that do not yet exist. […] Art and philosophy converge at this point. (Deleuze and Guattari 1994, p. 108)

Deleuze's claim that we 'lack an earth' can be referred to the deficiencies in our Promethean "sense" of the earth as the domain of the human, to be seized and measured—our present (Wiame 2015, p. 2). Yet, by naming this lack, Deleuze and Guattari call us beyond lamenting the loss of the earth—or people—demanding a new kind of relation therewith, where to resist means to project imaginative futures—to invent rather than remain in nostalgic paralysis. It is our supposition that to be 'true to the earth' is to think, or rather, to aesthetically and creatively engage an inhuman earth—the perspective of the cosmic rather than the human, the level of forces and intensities, which precede substantial forms, even that of the human subject.

This is the shard of hope that I want to trace at the end of the first passage, asking: *if* other molecular populations were possible, could slip into the first and give rise to a people yet to come—what *would (or could)* that be like? This time of production where forces are at their most deterritorialized gives us access to an underlying ontological truth that has hitherto been covered over by our own theorizing: the cosmic does not end with the sphere of capital but insinuates a plane of imma-nent, non-hierarchical relationality and connectivity, a world filled with discontinuities and oscillations, a deterritorialized earth of cosmic forces. Becoming worthy of what happens to us does not mean changing or controlling those events, but rather entering into them, becoming con-nected and engaged with them...relationally, intensively, and affectively. The event that we have to be worthy of is a unilateral uncompromis-ing intrusion, all the more profound because it is the inhuman itself; the earth as an assemblage of material processes (Gaia) (Stengers 2015, pp. 43–50), which can no longer be silenced or ignored.

And it seems that this is the task for thinking and learning, the peda-gogical imperative, that in one way or another lies before us—a common refrain presents itself, that what we need is an ontological shift in aware-ness, which I have entitled moving 'from the earth to the cosmic,' where the cosmic involves a transformation to *geological* [slow] time rather than anthropocentric historicity, and the recognition of the molecular forces that inform and transform us. As I have said, this would require two intertwined tasks: (1) developing an awareness to the level of force and intensity by which the cosmic arises and operates—what I call a sensitiv-ity to affect and immanence; and (2) the invention of practices and ways of being that allow for or precipitate this development.

According to Deleuze and Guattari, we have the tools at our disposal. What characterizes the modern age is a different ontological relation to materiality (to the earth), looking beyond the matter-form relation to the direct relation of material-forces, molecularised matter. Deleuze and Guattari envision the Earth as a plane of rhythmic, intensive vibration, which displaces the question of the origin of a people toward questions of pure relations, chance encounters, and perpetual motion (oscillators). The modern figure responsive to this terrain is not a founder, nor a creator, not even artist, but a cosmic artisan (Deleuze and Guattari 1987, p. 345), working from within the scrambling of terrestrial forces—including those we might consider "social" forces: machines, mass media, computers, weapons. Deleuze and Guattari define the artisan as "one who is determined in such a way as to follow a flow of matter, a machinic phylum. The artisan is the itinerant, the ambulant" (Deleuze and Guattari 1987, p. 409), who wanders and etches new paths. The cosmic artisan fabricates rather than replicates, her materials are flows and forces, which by necessity, testify to an always present overflow of the present, exposing the illusions of completeness, which buttress the stratifications of methods of control. We must become cosmic artisans. The questions are "how does this relate to the anthropocene?" Additionally, "what are the positive or effective political, social, environmental outcomes?"

Here Deleuze and Guattari's framework for geophilosophy is illuminative, and Greg Lambert (2005) does an excellent job of expressing the potential for a transformative politics in terms of geophilosophy's ability to identify the indices of the over-stratified earth characteristic of the Anthropocene in his essay, "What the Earth Thinks." Geophilosophy's appeal is that it creates a system of explanation that can be applied immanently and horizontally in order to make the relationality of different levels of being visible, and stratification, which is the capture and organization of forces, is indicative of geological as well as biological processes. Moreover, human beings create strata, through processes of coding and territorializing, at the meta-level of the socius. In other words, the coding of the earth is a fundamental activity that creates various social bodies and subjectivities, which, "in turn, [has created] the condition for the emergence of the great territorial machines that have distributed themselves across the surface of the earth" (Lambert 2005, p. 227). Each of these is like another level added to the mute and immanent continuum of the earth. Human societies are "mega-machines" (ibid.)—a language that fits well with an attempt to make sense of the processes that lend themselves to the epoch of the Anthropocene. Our current machinic assemblage is advanced Capitalism, and its specific characteristic is that it dismantles all those preceding it (Lambert 2005, pp. 227–229). And, though it operates through the deterritorialization of all flows, it is always for the purpose of recoding these in service of a greater degree of capture and stratification in order that *nothing escapes*—so much so that it seems to be an inescapable and inevitable fate (producing the overwhelming affect of fatalism). Therefore, geophilosophically speaking, our present condition is that of a crowded, bloated earth overburdened by territories, despotic forms of sovereignty, an Earth suffering from *too much* stratification.

Lambert observes that Deleuze and Guattari's political geology seeks to undo the totalizing underpinnings implicit in philosophies that rely on concepts of the absolute and universality—those that ultimately have been used in service of supporting the assumption of human domination over the earth and teleological progression that places human consciousness at its apex, and that leads to the fatalist assumption of Capitalism's universality. Geophilosophy provides a methodology for analyzing the construction of strata and accretions of power, exposing their inessentiality. This line of reasoning leads Lambert to the optimistic conclusion that rather than the culmination of a universal History, capitalism only produces *the illusion* of universality as its mode of capture and control (Lambert 2005, p. 229). Yet, this critique, the loosening of the universalizing illusion of Capital, is only visible to us from what Lambert calls the full body of the earth, by which he means "the absolute point of deterritorialization" (Lambert 2005, p. 230). I would like to add that this is where one must insist upon Deleuze and Guattari's ontological shift, from the earth to the cosmic, and the need for a commensurate shift in our own perspective. This cosmic perspective is akin to what I have called the sensitivity to immanence and reflects a register or plane *within* the earth comprised of molecular, inhuman forces. This emphasis on the necessity of a particularly holistic view, a cosmic perspective, which eliminates or surpasses the purely human realm of earthly existence as necessary for a properly conceived ethics, does not originate with Deleuze and Guattari. The Stoics, Marcus Aurelius in particular, do this when calling for meditative practices that minimize our human existence in light of the greater schema of material reality, of which we are just an infinitesimal part. Yet, Deleuze and Guattari's reconceptualization of the ontological as a ontogenetic process of becoming that operates below the common thresholds of substance and molar entities gives us a new standard by which we must judge what it means to be true to the earth (Sellars 1999; Sholtz 2019).

Moreover, through our recourse to geophilosophical analysis of processes of stratification, we can see that the very process that allows capitalism to spread across the earth is also the tendency that threatens to bring it to its limit—deterritorialization. Deleuze and Guattari are explicit about this even in the first volume of *Capitalism and Schizophrenia,* where they juxtapose the limited deterritorializations of Capitalism to the unlimited deterritorializations of desiring production that Capitalism seeks to dominate and control. The nature of Capitalism is to continually deterritorialize the socius, producing an awesome schizophrenic accumulation of energy or charge "against which it brings all its vast powers of repression to bear, but which nonetheless continues to act as capitalism's limit" (Deleuze and Guattari 1983, p. 34). A flow that might elude the code always haunts and threatens to expose the constructed nature of its universality.

As Lambert rightfully observes, the earth is that which resists all stratification (Lambert 2005, p. 235, also see Deleuze and Guattari 1987, p. 40). Because the earth, as resistance to stratification, operates within all strata, the potential for change lies then in our tuning into the rhythms of the earth; *there* are the processes of resistance that we seek, those that will allow for opening rather than foreclosing worlds. But, I reiterate, it is really a question of whether we are captured by these flows and affects *or begin to live and create from them,* to become adequate to perceiving ourselves as intertwined with these processes rather than having the Capitalist mega-machine devour, manipulate, and capture every flow and desire and us along with them.

This is the work of the cosmic artisan, to reframe the indices of modernity, the powers revealed through Capitalist capture and proliferation of the cosmic in an affirmative manner, and to produce new subjectivities that do not deny the present, but do not succumb to it either. The cosmic artisan intensifies and enlivens the event, connecting flows and traversing genres, in order that the bombardment of the miniscule, the mundane, or the machinic is transmuted into a vision of excessive beauty or intensity which provides the possibility of breaking open the configurations (and institutions) which have harnessed these molecular forces, those which operating as control mechanisms for setting and policing the limits. Cosmic artisans exist at the limit, as fabulators who counteractualise lines of flight, potentials that exist immanently, virtually, and intensively.

How do we become, these cosmic artisans? Who is worthy? What initiates such a transformation in awareness? This would require an ontological attunement that is resonant with the ontological messiness of the cosmic and inspires immersion and experimentation with these processes. But, how does one develop a new sensibility to cosmic immanence? Braidotti, who also identifies the need for a new kind of consciousness and critical thought that addresses the post-modern transformation of the politico-economy, makes the salient observation that inner, psychic or unconscious structures are hard to change by sheer volition (Braidotti 1994, p. 31). This is reminiscent of the paradox to which I referred earlier, that of developing a sensitivity to that which is beyond our normal levels of attentiveness, but it also speaks to the radically non-voluntaristic nature of what Deleuze and Guatarri want to accomplish. They are also interested in transformations of the subject that occur at the pre-subjective level of passive synthesis. Essentially, to effect these deep changes one must keep in mind the distinct levels between willful politics and unconscious desires, and develop strategies that are suited to each. Willful politics implies a *logos* based on reason and rational persuasion, while the latter, the realm of desires, expresses itself through non-signifying affect. The clear demarcation between these is rather like wishful thinking or desire itself. What passes as rational logos is built on a bed of lava, or even more dramatically, logos is mere façade, one more manipulation meant to engage and stimulate our deeper sensibilities—advertisers and war-mongers know this well.

Affect and Art: The Visceral and Visionary

In order to progress with our inquiry, it will be necessary to turn to realms more sensitive to the nuances of sensibility. That is, it is here that my philosophizing intersects with considerations of art, as a transformative potential that operates from the outside of philosophical thought, and the field of affect studies, which takes up the question of the nature of this being affected, this affective experience, explicitly. Affect theory offers a nuanced view of how our bodies are situated within a material environment and the way that non-signifying, non-semiotic, and often imperceptible forces work upon us, much of which is influenced by Deleuze's theoretical restoration of the autonomy of affect. Deleuze ascribes a radical power to affects; whereas concepts "lack the claw of absolute necessity... of an original violence to thought" (Deleuze 1995, p. 139). Affects epitomize "the claws of a strangeness or an enmity which alone would awaken thought from its natural stupor or eternal possibility" (ibid.).

At the same time, affect is the logos of postmodernity, and its invocation is marked with ambivalence, in much the same way that we have characterized the Anthropocene as the ambivalent site of an impending disaster *and* a call to invent a new future. As affect theorists have long insisted, affect is the level at which much of the information, disciplinary power, and the regulating forces of capitalism operate: "pre-individual affective capacities have been made central to the passage from formal subsumption to the real subsumption of life itself into capital" (Clough 2009, p. 221). Underscoring this contemporary shift toward affect, Colebrook provides a diagnosis of modern culture as "suffering from hyper-hypo-affective disorder" (Colebrook 2011, p. 45), which is exacerbated by the appropriation of affect through and by Capitalism, wherein we experience affect in terms of a diminishing intensity, all the while addicted to the consumption of more and more affects. The capacity to circulate affect becomes a matter of capitalist production, where bodily affect is mined for value and media is in the business of circulating and continuously modulating and intensifying affect. Food, sex, sociality are all marketed affectively, leading to 'affect fatigue' whereby the wider the extension of affective influx, the greater the diminishment of intensity. Thus Colebrook observes that we are in the grips of two catastrophic tendencies: "a loss of cognitive or analytic apparatuses in the face of a culture of affective immediacy," and yet a certain deadening of the human organism and its migration toward the generic, both of which are perpetuated by the tyranny of a relentless capitalism economy of consumption which routes affect for its own purposes.

What also becomes clear is that affect in and of itself is no panacea—it isn't the case that affects can "save" us from an over-intellectualized, over rationalized world, or that they will necessarily be agents of change in our perceptions or behaviors, because affect has already become the mode of exchange in our current economy. In fact, the problem is much deeper—the oversaturation of affect actually means that we have become impervious to its effects.[4] In order to think through these issues, Colebrook calls upon the work of Deleuze and Guattari, as thinkers who offer a "complex history of the relation between brain, body, intellect and affect" (Colebrook 2011, p. 50). While she is sympathetic to their work to uncover the power or force of affect and its centrality in human experience, she is also critical of the way that Deleuze's emphasis on affect has been reintegrated into discussions of affectivity, that is, of the assumption that the force of affect can be referred back to the affectivity of an organized living body (Colebrook 2011, p. 49). In order to see beyond is dilemma, we have to separate affect from affectivity in a more robust way. We need a concept of affect that would open a space for thinking beyond the immediacy of the "ready and easy responses craved by our habituated bodies" (Colebrook 2011, p. 50). We have to think the autonomy of the affect.

This more nuanced understanding of affect is certainly one that has it roots in the kind of autonomy that Deleuze ascribes to affect: "Affects [...] go beyond the strength of those who undergo them [...]" (Deleuze and Guattari 1994, p. 164). Returning to Colebrook's demand for an account of affect does not become reintegrated into the lived body and affectivity as such, I want to argue that we need to develop an even more radical account of affect's autonomy. Namely, that affect exists independently of living bodies altogether; affects are materially separate, active entities that act upon our bodies, a view that I believe is latent within Deleuze's account, but, because of our tendency to rely upon phenomenological description, is immediately lost.

In other words, affect must be perceived as not incumbent upon the affectivity of the subject but rather as an autonomous monument, comprised of circuits of force, which stand alone, outside of the body. Affect, understood thus, opens us to a different temporality than the affections that we feel through the lived body, and that this temporal disconnect can destroy the immediacy of affection that is often associated with affect, and, thus, would destroy the efficiency of an economy that systematically and seamlessly incorporates and neutralizes affect by creating a system of hyper-consumption which paradoxically anesthetizes the social body from the force of affect itself.

It is at this point that we must invoke the power of the artwork, as it presents an occasion to understand the nature of the affect as that which exists independent of our affective registers, and yet has a unique potential to disrupt and recalibrate our affectations. According to

Colebrook, "The power of art is not just to present this or that affect, but to bring us to an experience of any affect whatever [ ]—or that there *is* affect" (Colebrook 2004, p. 18). Deleuze emphasizes the particular double potentiality of artworks in *Logic of Sensation* and returns to this particular relation of affect and art in *What is Philosophy*, where he says explicitly: "It should be said of all art that, in relation to percepts or visions they give us, artists are the presenters of affects, the inventors and creators of affects. They not only create them in their work, they give them to us and make us become with them" (Deleuze and Guattari 1994, p. 166).

Experiencing the artwork's capacity to "create circuits of force beyond the viewer's own organic networks" opens up a space of delay (*an interval*)—frustrating immediate gratification. Posing this possibility of delay or interval becomes the occasion for thinking forces detached from the lived. Affect, rather than a response (the biological and internal model) must be considered from the perspective of that by which we are confronted and having an entirely other and external nature. "Affect becomes a genuine concept when it poses the possibility of thinking the delay or interval between the organism as a sensory-motor apparatus and the world that is (at least intellectually) mapped according to its own measure" (Colebrook 2011, p. 54). It is in this gap—between our lived bodies and the affect as a stand-alone entity, which cannot be reduced to the lived, that a space opens up for us to experience the inhuman, the forces of immanent being from out of which we are generated.

To Colebrook's demand for thinking the temporality of affect as an interval that breaks up the immediacy of our subjective experience, and thus our experience of the homogenous space of State philosophy and onslaught of Capitalist flows of affectivity, I would add that this also allows us to imagine affect in spatial terms, as a place in which inhuman forces can arise, or be illuminated. But rather than an empty space, or gap between spaces,[5] interval has to be thought as a temporal-spatial dimension that is already full, a crystallization and slowing down of the space that is already present, with its myriad relations, dynamisms, and forces, which would correspond to Deleuze's understanding of the minor as a way of occupying space and transforming political space from within the already instantiated major institutions and hegemonic formations.[6]

We encounter artworks as provocative, the combinatorial possibilities of which indicate the possibility of never before considered affects, which shock and confuse our ease of consumption. The feat of the artist is to straddle the line between chaos and order, to provide just enough consistency within the artwork for the myriad forces that are being captured to hold together, while allowing them the most freedom possible. Thus, artworks' framing of chaotic immanence allows for simultaneous thinking the carved out territory of the bloc of sensation and the transversality of the frame/sensation coupling as a kind of rhythmic bloc of sensation that interacts with its surroundings and provides a model of spatial interval (Sholtz 2015). These are studies of intensity that make visible or amplify these forces themselves, forming what could be considered a pulsating space by purposefully flirting with and precariously maintaining the tension between these two tendencies. Thus, these spaces of affect constitute an opening of immanence with which we can tarry to produce a sensitivity to this intensive and immanent realm that normally eludes us, or through which we clumsily pass unaware.

Yet, any naïve exuberance for merely producing *more* affects fails to account for oversaturated affective economy that has already routed and co-opted affect for its own purposes. In other words, the question is, "How to get out of the feedback loop of the human and the capitalist affect producing machine?" We must consider the kind of affects that must be generated in order to allow us to engage with this new concept of affect. What kind of activities, affects, and encounters can open a space whereby this sensitivity arises?

Intersection with Artistic Practice

In October 2015 while giving a keynote speech at the Moscow Biennale, Yanis Varoufakis, the former Greek finance minister, said, "Art must not be anodyne, culture cannot be decorative...[artists] should be feared by the powerful in our society, if you are not, you are not doing your job properly." Now, rather than interpreting this as a straightforward call for artists *to get political*, it strikes me that it holds a more profound message—as an implicit acknowledgement that art has a potential to open spaces of resistance and that it is uniquely poised to do so in a way that calls upon the artist in the mode of obligation. The exigency for the artist is amplified by a world that is practically cinched up by the overwhelming predominance of an all-encompassing capitalist economy that gobbles us, and our affects, up as quickly as they can be produced. Where and how can one escape from the singular economy of production if not in the intervals and spaces that artists uniquely open up. Moreover, art practices are the place in which the space of affect can be reflectively engaged. Of course, this is a potential of art, not its essence, a potential that becomes an imperative if one desires a different future.

My argument is that it becomes an imperative to produce affects that are themselves embodiments of delay or interval. I will focus on one example that can helps us understand this possibility of art practices to create or provoke intervals or delays in which forces of immanence overwhelm us.

SILENCE

I have in mind John Cage's explorations of the affect of silence. It may seem strange to speak of affects of silence, rather than a concept of silence, but this is exactly the precipice that must be traversed to shift toward an understanding of the autonomous power of affect. These affects, in particular, resist easy incorporation into our conceptual understanding, and they are unlike other affects that can be immediately connected to our own affective registers (as our tendency is to understand the products of art as reflecting our own anthropocentric language of affectivity, i.e. subjective emotion, feelings, expressions). I want to claim that these particular affects provoke an experience of interval or delay required for shattering the subjective paradigm and thus initiating us into a realm of inhuman force and immanence, which we have called an imperative for thought.

Cage is perhaps most well known for developing chance operations, which are meant to eliminate the subjective intention involved in the creation and highlight the aleatory as the main operator of the work. For instance, *Music of Changes*, which Cage expressly claims is "an object more inhuman than human" (Cage 2011, p. 36), imposes the aleatory by casting the ruins of the *I Ching* as a way of determining the structure of the composition. Indeed, Cage's methods of producing the aleatory in art were taken up in many other art registers, and set the tone for the development of performance art, as a medium that embraced the spontaneity of live action, minimally directive scores or instructions, and the unpredictability of audience reaction as the barriers between performer and spectator were challenged. But what is interesting is that Cage situates the aleatory in a larger framework beyond the orchestration of chance operations that disrupt intentional structure. What he suggests is that his method of chance operations was a stage along the way to exploring something more profound, the indeterminate, which is accessed by altogether abandoning structure, chance or anything otherwise. For this reason, Cage emphasizes the importance of the indeterminate with regard to performance. The purpose of indetermination is to bring about an unforeseen situation (Cage 2011, pp. 35–37), and, though chance operations to succeed in rendering the structure of a composition unknown from the beginning The performance itself is foreseeable as it follows the edicts that the chance-operations have determined. Maintaining that, "However, more essential than composing by means of chance operations, it seems to me, is composing in such a way that what one does is indeterminate of its performance" (Cage 2011, p. 69), Cage recounts his necessary progression from the intentional incorporation of the aleatory (chance operations) to a process that is itself aleatory (indeterminate).

Simultaneous to these experimental operations, Cage begins to develop a theory of silence, of which one only becomes aware once the structure and process of composition are disrupted. Traditionally, silence is seen to be the counterpart to sound, a mode of duration. Silence, then, is thought of in terms of the division of time-lengths and partitioning of sound and silence. But while Cage first attempts to make structure aleatory which eliminates "the presence of the mind as a ruling factor" (Cage 2011, p. 22), he is led to understand that structure is not necessary at all. In his subsequent work, he devises scores in which structure is no longer part of the composition, an activity characterized by process alone. It is in this context that he asks, what happens to silence, or the mind's perception of it?

Rather than a time-lapse between sounds, where there is a predetermined structure or organically developing one, "silence becomes something else—not silence at all, but sounds, the ambient sounds. The nature of these is unpredictable and changing. These sounds may be depended upon to exist. The world teems with them, and is, in fact, at no point free of them'"(Cage 2011, pp. 22–23). Cage insists that new music is nothing but sounds (Cage 2011, p. 7), which include those that are notated and those that are not. The non-notated are "silences, opening the doors of the music to the sounds that happen to be in the environment." Silence is not voided, empty space; it is an affect that holds open a space for the unintentional, ambient sounds that preexist us, that compose us, that exceed our activities: "inherent silence is equivalent to denial of the will" (Cage 2011, p. 53). Therefore, silence is a filled space, a space of plenitude, that eradicates the priority of our cognitive and affective circuits, and which opens an interval for that which arises independently therein—that is, concatenations of myriad forces of the external and yet immanent environment in which we are immersed.

The discovery of an unintentional silence, something that breaks free of cognitive determination (Cage 2011, p. 14), opens a space of materiality where forces arise. The composition becomes what arises in these spaces or intervals of silence—thus accomplishing two things: the eradication of the intentional subject and the rendering of the performance completely indeterminate, even more so given that the performers are something like inhuman, ambient forces. One could say that this study in silence brings about another affect, that of the indeterminate.

Perhaps in revisiting artworks such as this, with special attention to the affects that it was able to release, we can engage a new potential—a space in which humanity can become that which understands itself from a new conception of immanence and affect, to become a people sensitive to open, dynamic system of intensities, forces and multiplicities. This is not to become inhuman, but to think about the human, or being human, differently, as an open possibility constantly bombarded by and in tandem with myriad of forces and affective relations to other beings, human and otherwise. It is to inhere, to dwell even, in the same space— the interval, yet differently and with an alternate relation to these potential connections and minor voices. In this context, we would understand that *resistance is not loud*, it happens in the cracks between times and places. It is the silence of increasing intensity, the eventual release of an amplified force that tears through spaces, cultivating a new sensibility, as open vulnerability to this outside, for which art prepares the way.

CONCLUSION

The demands of revolutionary politics and recourse to aesthetic creativity do not always sit well with each other. My thinking attempts to straddle these unsettled boundaries. Before political efficacy, before institutional change, something more subtle, more ontological, must happen. This is what I have been calling for a new sensibility or transformation in our awareness. This new sensibility must be attuned to the normally imperceptible flows and forces that impinge upon us, the affective registers that enter into our reasoning and decision making—without attention to these, we are left with the disturbing perspective that nothing can change. It also necessitates a new relation to the earth and a more ontologically incisive perception of the human situation as embedded and horizontally interrelated to the earth and its existents. Being worthy of the event is to think the cosmic, which lets in a modicum of indeterminacy and freedom, this is the place of resistance: a gap, a wound, a space of deferral where fate does not prevail, where the future arises.

Deleuze says that "what we lack is "resistance to the present exist" (Deleuze and Guattari 1994, p. 108) and, in my view, this is a direct challenge to the fatalism and despair implicit in the attitude that there is no way to challenge state and corporate power, no way out of Capitalism. This position has led some to posit a kind of acceleration of Capitalist prerogatives as the only alternative—to drill and burn, however selectively, our way out of the world that we have irreparably changed/ damaged. Yet, as Deleuze explains, "absolute deterritorialisation is not defined as a giant accelerator; its absoluteness does not hinge on how fast it goes. It is actually possible to reach the absolute by way of phenomena of slowness and delay" (Deleuze and Guattari 1987, p. 56). In the spirit of Deleuze and Guattari, I want to suggest a different confrontation with speeds and slownesses. The interval, as an affective tarrying with slowness and delay, is a way of creating a sensitivity to immanence (the cosmic, earthly absolute) and all of its impermeable points and places that stratification fails to capture. In terms of the temporality of the Anthropocene, it is a response that neither pines for a nostalgic return to the past nor rushes toward a post-human apocalyptic future. Rather, this spatio-temporal delay draws our attention to our immanent present, while at the same time, *resisting* this present. This is a space not just for the earth/cosmic to arise, but for humans to recognize that our bodies, our affects, have already been captured in the great capitalistic mechano-sphere, and that our 'individuality' has been fashioned lock-step with the demands of the market and the flows of capital, in order that we may begin to redirect those flows and create a different future.

As the reversal to the beginning of this reflection on the Anthropocene, I would like to invoke the affect of hope. Hope, as a kind of optimism in the face of immense uncertainty and overwhelm-ingly oppressive conditions, is difficult to imagine, yet some are under-taking this task. In *Hope: New Philosophies for Change*, for instance, Mary Zournazi implores that re-enchanting life and politics through chance-taking is the way out of despair (Zournazi 2002, p. 274). Similarly (and quite reminiscent of Deleuze and Guattari's imperative to experiment), Stengers (2002) expounds the necessity of taking risks for moving to a politics of hope, as both crucial for generating the intensity and joy needed for making changes and for promoting a sense of com-munity and belonging. She characterizes risks as kinds of (revolutionary) events, inventions wherein lies hope for the future. If there is hope in all creative risks, we must create the space and the dispositions for such risks to be taken in order to reinvigorate our political imagination and to move beyond our reasons for despair. As we have seen, such a cre-ative political imagination reopens within the cracks and the fissures of the present, rather than appeal to a sterile utopian future. My claim is that this is what philosophy, what I, should be doing—creating inter-vals where thought and affect intertwine and the immanent forces of the earth can rise up within us. We must risk thinking differently. Opening ourselves to Deleuze's view of ontology of cosmic becoming and phi-losophy as a task of thinking that infinite movement is a risk—we risk our self-enclosure and our sense of stability and wholeness. But we also stand to gain—to gain a newfound awareness of the cosmic potential for the creative that exists as the flowing bedrock of our existence. There is hope because creativity is ontological. And, I would like to add, hope, or affirmation for that matter, it not a sign of naiveté. It is our communal responsibility to cultivate the kinds of affects that fundamentally move us forward rather than those that celebrate negativity and keep us mired in a repetition of the present. Hope allows us to take those risks.

NOTES

1. Most explicitly, Paul Patton's *Deleuze and the Political* (New York: Routledge, 2000), or more contemporarily, his *Deleuzian Concepts: Philosophy, Colonization, Politics* (Stanford: Stanford University Press, 2010); Nicholas Thoburn's *Deleuze, Marx and Politics* (New York: Routledge, 2003); Ian MacKenzie and Robert Porter's *Dramatizing the Political* (United Kingdom: Palgrave Macmillan, 2011); Nick Tampio's *Deleuze's Political Vision* (USA: Rowman & Littlefield, 2015).

2. "As opposed to the kind of exclusionary myths that create a people based on a common origin, blood or race, or even language, ... the kind of people that arises from this earth would have to be brought together out of their common dispersion, from the very fact that they are engaged in or produced by the deterritorialisations of dominant (molar) apparatuses" (Sholtz 2015, p. 242).

3. I am distinguishing my interpretation of Deleuze and Guattari's relation to postmodernity and contemporary capitalist society, from those who equate their ontology with capitalist deterritorialization, as if they were either merely offering descriptive analysis or, even more egregiously, embracing its inevitability and resolving themselves to the necessity of hyperbolized production, which, I believe, commits them to solutions relying merely on difference in degree rather than difference in kinds (of production). Deleuze and Guattari offer us, instead, the tools to recognize and analyze the powerful deterritorializing forces of capitalism (as limited deterritorializations which hinder and encumber the true libidinal forces of desiring production) in order that other modes of unlimited desiring production and deterritorialization can arise. Neither do I agree that Deleuze and Guattari's ontological position has theoretically exacerbated or buttressed neoliberal ideology. For instance, when Châtelet suggests Deleuze presents an affirmative ontology of chaos, one that correlates with and supports a political shift toward neoliberalism and its belief in the myth of auto-emergence that empowers the ideology of the sovereign power of free market (what he calls "seductive market-chaos" (Chatelet 2014, p. 60). Yet, to suggest that Deleuze and Guattari are advocates of pure chaos, of merely affirmation of difference and proliferation for its own sake is reductive. He offers more nuance than a mere repetition of the options presented by neoliberal adherence to the inevitability of the invisible hand market economy. They advocate something like attuning ourselves

to the chaosmotic plane out of which our experience and reality is generated; thus what is indeterminate is not chaotic, but virtual, and production is both a matter of blockage and delays as well as affirmation and speeds. Deleuze's ontology of becoming points toward relations of forces, a methodology for analyzing assemblages and recognizing consistencies, which exceed even the political and economic planes to force us to engage the outside of our human productions. This view of Deleuze and Guattari's philosophy suggests a cosmic potential that might combat the closures instituted by the globalizing forces of capitalism and the ever expanding technization of contemporary control societies.

4. In "Affect's Future: Rediscovering the Virtual in the Actual," for instance, Lawrence Grossberg (2010) posits that there are a great number of affective apparatuses to encounter and identify and that the failure to separate analytically those contexts has been a particular weakness of critical theory and cultural studies. These types of nuance with regard to affect are necessary to navigate the kind of contemporary environment that is being proposed.

5. As for Aristotle, interval (*diastēma*) is an empty space that is between bodies, necessarily without quality or mobility, and already implies a present relation between a body and its place (*topos*), rather than a place of durational emergence of bodies or form (Hill 2012, pp. 45–46).

6. Luce Irigaray develops a conception of interval through her counter-reading of Aristotle's understanding in "Place, Interval" (Irigaray 1993, 34–58). As opposed to Aristotle whose understands interval as an empty space between bodies indicating an immobile limit, she thinks the relationship between envelope and things as an open threshold, a place of passage and intersubjective becoming, which itself is predicated on the thought of difference that arises from thinking the space of interval. The development of the interval as a mobile or virtual place, that happens relationally, rather than as an already constituted presence provides an interesting parallel to Deleuze's ontology of becoming and theorization of the Event in particular. Yet, though her work is crucial for opening up a critical dialogue concerning the hegemony of the subject and helps us rethink the priority of form over matter, Irigaray's interval assumes the priority of sexual difference, while Deleuze's difference does not. Thus, from a Deleuzian perspective, Irigaray's difference is too tied to the prerogatives of subjects (sexuate beings), even while multiplying them, and does not go far enough toward the inhuman and pre-subjective outside.

Image 6.1   Dredge study (Mia Feuer, watercolor)

References

Benjamin, W. (2008). To the Planetarium. *The Work of Art in the Age of Its Technological Reproducibility and Other Writings on Media* (pp. 58–59). Cambridge: Harvard University Press. [Written Between 1923–1926, Originally Published 1928].

Braidotti, R. (1994). *Nomadic Subjects: Embodiment and Sexual Difference in Contemporary Feminist Theory.* New York: Columbia University Press.

Cage, J. (2011). *Silence: Lectures and Writings,* 50th Anniversary Edition. Middleton, CT: Wesleyan University Press [Original Publication 1961].

Châtelet, G. (2014). *To Live and Think Like Pigs: The Incitement of Envy and Boredom in Market Democracies* (R. Mackay, Trans.). Falmouth, UK: Urbanomic [Orig., *Vivre et penser comme porcs*, Exils, 1998].

Clough, P. (2009). The Affective Turn: Political Economy, Biomedia, and Bodies. In M. Gregg & G. J. Seigworth (Eds.), *The Affect Theory Reader* (pp. 206–228). Durham: Duke University Press.

Colebrook, C. (2004). The Sense of Space: On the Specificity of Affect in Deleuze and Guattari. *Postmodern Culture, 15*(1), 189–206.

Colebrook, C. (2011). Earth Felt the Wound: The Affective Divide. *Journal for Politics, Gender, and Culture, 8*(1), 45–58.

Deleuze, G. (1995). *Difference and Repetition* (P. Patton, Trans.). New York: Columbia University Press.

Deleuze, G. (n.d.). Lecture Transcripts on Spinoza's Concept of Affect (Released by Emilie and Julien Deleuze, Trans. Timothy S. Murphy). Lecture dt. 1978. Retrieved July 15, 2014, from https://www.webdeleuze.com/textes/14.

Deleuze, G., & Guattari, F. (1983). *Anti-Oedipus: Capitalism and Schizophrenia* (R. Hurley, M. Seem, & H. R. Lane, Trans.). Minneapolis: University of Minnesota Press.

Deleuze, G., & Guattari, F. (1987). *A Thousand Plateaus: Capitalism and Schizophrenia* (B. Massumi, Trans.). Minneapolis: University of Minnesota Press.

Deleuze, G., & Guattari, F. (1994). *What Is Philosophy?* (H. Tomlison & G. Burchill, Trans.). London: Verso.

Duke, H. (2016). Assembling the Mecanosphere: Monod, Althusser, Deleuze and Guattari. *Deleuze Studies, 10*(4), 514–530.

Flaxman, G. (2012). *Gilles Deleuze and the Fabulation of Philosophy*. Minneapolis: University of Minnesota Press.

Grossberg, L. (2010). Affect's Future: Rediscovering the Virtual in the Actual. In M. Gregg & G. J. Seigworth (Eds.), *The Affect Theory Reader* (pp. 309–338). Durham: Duke University Press.

Hill, R. (2012). *The Interval: Relation and Becoming in Irigaray, Aristotle, and Bergson*. New York: Fordham University Press.

Irigaray, L. (1993). Place, Interval. *The Ethics of Sexual Difference* (pp. 34–58). Ithaca, USA: Cornell University Press.

Lambert, G. (2005). What the Earth Thinks. In I. Buchanan & G. Lambert (Eds.), *Deleuze and Space* (pp. 220–239). Edinburgh: Edinburgh University Press.

Saldanha, A. (2015). Mechanosphere: Man, Earth, Capital. In J. Roffe & H. Stark (Eds.), *Deleuze and the Non/Human* (pp. 197–216). Basingstoke: Palgrave Macmillan.

Sellars, J. (1999). The Point of View of the Cosmos: Deleuze, Romanticism, Stoicism. *Pli, 8*, 1–24.

Sholtz, J. (2015). *The Invention of a People. Heidegger and Deleuze on Art and the Political.* Edinburgh: Edinburg University Press.

Sholtz, J. (2019). Deleuzian Exercises and the Inversion of Stoicism. In K. Lampe & J. Sholtz (Eds.), *Contemporary French Reception of Stoicism.* London: Bloomsbury Press (Forthcoming).

Stengers, I. (2002). A "Cosmo-Politics"—Risk, Hope, Change. In M. Zournazi (Ed.), *Hope: New Philosophies for Change* (pp. 244–272). Abingdon, UK: Routledge.

Stengers, I. (2015). *In Catastrophic Times: Resisting the Coming Barabarism* (A. Goffey, Trans.). London: Open Humanities Press.

Vitousek, P. M., Aber, J., Howarth, R. W., Likens, G. E., Matson, P. A., Schindler, D. W., et al. (1997). Human Alteration of the Global Nitrogen Cycle: Causes and Consequences. *Issues in Ecology,* No. 1. http://www.esa.org/esa/documents/2013/03/issues-in-ecology-issue-1.pdf.

Wiame, A. (2015). Reinventing Our Links to the Earth: Geophilosophy, Conceptual Mapping and Speculative Fabulation. (Unpublished, Read at Deleuze and Guattari and Africa Conference, University of Cape Town, 15–16 July 2015). http://deleuzeguattari.co.za/.

Zalasiewisc, J., & Schwagerl, C. (2015, April 29). The Anthropocene with Jan Zalasiewisc and Christian Schwagerl. Retrieved November 23, 2015, from https://www.youtube.com/watch?v=xP9P2i5jx-4.

Zournazi, M. (2002). *Hope: New philosophies for Change.* New York: Routledge.

CHAPTER 7

Sounding the Anthropocene

Mickey Vallee

INTRODUCTION: SOUNDING

Sounding is not the same as sonification, the latter of which refers to contemporary practices that organize data into sound (a transduction). For instance, when John Luther Adams sonifies the collective sound of Alaska's geophysical movements in *The Place Where You Go To Listen* at the University of Alaska in Fairbanks, synthesizers and lights connected to seismic readers across the State perform an immersive, ongoing, and live composition whose composer is a vast, expansive, and complex landscape ecology. His installation resonates with and is composed alongside the dynamic movements of the earth, and in general is anchored in the place which gives it its raw material, his music specifically tied to the place it is in: 'My music is going inexorably from being about place to becoming place' (qtd. in Ross 2012, p. 8).

Similarly, sounding has little to do with sound, and less so with the physiology of hearing. So we have little recourse to sounding so long as we take sound as limited to human hearing (which picks up 20 Hz–20 kHz, below which are infrasonic vibrations and above which are supersonic vibrations). The purpose of sounding is not to pursue sound as an object of knowledge. The collective research towards sound

M. Vallee (✉)
Athabasca University, Athabasca, Canada

© The Author(s) 2018
j. jagodzinski (ed.), *Interrogating the Anthropocene*, Palgrave Studies in Educational Futures, https://doi.org/10.1007/978-3-319-78747-3_7

confirms that it is pursued with the interest of transducing information for the benefit of widening perception for inclusion: Alexander Graham Bell's fascination with sounding circulated around Deaf culture (Shulman 2008); Norbert Wiener's early experiments on voice were intended to help those with voice loss (Mills 2011).

To approach the practice of sounding is to question the limits of visibility while accounting for the embodied practices of individuals who collectively work towards sounding. Sounding is decidedly practical. Two perspectives help in this matter of differentiating yet complementary definitions. Martin Heidegger's definition of sounding is that which hovers at the limits of representation, constituting 'the character of strife (earth—world)' (qtd. in Smith 2013, p. 87). Meanwhile, more recently, Julian Henriques (2008) has described sounding as a 'kinetic activity, a social and cultural practice, a making and becoming' (p. 219). For our purposes here, these two definitions are entirely compatible. That is, Heidegger's definition of sounding is intended to capture the stress between revealing and concealment (that an object's sound gives it a contour, shape, and extension, all the while remaining enchained to the evanescence of sound). While the kinetic, social, and cultural practices that Henriques describes are useful for approaching the collective forces necessary for extracting imperceptible vibrations. The central issue around sounding is thus methodological, and sounding is methodological: it is a *meta* and a *hodos*, a way.

Sounding, if we account for the tension it produces between revelation and concealment, along with the ongoing labour to suspend this tension for human interest, is ontological as well as embodied, deserving of the term 'onto-kinetic sounding'.

How, then, do we think through the sound in sounding? Obviously there must be some correlation between the two, if either is to gain any measure of value from the other. Although it is tempting to suggest that sounding has always relied on the frame proposed (onto-kinetic), I suggest here that sounding beholds a unique relationship with emerging sound technologies, which can capture a higher-resolution larger data set of imperceptible vibrations through a coded and technologically bound interface. Contemporary scientific research uses sounding as a method for detecting changes in biodiversity and the environment. In sounding, animals are defined by their acoustic properties over and above their visual phenotypes.

The onto-kinetic model of sounding thus propagates the following paradox: emerging sound technologies are capable of high definition, closer rate analysis, and require less human interference in the collection and analysis of data in doing so. There is a non-anthropocentric condition in sounding, insofar as actual sound waves are concerned, but a renewed anthropocentric condition insofar as much sounding research is intended to reduce biodiversity loss. All this research depends on machines that listen to and analyze data. And the purpose of much of this research is to become increasingly reliant on technologies that listen in place of the human ear, because they hear what ears cannot. The movement towards autonomization is especially important to non-anthropocentric research, which decenters the exceptional listener, and transforms it into something programmable and autonomous. Sounding, then, only disposes of the sound that can be heard by human ears.

This chapter addresses some of these soundings of the Anthropocene through a discussion of scientific research, including the degree to which the public is becoming involved in research projects. It is the intent of the chapter to work through the concept of sounding by aligning the Anthropocene with sounding in the following ways: by constructing both as tensions on the horizon between world and earth, as Heidegger described sounding, while accounting for the kinetic energy taken up to curb the Anthropocene's effects. With this, the argument is forwarded that the dystopian shadow of the Anthropocene might be neutralized by the turn towards data analysis and the measure of global populations in real time. Thus, while it might be assumed that the Anthropocene delivers (a) dystopian fears or (b) a new mode of global self-governance/ control, the concept of sounding resounds as a means of anticipation, hope, and the politics of action for embodied transformation. Sounding requires three components: (1) the global concept through which sounding is theorized along the character of a strife between earth and world (*Anthropocene*); (2) the scientific practice of sounding as a kinetic transduction of imperceptible vibrations (*bioacoustics*); and (3) the collective response to the necessity for sounding on global and local time-frames (*citizen science*).

ANTHROPOCENE

The Anthropocene is generally understood as a heuristic device that describes the human species as a geologic force powerful enough to change the global environment, and was introduced by Paul J. Crutzen (2002) in the journal *Nature* (though the term has yet to be 'officially announced', a peer-review journal, *The Anthropocene Review*, is already devoted to the subject). There are many contrasting and conflicting definitions of the Anthropocene, along with some warnings that the concept of Anthropocene may reiterate the 'environmentality' involved in the production of responsible citizenship (see Gabrys 2014). Either way, we might think of Anthropocene as a strategy. This is reinforced by Simon Dalby's recent assessment, for instance, which argues that the term Anthropocene varies between a well-defined geological heuristic tool, and a strategy for 'reimagining humanity's place in the cosmos' (Dalby 2015, p. 8), as well as a prelude to the 'possibility' of 'a period when humanity takes seriously ... a "sustainable earth"' (p. 10). Speaking to this ambivalence between objectivity and subjectivity, between fact and imagination, between science and affect, is partly the aim of this chapter. Though by no means do I intend to present evidence for or against the Anthropocene, as though the empirical evidence is intended to gather up and provide a context of justification for a new master narrative. Indeed, if the Anthropocene happened (implying that it *is* happening and *will continue* to happen), it is less the theories that will persuade us than the facts they are theorized upon. This is not intended to diminish the impact of philosophical thought on the Anthropocene: as the chapter demonstrates, studying under the context of Anthropocene requires the boldly objective data collection and analytic skills of a scientist, but the subjective reticent care of an artist.

As a term, Anthropocene may be more or less approached as an immersive concept, which we can base on new technologies and a new technologically based attunement to imperceptible forces: in one sense, the human species as a geologic force, and more specifically the vibrations of those changes that cannot be heard in the limited range of human hearing. We might refer to this intertwining of human, technological, and natural encounters as a 'becoming-imperceptible', a term belonging to Deleuze and Guattari (1987) and described recently by Rosi Braidotti (2013) as 'the point of evacuation or evanescence of the bounded selves and their merger into the milieu, the middle grounds, the radical immanence of the earth itself and its cosmic resonance' (p. 137). Deleuze and Guattari describe it earlier as crossing the threshold of distinguishing between categories of life forms towards vitality, or a 'plane of consistency'. In terms of visibility, however, the Anthropocene best serves the scientific community and the citizens that science serves, causing individuals and communities to think along the ethical lines of serving the Anthropocene.

If attending to the senses relies on the work of our intuition, it means getting more into the 'rhythm' of the empirical world, of becoming radically empirical, rather than agonizing over how to sort it all out. Getting into the rhythm is becoming attuned to the earth's vibrancy. And vibrancy is a sub-concept of sounding, because sounding describes only a certain component of the vibrations we are moved by, touched by, in what we hear and what we don't hear, like being in the presence of a bat swarm, whose millions of ultrasonic chirps coalesce into a palpable and gut-deep vibration when in its presence, but alone each singular bat is entirely imperceptible without a recording device for ultrasonic transduction (Mason and Hope 2014, p. 116). Sounding accounts for how transformation occurs through repetition, and implies that repetition cannot be articulated meaningfully (i.e. rhythmically) without the articulation of difference. This complicates the possibility for sounding, especially as sound is captured, transduced, and made sense of in the context of such scientific fields as bioacoustics.

BIOACOUSTICS

Bioacoustics has a longstanding Romantic association, reflecting a longstanding quasi-Romantic relationship with the 'sounds of nature', from Ludwig van Beethoven's strolls through nature, to Olivier Messaien's synaesthetic representations of birds, to R. Murray Schaefer and the World Soundscape Project's archives at Simon Fraser University, to the recent rise in bioacoustic instruments and the popularization of sound recording as a pastime for people interested in capturing for analysis as well as for cultural production the sounds of bats, birds, and plants: South Texas Wintering Birds, Mourning Warbler Song Mapper, Noise Map, Dog's Emotional Calls, Wolf Howls, New Forest Cicada, city nighthawks, bat echoes, human laughter, whale song, 'noisetube', the conversion of the Earth's ionosphere into sonic data, FrogWatch USA, 'sound around you project', plant phenology, and some of the most interesting developments in transducing the sounds of deep space (sounds which human ears cannot hear), as well as the sonic vibrations emitted by plants in times of stress such as times of drought (see scistarter.com for a description of these and other projects). Over time, through simple Apps and through technologies that are made and deliverable and whose function requires no more than a quick YouTube tutorial are being made cheap and readily available for analysis through spectrogram or wave formats. Bioacoustics is primarily concerned with communication, focusing on the sounds emitted from the vitality of living organisms which are studied in their natural contexts, and how sounds communicate with other sounds, which implies an interest that is not limited to animal sounds, but geologic, botanical, elemental, human, and non-human sounds (though animal sounds appear to remain the main focus, such as dolphins, bats, elephants, whales, and so on).

Michael Gallagher (2015) refers to this type of field recording as 'the nature style' where bioacoustics devices are capable of capturing vibrations imperceptible to normal human hearing (mice emit ultrasonic mating calls, plants emit infrasonic and ultrasonic vibrations in times of distress, environments vibrate with the seismic activity of the earth, for instance). Such a technique attempts to erase the human perspective, despite the fact that human activity is intrinsically involved in locating environments untouched by humans (one must fly there, drive there, step through plants to get there). To remove the human-hearing element, then, requires much human action and interaction.

The question of bioacoustics then is that of performance, the performance of non-human entities communicating with other non-human entities for the purposes of human capture and analysis. This is a long-standing practice in the Canadian context, from R. Murray Schafer, to the World Soundscape Project, and has a long history of landscape painting, a sensorial phenomenon. But Bioacoustics research is based less in the aesthetic consideration of the 'nature which moves itself' than it is of using the aesthetic (the 'sensoria' of the Anthropocene) as a means of rupturing and changing our ethics and our morality. Bioacoustics seeks to repair the damage of aesthetic distantiation as well as corporate extraction, both of which belong to the same colonialist enterprise.

An interesting development here is the increasing use of ultrasonic devices (either as an app or a piece of hardware that transduces sounds that are imperceptible to human hearing, such as bat chirps, distress signals in rats, and botanical communication), currently a hobby and pastime that is rising in popularity in the UK, in Canada, and in the United States, with such companies as *Wildlife Acoustics* selling handheld devices and smartphone extensions specifically for monitoring bats, bird and land animals, and marine life. Ultrasonic devices also transduce imperceptible vibrations into perceptible sounds in live time, allowing the recordist to analyze frequencies with bat movements. The ultrasonic devices are also equipped, in certain more expensive cases, to record and slow down bat chirping. Measuring the beats and duration of wing sound, mid-flight body collapse, and echolocation chirping (which bats use to locate food sources), is intended for the UK's National Bat Monitoring Programme to understand which bats are residing where, how their flight patterns are working, and where they are more and least active.

The abstraction of data from the natural world has a very hard-wearing ethical dimension, with implications for philosophies of Anthropocene. Even with the immensity of data accrued from citizen scientists, there there's still a great deal of scepticism in the scientific community about its efficacy. For instance, some criticize that the eager, community-oriented, and naturalistic images of citizen scientists make some researchers wary about 'amateurs' or 'voluntary biologists' who claim to contribute to a further understanding of environmental issues (Greenwood 2007). But, an opposing argument has stated that as more people are drawn towards citizen science, those community-oriented and knowledge building and philosophically necessary components of the Anthropocene and are becoming newly defined components belonging to scientific knowledge. For researchers the term 'big data' suffers from a 'trough of disillusionment' (Sicular 2013), as the capacity for data storage has reached from bit to byte to kilobyte to megabyte to gigabyte to terabyte to petabyte to Exabyte to zettabyte to Yottabyte (or 2 to the power of 80 bytes, which Rob Kitchin describes as 'too big to imagine' (2014, p. 70)), and terms for larger amounts are going to be needed in the next decade. This massive amount of data makes up for two potentials in data collection: one, on the side of the Latin term for data, *datum*, describing the elements that have the potential to be extracted from phenomena, versus *captum*, which describes those elements which *are actually captured* from phenomena. The former is potential, the latter actual.

CITIZEN SCIENCE

A principal axiom of citizen science is that of participation, where furthering knowledge comes as its own reward, and where the other side of participation (passive observation) is stood on the defensive; Citizen science rewards its scientists with the great outdoors, solitude, and the vicissitudes of nature (Lawrence 2006, p. 292), while proffering a set of habits that involve a fissure of the gap between everyday life, knowledge, and nature. But citizen science is also an experience belonging to cultural capital (Bourdieu 1984), as it is a field comprised of those who consume the narratives of the Anthropocene, and who, through these narratives, produce their own exclusions. By acting ethically, they are excluded from the 'rest' of the human race—the trip from perpetrator to bystander is by way of participation. The citizen scientist is privileged to many responsibilities: to their community, to research teams, to government, as well as to social media. Anna Lawrence's (2006) figure of dynamic interactions in Voluntary Biological Monitoring depicts the interrelations between action, experience, data, and policy. The citizen scientist is thus presented with a range of responsibilities that are embodied and cognitive: to understand the world as an interconnected mesh (Morton 2010); to understand that in every passing present looms an imitable and potentially catastrophic future; to become creative in their means of data collection and data sharing; a political awakening to the rhizome of human and non-human agents (Latour 2005); and a commitment to immersing themselves in a field.

But it's also often assumed that 'citizen science' rests on a 'deficit model'; that people are not active without contributing to recognized knowledge building and community building (Lorimer 2014), which implies a status game in the new 'inclusionary economy'. Such a perspective reflects the earliest developments in sustainable ethics (Irwin 1995): that a person's inner worth is validated by their 'new relation' with a sustainable world. This is an old linkage between self-worth and unsustainable practices that are treated as though they are pathologies to be corrected through expert knowledge, but this would be a radical bio-political reading of citizen science. Citizen scientists are encouraged to use technology in order to better scientific knowledge, have broader nationally expanding data collection archives, and contribute towards a better global future. Given that technologies are *relatively affordable* to the general public, the benefits of citizen science are brought within the orbit of science and knowledge mobilization, as well as multiple and immersive modes of data collection.

Certainly, the 'citizen science' culture and its 'inclusionary economy' place pressure on everyday people to live according to their duty to nature. Those who dispose of their recycling, drive excessively, shop at WalMart, or let balloons go into the sky as part of a memorial service (all activities the author has recently taken part in), are not only ignorant but are earth-citizens in utero. In particular, 'participatory citizens' engage in all sorts of earth-bound activities, taking photographs while sea-diving (Latimer 2013, p. 88), sharing their data through social networks (Lievrouw 2010), forming ties with the scientific community in traditional peer-review publications and credited for contributory, collaborative, and co-created knowledge about noise pollution, monarch butterflies, volcanoes, soil samples, marine phytoplankton, redwood forest plants (Connors et al. 2012, p. 1273), as well as customized citizen science projects to suit a range of activities: at the zoo, at home, at night, school, sports stadiums, the beach, emergency response, online, in oceans, streams, rivers, lakes, in snow, rain, the car, on a hike, on a walk, run, or while fishing (see scistarter.com).

Though, some argue that it is impossible for citizen scientists to contribute to knowledge building so much as they are co-producers in conservation habits (Cornwell and Campbell 2012), many of these are being incorporated into the traditional university system of research which creates more transparent ties between the research process and research dissemination. This list of actions and benefits is intended to transform subjectivity from a self-reflexive entity into a worldly one. However, to reiterate, any relationship between Anthropocene and becoming-imperceptible is wrought with difficulties and may not work simply by imploring people to become more active in their everyday travels. The manner in which Anthropocene is conceived, is that of a 'massive concept' devised through the technical and diverse writings from 'biogeographers' and 'environmental historians' that link human mobility through intense networks that crumble traditional global boundaries, or term it an 'homogecene' (Olden 2006) that reduces biodiversity and increases homogeneity. If individuals are expected to converge under the rubric of eliminated homogeneity, is it possible to do so while still thinking in terms of the collective? Anthropocene is conceived as a 'massive consequences' rubric, brought to light in images of environmental catastrophe—or, 'epiphanies of catastrophe'. The Anthropocene has a 'vibrating body', then, perceptible in terms of data analysis and grounded in scientific fact.

A central component to citizen science, and one that belongs to Deleuze & Guattari's notion of the 'plane of consistency', is the transformation of natural immersion. A main site where the keys to transformation in the biosphere are found is in the birdcall, a longstanding historical practice for amateur and professional ornithologists (and the separation itself is something of a new division). In 1749 Professor Johannes Leche from the Turku Academy studied with a team of researchers made up of interested members of the public, throughout dozens of societies and university-affiliated research groups (see Greenwood 2007). The study of bird migration, habitat and population has held the scientific imagination in terms of a reliable heuristic device for gaining a handle on environmental change. The earliest research groups were, thus, interested in transformation. The same goes for any modality of transformational thought, which might take into account the manner in which clashes of culture occur *within* the differences of the similar. Tocqueville had written, for instance: 'As for me, ... I don't know when this long voyage will end; I'm tired of repeatedly mistaking misleading vapors for the shore, and I often ask myself whether the *terra firma* that we have sought for so long really exists, or whether our destiny might not rather be to ply the seas eternally!' (qtd. in Runia 2014, p. 179). And what we face now is a situation where the potential (the full range of *datum*) are closer to actualization than in the past, rendering the space of data storage a massive assemblage of human, non-human, more-than-human, and post-human actors.

CONCLUSION: WORLD/EARTH

The need to address places as intersections between time, space, materiality, memory, and bioacoustics reveals those vibratory intersections that are otherwise imperceptible to the normal sensorium, but come about instead through technologies of transduction. Rather than dismiss immersion, I would rather explore further the concept of immersion as it is related to transduction in such a way that is parallel to the terms *datum* and *captum*. Transduction is a process of making sense of vibration, and the turning points between these activities. *Sound* is a wholly unsatisfying reduction of this world. Against this, I argue that the most misleading term amongst us in sounding is the tired notion of 'immersion': that sound *surrounds* us, while the visual *directs our attention*, that sound is *fleeting*, while the visual is *permanent*. Sounding represents the bundled up participation of pasts as they flow in the present towards the future, but the question of immersion prioritizes the temporal figuration of sound over and above the *places* that bioacoustics and field recording give us as potential. To suggest that sounding is simply immersive, then this reverses the flow of our scientific capacity to measure and analyze *for the betterment of community* and *for an increased understanding of culture/ nature relations*. Soundings, I argue, are not immersive qualities: they are directives, they are orientations, they are the manifestation of *associations*.

Sounding diminishes the dystopian rhetoric of the Anthropocene. While the Anthropocene certainly invites a level of anthro-centric regret, this chapter has attempted to introduce the possibility for transformation through sounding. There is thus no argument over whether the Anthropocene is or is not real; it is difficult to contest that humans have caused major environmental changes. And while we might think that the new age of 'environmentality', or self-governance in the image of the responsible citizen (see Gabrys 2014), is an example of ecologically driven social control, it also represents a 'crisis moment' for interdisciplinary collaborations. That is, the coming together of technological innovation and community involvement is part of what drives towards new scientific collaborations and interdisciplinary and transdisciplinary partnerships. Such partnerships offer the opportunity to theorize new decentering methods of scientific analysis that place human exceptionalism at bay, yet require human exceptionalism in order to properly facilitate its own decentering.

To conclude, I suspect that the problems outlined here come from the images of the imperceptible, involving immersion and transduction, datum and captum, sounding and sound, apparatus and capture. As Gilles Deleuze writes in his postscript to the society of control, the suspicion is that we are post-disciplinary, that we no longer 'contain bodies' so much as we 'set them free', that we wish to learn about multitudes of species in flight, in massive swarms, with which we are fascinated, their modulations and transformations holding us captive; as he writes with Félix Guattari in *A Thousand Plateaus*, we are in a time dominated less by discipline than by control, less by the bird and more by the insect: 'the reign of birds seems to have been replaced by the age of insects, with its much more molecular vibrations, chirring, rustling, buzzing, clicking, scratching, and scraping. [...] The insect is closer, better able to make audible the truth that all becomings are molecular [...]. The molecular has the capacity to make the elementary communicate with the cosmic' (Deleuze and Guattari 1987, p. 308). Such a position stands in contrast to the 'bird-watcher': here the gaze captures the bird in isolation, catalogues it, isolates it, captures its warble. Here one doesn't hear so much as they feel through a big sensorium: immersion and transduction. This marks the threshold of the 'becoming-imperceptible', beyond animal, beyond the minor, beyond the molecular, it is enabling, it is productive. The temptation to depict the citizen scientist as an interpellated subject of neoliberal enjoyment is strong, but it remains too easy of an answer.

And where we currently reside in the ambivalences between nature and culture, here we prefer to vibrate between them. Perhaps, to see that sounding really means the fate of all sound, that needs to come together as one political issue, one polis, one vibration.

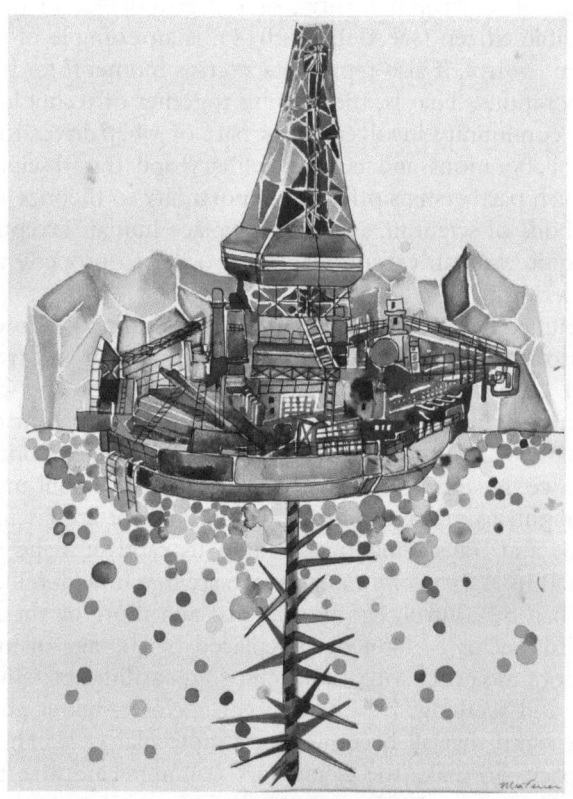

Image 7.1   Off shore study (Mia Feuer, watercolor)

REFERENCES

Bourdieu, P. (1984). *Distinction: A Social Critique of the Judgement of Taste.* Cambridge, MA: Harvard University Press.

Braidotti, R. (2013). *The Posthuman.* Cambridge: Polity Press.

Connors, J. P., Lei, S., & Kelly, M. (2012). Citizen Science in the Age of Neogeography: Utilizing Volunteered Geographic Information for Environmental Monitoring. *Annals of the Association of American Geographers, 102*(6), 1267–1289.

Cornwell, M. L., & Campbell, L. M. (2012). Co-producing Conservation and Knowledge: Citizen-Based Sea Turtle Monitoring in North Carolina, USA. *Social Studies of Science, 42*(1), 1101–1120.

Crutzen, P. J. (2002). Geology of Mankind. *Nature, 415,* 23.

Dalby, S. (2015). Anthropocene Formations: Environmental Security, Geopolitics and Disaster. *Theory, Culture & Society, 34*(2–3), 233–252.

Deleuze, G., & Guattari, F. (1987). *Capitalism and Schizophrenia: A Thousand Plateaus.* Minneapolis: University of Minnesota Press.

Gabrys, J. (2014). Programming Environments: Environmentality and Citizen Sensing in the Smart City. *Environment and Planning D: Society and Space, 32,* 30–48.

Gallagher, M. (2015). Field Recording and the Sounding of Spaces. *Environment and Planning D: Society and Space, 33*(3), 560–576.

Greenwood, J. J. D. (2007). Citizens, Science and Bird Conservation. *Journal of Ornithology, 148*(suppl. 1), S77–S124.

Henriques, J. (2008). *Sonic Bodies: Reggae Sound Systems, Performance Techniques, and Ways of Knowing.* London: Bloomsbury.

Irwin, A. (1995). *Citizen Science: A Study of People, Expertise and Sustainable Development.* London: Routledge.

Kitchin, R. (2014). *The Data Revolution: Big Data, Open Data, Data Infrastructures, and Their Consequences.* London: Sage.

Latimer, J. (2013). Being Alongside: Rethinking Relations Amongst Different Kinds. *Theory, Culture & Society, 30*(7/8), 77–104.

Latour, B. (2005). *Reassembling the Social: An Introduction to Actor-Network-Theory.* Oxford: Oxford University Press.

Lawrence, A. (2006). 'No Personal Motive?' Volunteers, Biodiversity, and the False Dichotomies of Participation. *Ethics Place and Environment, 9*(3), 279–298.

Lievrouw, L. A. (2010). Social Media and the Production of Knowledge: A Return to Little Science? *Social Epistemology, 24*(3), 219–237.

Lorimer, J. (2014). Multinatural Geographies for the Anthropocene. *Progress in Human Geography, 36*(5), 593–612.

Mason, V., & Hope, P. R. (2014). Echoes in the Dark: Technological Encounters with Bats. *Journal of Rural Studies, 33,* 107–118.

Mills, M. (2011). On Disability and Cybernetics: Helen Keller, Norbert Wiener, and the Hearing Glove. *Differences, 22*(2–3), 74–111.

Morton, T. (2010). *The Ecological Thought.* Cambridge: Harvard University Press.

Olden, J. D. (2006). Biotic Homogenization: A New Research Agenda for Conservation Biogeography. *Journal of Biogeography, 33*(12), 2027–2039.

Ross, A. (2012). Song of the Earth. In B. Herzogenrath (Ed.), *The Farthest Place: The Music of John Luther Adams* (pp. 13–22). Boston: Northeastern University Press.

Shulman, S. (2008). *The Telephone Gambit: Chasing Alexander Graham Bell's Secret.* New York and London: W. W. Norton.

Sicular, S. (2013, January 22). Big Data Is Falling into the Trough of Disillusionment. *Gartner.*

Smith, D. N. (2013). *Sounding/Silence: Martin Heidegger at the Limits of Poetics.* New York: Fordham University Press.

PART III

Media and Artistic Responses

Image III.1   Birds (Mia Feuer, watercolor)

CHAPTER 8

Geoartistry: Invoking the Postanthropocene via Other-Than-Human Art

David Fancy

When we describe a landscape, an insect or a flower as 'beautiful', such statements usually carry with them the understanding that the entities in question bear a 'natural' beauty unmarked by human intention and are therefore not typically understood to embody 'artistic' kinds of beauty. The theologically oriented among us might see the 'hand of God' at work in these 'creations', but regardless of how we attribute the causalities informing the landscape, insect or flower, they are still not traditionally understood to be 'beautiful' in the sense usually ascribed to creations generated by human hands. Deleuze and Guattari, however, invite an exploration of the artistic capacities of other-than-human entities when they suggest in their early collaboration, *A Thousand Plateaus* (1987), that 'Messiaen is right in saying that many birds are not only virtuosos but artists' (316–317) given these creatures' demonstrated capacity for activities such as the expression of varying territorial songs or the construction of complex dwellings. As I will outline further below, Deleuze and Guattari assert that central to this other-than-human artistic expression is an understanding that the world is constituted by onto-genetic and self-producing systems rather than by traditional identitarian formations

D. Fancy (✉)
Department of Dramatic Arts, Brock University, St. Catharines, ON, Canada

© The Author(s) 2018   217
j. jagodzinski (ed.), *Interrogating the Anthropocene*, Palgrave Studies in Educational Futures, https://doi.org/10.1007/978-3-319-78747-3_8

issuing forth from transcendent evolutive lineages. They suggest that a *distributed* 'autonomy of expression' (317) is a key element of both such onto-genetic and machinic processes, as well as being central to the capacity for other-than-human artistic expression not anchored in traditional notions of human agency or intentionality. Deleuze and Guattari's *avant la lettre* posthumanist claims resonate in our current intellectual milieu in which the collapse of the distinction between human and natural history is active and ongoing. As such, it comes as little surprise perhaps that discussions of other activities traditionally understood to be the province of the human such as the artistic are also up for reconsideration.

With regard to the matter at hand in this volume, namely the question of the 'Anthropocene', understanding the capacity for artistry of other-than-human entities, as well as the capacity for other-than-human entities to *experience* artistic production, could serve as a significant aspect of humans' ability to imagine and negotiate pathways towards postanthropocenic[1] futures. Indeed, the editors of a volume of *Deleuze Studies* dedicated to the Anthropocene remind us that 'it is overwhelmingly clear that the Anthropocene will have profound impact on philosophy', including considerations about the nature of artistic production, thereby 'shocking [philosophy] out of anthropocentric complacency and the accustomed ways of thinking which situate the human apart from nature' (Saldanha and Stark, p. 431). But are we not already in a realm of unresolvable contradiction, suggesting that humans themselves need to think the postanthropocenic and that human thought needs to come to terms with other-than-human forms of artistic creation? Saldanha and Stark continue by reminding us that 'a critique of human exceptionalism and humanist essentialism does not necessarily mean dissolving the specificity of the human into a free-flowing, all-encompassing and chaoid Life' (434). Rather, we have no choice from a Deleuzo–Guattarian perspective but to 'agree with the humanist tradition that it is incumbent on thought to examine reality *from within* the intractabilities and ambiguities of the human perspective' (ibid., author's emphasis). They amplify their position by echoing Elizabeth Grosz' suggestion that the 'humanities requires a recognition not only of the politics of the human that informs and produces knowledge of the human; but also of the ontological forces of the nonhuman that press the human from both within and outside' (431). As such, with a view to contributing to thinking beyond the supremacy logics that have accelerated what has become the planetary calamity of the Anthropocene, this chapter will posit a range of artistic production as well as *reception*—conceptual terrain not extensively covered by Deleuze and Guattari—outside the realm of what is traditionally understood to be 'artistic' by engaging with examples of landscapes, of cloud formations, of flora and of various other other-than-human animal activities. In such thinking, multiple phenomena will be understood to generate and also *experience in turn* artistic production according to the range of capacities for perception and expression afforded them by the assemblages of which they are constituted.

In their final collaborative text, *What is Philosophy*, Deleuze and Guattari articulate the territories of the three central human pursuits of science, art and philosophy, and argue for the existence of a *geophilosophy* that ties the activity of thought in various ways into the wider processes of territorialization of the earth (85–113). Recognized as a human activity drawing from and gathering bundles of preindividual processes, this radically constructivist and expanded understanding of philosophy situates speculative thought in much broader cosmic and ontological dynamics. As such, geophilosophy speaks to a mode of thinking that both emulates and affirms the multiplicitous complexity of becomings of natural reality. This insight is borne from each thinker's commitment to postidentitarian, differential and processual forms of thinking that seek to work outside inherited models of thought anchored in binarist conceptions of substance, ones bound by restrictive logics of recognition and representation (Deleuze 1994, 2004; Guattari 1995). In such a vision, philosophy's role is to generate new concepts, to 'bring forth events' (Deleuze and Guattari 1994, 197) and to '[lay] out a plane of immanence that, through the action of conceptual personae, takes events to infinity'. Similarly, they suggest that art, in a world of multiplicity and becoming, is marked by a similar aspiration 'to create the finite that restores the infinite: it lays out a composition that, in turn, through the action of aesthetic figures, bears monuments or composite sensations' (ibid.). Art can do this, they argue, because the artist extracts percepts—the essences of perceptions—from the quotidian perceptions they are always already bathed in. Artists also simultaneously extract affects from experienced affections—those shifts initiated by the engagement with percepts—and bundle affects and percepts together within the materials and forms in which they then express themselves: words and paper, clay and figures, bodies and performances, and so forth. Art's drive 'to raise lived perceptions to the percept and live affections to the affect' (170) takes place, as Deleuze and Guattari suggest is the case with geophilosophy, 'in the relationship of territory and the earth' (85) and, like this geophilosophy, art can be understood to link 'the cry of humanity and the earth's song' (176). I suggest that such instances can be described, as with Deleuze and Guattari's philosophy's turn earthward, as *geoartistry*, or as being *geoartistic*. Both a concept and an unlimited series of artistic practices—phenomena that instantiate events in thought and precipitate the finite that restores the infinite through specific moments of creation—*geoartistry*, like philosophy can help serve to '*summon [...] forth a new earth*' (99).

It is clear that the risks of articulating geoartistry could readily be infiltrated by a recuperative anthropocentric gesture of understanding increased complexity of relationalities between other-than-human entities *solely* for the purposes of sustaining the planet's ecosystem for continued human viability as a species. In order to keep this (inevitable?) motivation somewhat in check throughout the ensuing explorations, let me first invoke what Guattari describes as the 'ethico-aesthetic' aspects of artistic production and suggest how it can pertain to a wider than human spectrum of artistic activity (1995). Ethico-aesthetics are informed by the ways in which the Deleuzian and then Deleuzo–Guattarian ontological projects draw significantly on the Spinozan project of ethics characterized not by suggested adoption of transcendent moral anchors (a God, Reason or other principle) posited to be outside the systems they sustain, but instead immanent to or inhering within the system they support and sustain. An immediate implication of such an immanentist (rather than transcendent) position is that prescriptive moral projects are repositioned instead into ethical ones responsive to the situatedness of the interactions of bodies and forces in the world. Central to this understanding is that bodies are not fixed and bound locations:

> That important thing to understand life, each living individual not as a form, or of a development of a form, but as a complex relation between differential velocities, between deceleration and acceleration of particles. A composition of speeds and slownesses of a plane of immanence. (Deleuze 1988, p. 123)

Indeed, Deleuze writes that for Spinoza, 'a body, of whatever kind' is simultaneously by a 'kinetic component' composed of 'an infinite number of particles' constituted by 'relations of motions and rest, of speeds and slownesses between particles' and a 'dynamic component' by which 'a body affects other bodies, or is affected by other bodies' (ibid.). This constitution and co-constitution of bodies via singular composition as well as mutual engagement with other bodies invites a recognition of the situated nature of ethical engagement. Instead of depending on an arbiter of justice external to bodies to legislate what is or is not moral behaviour, Spinoza suggests instead that which is ethical is that which expands a body's capacity to act is ethical and joyful, that which restricts it is unethical. In his reading of Spinoza, Deleuze directs the reader to understand the extra-individual nature of 'affections' constitutive of human or any other perceptual experience in that affections 'involve both the nature of the affected body and that of the affecting external body' (49).

It is then the transmissional and shared aspect of the affections that makes them constitutive of a variety of bodies simultaneously, and therefore cannot be reduced to the property of one specific body but instead circulate between bodies. This circulation of constituting and constitutive affections cause 'transitions' and 'passages' to be experienced between varying states—durational feelings called affects—that allow bodies 'to pass to a greater or lesser perfection' (48). This increase or decrease of perfection is oriented not towards a perfection exterior to the body with which the body is being compared or adjudicated, but instead towards the functionality and expression of the coherences inhering within and constituting the body itself. From the perspective of this kind of sensitive ethical libertarianism, if I were to disparage or minoritize a person because of set of attributes affiliated with a particular identity category, such as gender or ability, my generation in them of an 'unjoyful' affect would be understood to be unethical not because of the application of a prescriptive external regulation that insists that insulting is wrong because it is immoral to do so, but because of a situated ethical matrix that reduce the perfection of the targeted individual in that particular moment. While external edicts can serve as powerful cultural repositories of the discoveries of what reduces perfection ('Thou shalt not kill'), to hypostasize these as external to the many situated encounters that produced them is to foreclose the possibilities of bodies responding to the specificities of individual and situated instances. Deleuze describes how for Spinoza productive encounters between bodies that mutually enhance their singular velocities and coherences generate shared or joyful encounters:

> when we encounter a body that agrees with ours, has the effect of affecting us with joy, this joy (increase of our power of acting) induces us to form the common notion of these two bodies. (Deleuze 1988, pp. 118–119)

Given the *creativity* involved in the work or curating such joyful qualities, Deleuze asserts following Spinoza that, 'The common notions are an Art, the art of the *Ethics* itself: organizing good encounters, composing actual relations, forming powers, experimenting' (119).

The capitalization of 'Art' here invokes a register in which cultivating ethical relations between bodies can be understood to involve a great deal of skill and nuance. Let me turn now to the writing that Deleuze and Guattari undertake in a section of *A Thousand Plateaus* entitled '1837: Of the Refrain' in which they discuss the emergence of artistic activity proper from the work of territoriality undertaken by various bodies operating via variously ethical encounters. Their machinic articulation of the unfolding of complex relationships of assemblages is a refinement and extension of elements of Deleuze's earlier work on Spinoza. In their discussion of the refrain (*ritournelle*), Deleuze and Guattari describe how the use of music, or any other form of expression, can create 'a circle around [an] uncertain a fragile centre' (311), a tempo-spatiality perimeter that constructs a discernable location distinct from the relative chaos that surrounds it. Such spatio-temporalities are not cut off from the spaces and velocities that surround them. Instead, the borders are selectively porous in that they continuously reintroduce the potentiality of chaos, but on their own terms. In other words, inviting chaos rather than being subject to it: 'One opens the circle not on the side where the old forces of chaos press against it but in another region, one created by the circle itself' (ibid.). This opening of the body for encounters with affections that surround it is an exercise of seeking to activate the potentialities by which it is encircled:

> As though the circle tended on its own to open onto a future, as a function of the working forces it shelters. This time, it is in order to join with forces

of the future, cosmic forces. One launches forth, hazards an improvisation. But to improvise is to join the with the World, to join with it. (ibid.)

The role of this 'refrain' is, Deleuze and Guattari indicate essentially territorial: 'Bird songs: the bird seeks to mark its territory' (ibid., p. 312). These 'territorial assemblages' are however ones that are clearly open to their own deterritorialization and evolution through the improvised capacity for expansive inclusion of elements outside initially contained by the portion of the process of the refrain that circumscribes spatio-temporal parameters of 'chaos, terrestrial forces, cosmic forces' (ibid.). Milieus are the 'blocks of space time' constituted by the territorializing act refrain (ibid., p. 314) that 'pass into one another' opening them to chaos and potentials for change due to the rhythmic states that exist in the transition between milieu (313). Milieus become actual territories 'when milieu components cease to be directional, becoming dimensional instead, when they cease to become functional to become expressive' (315). They provide the example of 'colour in birds or fish':

> Colour is a membrane state associated with interior hormone states, but it remains functional and transitory as long as it is tied to an action (sexuality, aggressiveness, flight). It becomes expressive on the other hand, when it acquires temporal constancy and a spatial range that make it a territorial, or rather territorializing, mark: a signature. (ibid.)

They counter Lorenz' claim that aggression is the only basis for territory (ibid.), suggesting that while aggression is a manifestation of territoriality, it cannot solely explain the work of territorialization itself, and that its source must be sought elsewhere: 'precisely in the becoming expressive of rhythm or melody, in the emergence of proper qualities (colour, odor, sound, silhouette)' (316). They ask, 'can this becoming, this emergence, be called Art?' which, if answered affirmatively 'would make the territory a result of art' (ibid.), with such a property being 'fundamentally artistic because art is fundamentally *poster, placard*' (ibid.). From this they derive the insight that 'what is called *art brut* is not at all pathological or primitive; it is merely this constitution, this freeing, of matters of expression in the movement of territoriality: the base or ground of art' (ibid.). The implications here are significant: they suggest, as 'from this standpoint art is not the privilege of human beings' (ibid., p. 317).

A further and essential step occurs, they argue, when the post or placard, the 'signature' of the becoming expression through the process of territorialization is transformed, when territorializations are, 'No longer a signature, but a style' (318). This moment takes place when the 'expressive qualities entertain variable or constant relations with one another (this is what matters of expression *do*)' (ibid. emphasis authors'). In other words, the expressive qualities 'no longer constitute placards that mark a territory but [rather instead] motifs and counterpoints that express the relation of the territory to interior impulses or exterior circumstances' (ibid.). According to Deleuze and Guattari's onto-genetic model, it is clear that all phenomena are creative, and as such engage with a kind of artistry that produces the *art brut* of territorial expression. However, a distinction can be made, they suggest, namely that 'what objectively distinguishes a musician bird from a non-musician bird is precisely this aptitude for motifs and counterpoints' (ibid.). These qualities, regardless if they are 'variable, or even when they are constant' serve to 'make matters of expression something other than a poster': with these new variations they become a *style* 'since they articulate rhythm and harmonize melody' (ibid.). The relationship these dynamics entertain with the later discourses of ethico-aesthetics is clear in the use of the Spinozan language to describe the bird's relationalities with the world about it:

> We can say that the musician bird goes from sadness to joy or that it greets the rising sun or endangers itself in order to sing or sings better than another. (ibid.)

They are quick to assert that: 'None of these formulations carries the slightest risk of anthropomorphism, or implies the slightest interpretation' (ibid.), instead they affirm that there is 'a kind of geomorphism' (ibid.) or modulation of relationalities between a wide range of bodies at work in such instances. Key to the *artistic* quality of such encounters and interrelationships is the emergence of novel and original expression generated by the interplay of components interacting via a complex non-binary dialecticism:

> The relation to joy and sadness, the sun, danger, perfection, is given in the motif and the counterpoint, even if the term of each of these relations is not given. In the motif and the counterpoint, the sun, joy or sadness, danger, become sonorous, rhythmic, or melodic. (ibid., p. 319)

Deleuze and Guattari describe these complex interrelationships, borne from exercises of territoriality and emerging further through more differentiated counterpoints, as a 'machinic opera' (330)—one that is machinic because of 'this synthesis of heterogeneities as such'[2]—that binds 'together orders, species, and heterogeneous qualities' into an expressed machinic 'statement' or moment of 'enunciation' (330–331). They provide a recurring example of the stagemaker bird that combines a variety of different registers of expression to generate sounds, movements, and colours in expressive collaboration with other aspects of the machinic assemblage surrounding it that produce:

> a consolidation that 'consists' in species-specific sounds, sounds of other species, leaf hue, throat colour: the stagemaker's machinic statement or assemblage of enunciation (331)

Essential to the argument is the understanding that this notion of the assemblage of enunciation, a gathered and yet distributed enunciation that evidences the 'autonomy of expression' central to their claims, 'cannot be arrived at in terms of behavior'. Such an approach territorializes the agency of the expression into one individuated bound form, 'rather [than] only in terms of assemblage' (332). The refrain and its forces of territorialization, counterpoint, expression, and style, *work through* the stagemaker rather than the bird being the autonomous source of the expression:

> the animal is instead prey to "musical rhythms" and "melodic rhythmic themes" explainable neither as the encoding of a recorded phonograph disk nor by the movements of performance that effectuate them and adapt them to circumstances. (332)

They assert that the 'opposite is even true', namely that 'the melodic of rhythmic themes precede their performance and recording' (ibid.). It is in fact any animal's, human or otherwise, capacity to move from what is 'innate' in combination with the rest of the machinic assemblages with which it intersects, stretching from processes in the 'intra-assemblage'—within the individuated animal body—'across all the interassemblages and reaches all the way to the gates of the Cosmos' (333). In other words, these creative and artistic forms of expression open out onto the affective and perceptive dynamics not only constitutive of the 'socius'—'interassemblages of courtship and gregariousness' (333)—but to those wider inter-assemblages that constitute the nature of onto-genetic production at its widest possible scale. The simultaneously territorializing and expressive nature of the stagemaker's song, run through as it is by the rhythmic transitionality central to the refrain, expose the drive of the autonomous expression gathered in the assemblage, namely, as Deleuze and Guattari will describe a key element of the artistic function later in their *What is Philosophy?*, as being the drive to 'create the finite that restores the infinite' (197).

STYLE, SOCIUS, COSMOS

An aspect of the thinking here has to be the need to articulate as best as possible—considering the interspecies or inter-entity ventriloquism that humans writing about other-than-humans necessarily involves—the experience of geoartistry in and of itself from the perspective of the other-than-human entity/ies at hand, and how the experience may be joyful for them in the sense that the entity/ies' potentials are expanded from the experience. This can serve to provide different examples than Deleuze and Guattari who tend to focus on the other-than-human *animal* in their thinking about examples of the autonomous expression of the artistic style that opens out onto territory, socius, and cosmos. Additionally, such an undertaking can proceed via the delicate task of imagining what percepts and affects other-than-human and other-than-animal entities more widely might extract in their experience of the monumentality of the work of the artist as a means of contributing to postanthropocenic presents and futures.

How would one, for example, go about describing plant expression as 'artistic?' In 1973, Dorothy Retallack published a book entitled *The Sound and Music of Plants* outlining her experiments of how music affects plants that appeared to demonstrate a strong relationship between types of music, volume and plant growth. Perhaps reflecting Retallack's own unconscious bias of a certain musical conservatism, plants exposed to 'soothing' music thrived, those exposed to the generally discordant and atonal music of modernists Schönberg and Webern failed to thrive, while plants subjected to Led Zeppelin and Jimi Hendrix died. More recent research by K. Creath and G. Schwartz (2004) has suggested that 'sound vibrations [...] directly affect living biologic systems' (113) and in a literature review A.R. Chowdury and A. Gupta (2015) summarize a wide range of global research by stating that 'specific audio frequencies in the form of music facilitated the germination and growth of plants, irrespective of the music genre' (33). The Damanhur group has been working since the 1970s to find ways of translating the subtle shifts in electromagnetic patterns in plant leaves and roots into music audible for humans, and currently promote a Musical Instrument Digital Interface (MIDI) instrument that amplifies such bio-signals into musical 'conversation' between plants that appear to increase their vitality and growth.

Switching to the language of Deleuze and Guattari's discussion of the refrain, how does the plant experience its interactions with the intra-assemblage of its own territorialized intra-assemblage, and the wider and multiple inter-assemblages of which it is part? From the initial perspective of the plant as a body that announces its territoriality, the xylem and phloem vessels and tubes of the stem serve to circulate water and nutrients extracted by root systems and the photosynthetic processes of the leaves in order to sustain the coherence of the plant's intra-assemblage. The fragrance and colour of the plant flower, generated in a complex relationship between the pistil, stamen, sepals, and petals, guarantees the plant's reproductive continuance from a simple biological perspective. The flower, in its function as the main placard signalling an invitation to pollinators and other inter-assemblage factors, is the 'base or ground of art' in its 'constitution, [its] freeing, of matters of expression in the movement of territoriality' (1987, p. 316).

Following Deleuze and Guattari's progression then, is it possible to suggest that there is an opportunity in the extended refrain of the territoriality of a plant's signalling for a shift from expressive 'signature' to artistic 'style'? There are at least two ways of proceeding to answer this question. The first involves understanding that the examples of other-than-human animals that have artistic capacities have such capacities as a result of organismic complexity. In the case of the stagemaker bird for example, such complexity permits them to inhabit spatio-temporality in such a way that affords the types of contrapuntal and improvisatory interventions that Deleuze and Guattari suggest elevates signature to style. Birds move quickly, have more complex neuronal pathways—and thus are more liable to have the capacities for articulating creativities beyond the art brut phase of expressive territoriality. It might seem to follow that it is more likely that the intra-assemblages of birds can harness the autonomous capacity for expression so as to be able to express style than plants can. Despite their inherent organismic complexities, plants may be more likely, at least from a human philosophical perspective that privileges individuated expression, to need the collaboration of inter-assemblage arrangements to reach the dynamics of counterpoint and deterritorialization of motif understood to progress from signature to style.

But is not this privileging of individuated expression, with all of its anthropocentric echoes of human agency and autonomy, particularly limited from a machinic postanthropocenic perspective that foregrounds *distributed* expression? Indeed, would it not follow that, from the perspective of distributed machinism, geoartistry in its more complex manifestations of *style* is liable to occur in the plant's relationship with water, light, soils, wind, geological, macro-meteorological and other factors, and over a much more extended period of time? Following this logic, we might assert that in less 'complex' organisms, autonomy of expression generates geoartistry in the accumulated complexity of the inter-assemblage over months, seasons, and millennia, rather than in the individuated 'behaviour' of the stagemaker bird more intelligible within an anthropological ethological register. The body of the meadow, the watershed, the hemisphere, the earth, serves as the context for the concatenation of signatures from a range of smaller bodies contained within these larger inter-assemblages. Dynamic counterpoint—a central element of Deleuze and Guattari's understanding of the progression from signature to style—between the plant and clouds generated by orographic precipitation from humidity rising up a mountainside contributes to the particular distributed expression of the ecosystem of the mountainside. As additional macro factors such as tectonic plates, volcanic eruptions, carbon emissions intervene in the amount of light and precipitation on the mountainside, the plant responds by improvising over multiple iterations and generations of 'itselves' to find distributed and different expression in order to move through and meet the trauma and challenge of its own tribulations in a changing world. As such, the wider inter-assemblage is the main vehicle inviting the reintroduction of the potentiality of chaos into spatio-temporal perimeter of the inter-assemblage—as per Deleuze and Guattari's logic of the refrain—as well as the many smaller bodies and intra-assemblages that compose it. If it were possible to view these processes in an accelerated fashion, the capacity for autonomous expression harnessed by the aggregate complexity of the inter-assemblage of the mountainside might start to look, from a human perspective, more like a style than a signature.

If, then, organismic complexity is key to the capacity for counterpoint and improvisation that marks a shift from the *art brut* signature of territorial expression to the style associated with human artistic expression, and if complexity can be understood to intensify via the passing of time and the expansion of the scope of inter-assemblages, then geoartistry, even in its most complex forms, can be understood to inhere in every aspect of the natural world. Time (or indeed the ability for simultaneous multiple responses of and in velocity) plus breadth of inter-assemblage (or indeed complexity of intra-assemblage) increases the extensive possibility of geoartistry inherent in the intensive potentials of an immanent and machinic universe. It is thus that, when Deleuze and Guattari wonder, 'Not only does art not wait for human beings to begin, but we may ask if art ever appears among human beings, except under artificial and belated conditions,' (1987, p. 320) it is not only in other-than-human animals that artistic precedence can be traced, but in all phenomena on the earth. The planet as expression of geoartistry in and of which human artistic activity is a particular extension and expression; the planet as assemblage experiences Spinozan joy in its unfolding expression.

A second and complementary way of looking at the question: 'is the plant engaging in artistic creation', is to imagine the experience of the plant, as well as of the inter-assemblages of which it is part, and in so doing explore the question of ethico-aesthetics as it pertains to the experience of the plant and the wider inter-assemblage itself (or indeed 'themselves' if we attempt to think in multiplicity here). This constitutes the more difficult speculative part of the argument where thought, as Claire Colebrook has recently written about philosophy and the Anthropocene, 'finds itself through a constant process of self-erasure' (2016, 542). Let me return then to the language Deleuze and Guattari use in *What is Philosophy?* to speak to the experience of what artists do, what art is, and how art affects, and explore to what extent it can be traced onto the other-than-human assemblages and inter-assemblages from which creative expression is always already emerging.

As I outline above, they note that art wrests 'the something lasting' from perceptions and affections, the extraction and curation of particular duration elements of what artists perceive in the world and the affections and shifts of experience that are generated in encountering the world, and the work of art which is its intensification. They describe these extracted elements as 'percepts' and 'affects' that are bundled together via the media an artist is using (paint, words, movement, etc.) in a fashion that is durational, that lasts. In a clear indication of the extra-human nature of the sources of the artistic, affects are described as the 'non-human becomings of man' (169). Given the work of art's capacity to sustain and generate the associated perceptions and affections in those encountering the artistic expression, Deleuze and Guattari use the term 'monument' to describe the work of art, although not monumentality as defined by its fixity and finitude. Instead the monument of the work of art is determined by its capacity to continue, via the 'block of sensations' (164) that are the bundled percepts and affects, 'to create the finite that restores the infinite: it lays out a composition that, in turn, through the action of aesthetic figures, bears monuments or composite sensations' (197). Central then to the postanthropocenic understanding of art is to be able to, at the very least, speculate and imagine the experience of the other-than-human entity's relationship to what might be constituted as 'monumental' from their perspective.

WALKING TO THE EDGE OF THE FIELD

An impure critical and philosophical approach fused with speculative prose is one methodology for achieving these ends. For example: I live on a small acreage surrounded by fields and forests in the country in the Niagara region of south-western Ontario. Every year towards mid-June, fireflies (beetles of the *Lampyridae* family of the *Coleoptera* order) generating bioluminescence via sacs of fluid in their abdomen, appear at dusk and into the evening in the forests and fields on and around my house. The biologically established reasons for this bioluminescence are largely that it represents an evolutionarily differentiated tactic for attracting mates, or indeed for attracting other insects to prey upon. From a geoartistic perspective, the *art brut* expressed by the insect's territorialization is a signature of the insect's presence. From the insects' perspective, what can we imagine their relationship might be with the perceptual and affective realities of the experience of flying about communicating with each other for various reasons via this light? And what of the inter-assemblage realities of the other insect's, plants and bats' perceptual and affective realities in the face of the light of the Lampyridae? Is a possibility of style inherent here? One evening last summer, a new moon appeared from behind large cumulonimbus clouds over the field shortly after dusk, and the flight patterns of the insects changed as their velocity increased as well as their height off the ground, with a seeming shift from more randomized patterns of flight and insect arrangements in and above the field, to a more uniform vertical ascent and descent type of flight that rose to 20 feet above the ground, then back down again. The moon disappeared behind a cloud and after about ten minutes the flight pattern shared by about 50% of the hundreds of fireflies that I could perceive over about three acres from a slightly elevated vantage point returned to lower-flying and to a more randomized pattern of flight. After twenty minutes the moon reappeared, this time somewhat higher in the sky and again, after about ten minutes, over half the insects in the field had regained the elevated flight pattern with the vertical ascents and descents. After approximately thirty minutes of this, small clusters of 10–15 insects congregated in eight or nine spots throughout the field and began blinking on and off in largely shared patterns. The pattern of the bioluminescence was not homogenous within each group, or uniform across the groups I could keep in my field of view at one time, but again at least half the insects in each of these smaller clusters would, every minute or two, activate and deactivate their bioluminescence simultaneously every three or four seconds. The experience of watching this was of a gradual and gathering coherence of the pulsing of the insects across my field of view.

This activity, described from a behaviourist perspective, might include references to rhythmical entrainment of bioluminescence for the purposes of reducing exposure to predation, or indeed heightened flight patterns given increased visibility for the insects across visible light spectrum permitting more extended departures from the safer confines of the long grass. These or similar reasons might well be part of the overall reality of the insects' experiences in the moment, but what of the inter-assemblage relationalities between the insects and the curiously bright echoing of their own luminescence by the moon that could serve as an invitation to an intensified experiential sensibility among the insects? Might this invitation and the shift of movement it seemed to invoke be one among a number of affections induced by the perceptions generated by the inter-assemblage counterpoint of firefly-moon? As suggested above, we might not be able to assert that the complexity of either of the organisms—firefly or moon—has the capacity to engender the reflexivity necessary to create art with the self-consciousness (even if it is informed by energies beyond consciousness) usually understood to be necessary for human artistic work. Nonetheless, is there something about the time plus scope of inter-assemblage arrangements that extracts, for these insects if not for the moon, an experience of monumentality, of elevated intensity and extensity leaving percepts and affects together in sufficiently indelible form to induce a monumentality of this kind every time a moon rises on certain occasions above a field of fireflies? Is there a joy of the fireflies—or rather the complex inter-assemblage of field, moon, fireflies, cloud, wind, grass, tree, bat—in these moments?

Indeed, these speculations raise more questions than answers. We might wonder, for example, if there is something anthropocentric about the distinction between the signature—that which marks territory—and style as that which expresses innovation drawn by Deleuze and Guattari. Is that which is signature experienced as style in the situatedness of experience of other-than-human entities? What does it matter? By implicitly privileging the artistic and by focusing on its higher-order processualism—the distinction between imitative verses artistic birds—is there not a risk of providing the opportunity for anthropocentric thinking and its attendant forms of supremacy logics to be reterritorialized via assertions of the inherent 'advanced' neuronal complexity of the human *vis-à-vis* other animals and entities? On the other hand, are human realities and singularities suppressed in their value in the face of such considerations? One proviso to be considered here in relation to this latter concern would be that the existence of other-than-human creative and artistic events do not necessarily evacuate the human artistic expression of its singularity, nuance, and potentialities, but simply refuse to understand such human expressions as having supremacy over other kinds of non-human becoming. Saldanha and Stark (2016) remind us that, 'Deleuze's anti-humanism is quite different from recent anti-Kantian developments like speculative realism, which aim to replace the humanist and critical legacies with serene indifference, and even explicit nihilism' (435), and also suggests that Deleuze would have little time 'for those radical ecologists who argue the human species is but a skin disease of the Gaia hyper-organism' (435).

Regardless of the remaining amount of unthought terrain, perhaps what is clear from these speculations here is that ultimately it behooves those humans attempting to move beyond the anthropocentric histories of reductive materialization to articulate the ways in which other-than-human entities might 'experience' creative and artistic phenomena given that these can be central to each entity's expansion or contraction of abilities and potential, of their capacity for joy. In other words, understanding with nuance the experiences of beauty generated and taken up by other-than-human entities can help humans attempt to understand these entities' 'value' on their own terms. This is most especially significant at a time when, in a perhaps valiant but ultimately misguided attempt, certain green economists are financializing every natural process as a means of defending it: we have to save the bees because they contribute so many billions of dollars to the economy, we need to save wetlands because they filter massive amounts of water and this would otherwise cost 'us' billions to do (Balmford et al. 2002; Costanza et al. 2017). In a time when such strategies—ones that are a manifestation of the reductive numerations and objectifications that have led to the deleterious planetary effects of the Anthropocene in the first place—are being undertaken as a latch ditch effort to preserve 'natural systems' from human encroachment and destruction, continuing to find ways to describe other-than-human entities' qualities and capacities for artistic expression would seem to and important part of the collective and deeply political work of imagining ways in which humans and other-than-human bodies can live together.

NOTES

1. The absence of a hyphen between 'post' and 'anthropocenic' is, following a similar absence of hyphen in 'postcolonialism', is not necessarily meant to designate a time temporally 'after' the Anthropocene (although postanthropocenic thought can certainly be guided by wishing to imagine/ achieve this end), but rather instead invokes an engagement and response *to* the Anthropocene even as it is unfolding, indeed, from the very first moments of its manifestation.
2. They note that 'a machine is like a set of cutting edges that insert themselves into the assemblage undergoing deterritorialization, and draw variations and mutations of it' (333), and that '*Machines are always singular keys that open or close an assemblage, territory*' (emphasis authors') (334).

Image 8.1   Turret (Mia Feuer, watercolor)

REFERENCES

Balmford, A., Bruner, A., Cooper, P., Costanza, R., Farber, S., Green, R. E., et al. (2002). Economic Reasons for Conserving Wild Nature. *Science, 297*(5583), 950–953.

Chowdury. A. R., & Gupta, A. (2015). Effect of Music on Plants—An Overview. *International Journal of Integrative Sciences, Innovation and Technology, 6*(6), 30–34.

Colebrook, C. (2016). "A Grandiose Time of Coexistence": Stratigraphy of the Athropocene. *Deleuze Studies, 10*(4), 440–454.

Costanza, R., de Groot, R., Sutton, P., van der Ploeg, S., Anderson, S. J., Kubiszewski, I., et al. (2017). The Value of the World's Ecosystem Services and Natural Capital. In P. Newell (Ed.), *The Globalization and Environment Reader* (pp. 117–133). Oxford: Wiley.

Creath, K., & Schwartz, G. (2004). Measuring Effects of Music, Noise, and Healing Energy Using a Seed Germination Bioassay. *Journal of Alternative and Complementary Medicine, 10*(1), 113–122.

Damanhur Group. (2011, October 24). https://www.youtube.com/watch?v= aZaokNmQ4eY.

Deleuze, G. (1988). *Spinoza: Practical Philosophy*. San Francisco: City Lights Books.

Deleuze, G. (1994). *Difference and Repetition*. New York: Columbia University Press.

Deleuze, G. (2004). *The Logic of Sense*. London: Continuum.

Deleuze, G., & Guattari, F. (1987). *A Thousand Plateaus: Capitalism and Schizophrenia*. Minneapolis: University of Minnesota.

Deleuze, G., & Guattari, F. (1994). *What Is Philosophy?* New York: Columbia University Press.

Guattari, F. (1995). *Chaosmosis. An Ethico-Aesthetic Paradigm*. Sydney: Power Publications.

Retallack, D. (1973). *The Sound and Music of Plants*. Camarillo, CA: Devorss.

Saldhana, A., & Stark, H. (2016). A New Earth: Deleuze and Guattari in the Anthropocene. *Deleuze Studies, 10*(4), 427–439.

"Like Watching a Movie": Notes on the Possibilities of Art in the Anthropocene

Bradley Necyk and Daniel Harvey

INTRODUCTION

On May 7th, 2016, the *St. Albert Gazette* ran a cover story on the forest fire in northern Alberta, Canada that caused the evacuation of the town of Fort McMurray and environs. Below a photograph of a billowing pyrocumulus cloud blocking out the sky, the header reads: "It was like watching a movie." Halfway through the article, the evacuee expands on his description: "'It was like watching a movie,' he said. 'Smoke everywhere, you're seeing people walking down the road with their kids and a gas can, looking for gas... ambulances and fire trucks going everywhere'" (Paterson 2016). The fire was like a movie, then, and a specific type of movie at that. The language resonates with the genres of disaster, of science fiction, of apocalypse and dystopia, in which the trappings of civilization have been ripped away, and individuals are left on their own to seek out the barest of necessities: fuel, family members, and so on. From this spare description, the conventions of the film seem readily apparent, part *The Road* and part *Armageddon*.

B. Necyk (✉)
Psychiatry, University of Alberta, Edmonton, Canada

D. Harvey
English and Film Studies, University of Alberta, Edmonton, Canada

© The Author(s) 2018
j. jagodzinski (ed.), *Interrogating the Anthropocene*, Palgrave Studies in Educational Futures, https://doi.org/10.1007/978-3-319-78747-3_9

And of course, for those 80,000 individuals forced to flee, this was a kind of apocalypse, as homes and possession were destroyed, by fire or water or smoke or the pollution that comes from modern subdivisions going up in flames. "The Beast," as the Canadian media nicknamed the fire, was active around the town for almost a month, and although it retreated northeast by mid-June, it was still burning nearly a year later, having overwintered below ground in the peat moss and dirt of the boreal forest (Quinlan 2017, n.p.). At its height, the fire consumed 1.5 million acres of land. All told, the fire's damages have been estimated at nearly 10 billion dollars Canadian in direct and indirect costs, and will likely go higher (Weber 2017, n.p.); the Fort McMurray forest fire is the most expensive natural disaster in Canada's history.

We are struck by the phrasing of this description of the fire; "like watching a movie" captures, we thought, something important about the ways in which we as a society have been responding to natural disasters within an era of pandemic climate change. For all its flat banality, the phrase signals a certain way of seeing, and hence of relating to, the catastrophic impacts of what has come in the last decade to be called the "Anthropocene." That is, the genre that it implies—contemporary action films of various types—indicates a form of understanding, one predicated on a specific duration, descriptive arc, character types, mode of consumption, and narrative closure. This phrase contains latent within it an aesthetic and ethics of the spectacular: apolitical (in the sense of maintaining the status quo); idealist (cut off from the material histories that made such an event more likely); individualized (rather than social or structural); and atemporal (existing as a kind of nonevent outside of time, with no history or future). Framing an extreme event such as this outside of its climatic context (global warming increases the likelihood and severity of wildfires), framing it as spectacle, makes it all the easier to ignore the decidedly nonspectacular causes and effects of that context. This is a serious problem, if we are to confront the changing conditions of life in the Anthropocene: the spectacle-events of climate change and global warming captivate our attention, making it more difficult to focus on all those effects that are less visually arresting, or even visible.

Why connect this natural disaster, among so many, to issues of representation raised by and in this new epoch? Its location, for one. Fort McMurray sits roughly 400 kilometers north of Edmonton, the province of Alberta's capital city, and acts as key site for the development and expansion of the Athabasca tar sands; these deposits are the largest in Canada, containing roughly 166 billion barrels worth of extractable oil in the form of crude oil and bitumen (Alberta Energy Regulator 2016, p. 4). Alberta's tar sands development are regularly singled out by environmental groups within North America and globally as a major source of air and water pollution, and a significant contributor to process of global warming. According to a 2013 report in *Scientific American*, each barrel of oil produced out of the tar sands emits between 79 and 116 kilograms of greenhouse gas (depending on the type of extraction process used), "roughly 14% more...than the average oil used in the United States" (n.p.). Fort McMurray's population, economy, and demographics have therefore been primarily driven by tar sands development and extraction. It was perhaps unsurprising, then, that the fire instigated a public debate about its links to climate change, and the culpability of the city in contributing to global warming—and hence, to the increased likelihood of disasters like the fire. On the one side of the debate stood those who saw any such links as disingenuous at best, and politicizing a disaster at worst; on the other, those convinced that the tar sands projects directly contribute to global warming, which in turn leads to more common, and more intense, forest fires.

Encapsulated in this throw-away phrase, then, we can discern a larger problem of representation, one highlighted in recent theoretical works on the Anthropocene. Rob Nixon's *Slow Violence and the Environmentalism of the Poor*, and Donna Haraway's *Staying with the Trouble*, seem particularly resonant here. The introduction to Nixon's work lays out this problematic with a sharply pointed question:

> in an age when the media venerate the spectacular...how can we convert into image and narrative the disasters that are slow making and long in the making...that are anonymous and star nobody...that are attritional and of indifferent interest to the sensation-driven technologies of our image-world? (2011, p. 3)

While Nixon poses this question in the context of postcolonial literatures and environmental activism, to tease out the interrelations between them, this question seems especially pertinent in the context of the current study. The view of Fort McMurray-as-spectacle implied by comparing it to a movie brackets out everything that doesn't fit within a filmic frame, eliding spatial and temporal scales, the history of tar sands production and of Fort McMurray itself, along with the material impacts left behind by the fire: the pollution caused by the burning of modern homes; the impacts on infrastructures, both physical and social; the cardiovascular issues and instances of PTSD now cropping up in residents and first responders alike; and so on. None of these fit within the frame of the film, either too large or small in scope and importance for an action film. Indeed, the simile of "like watching" seems to emphasize the issues with such framing, at once creating an equivalence between film and event to reduce the latter's importance—exactly like a movie, and only a movie, untethered from its larger import or meaning—while at the same time suggesting a certain passivity.

As Haraway has commented in her recent work, the scale of the stories we tell ourselves matters: too big, and they can overwhelm us, breeding fatalism and nihilistic self-destruction; too small, and they seem unimportant, and can just as easily breed either acceptance or ignorance, neither of which seem likely to generate action. This simile diminishes the story of the Fort McMurray fire in both ways, shutting down possibilities for thought and action by positioning its representation within a story that is too big (the generic form of the apocalyptic disaster narrative, in which the end, after all, is already determined in advance), and too small (only a spectacle happening to others, nothing to get too worried about), in which participants in the larger story of the Anthropocene (that is, all humans) become passive viewers cut off from any specific Anthropocene event. Further, it becomes a once in a lifetime event for those directly affected, something that took place outside of their everyday, 'real' life, and can easily be turned into a story deployed again and again rather than something with which to grapple.

This mapping of event onto film speaks to an inability or refusal to cognitively map our current situation within a story that is large enough to impel action, yet small enough to dispel paralysis. As Haraway puts it (consciously echoing the words of anthropologist Marilyn Strathern), "it matters what matters we use to think other matters with; it matters what stories we tell to tell other stories" (Haraway 2013, n.p.). The types of stories we tell ourselves, the materials we use to build our stories, and how we build them, impact the new types of stories that we can be able to tell, and the stories, materials, and generic forms made available through this spectacle's apocalyptic action seem poorly designed to generate new stories that will lead to greater flourishing.

What, then, is to be done? How can we combat the spatio-temporal closure Nixon identifies in the contemporary spectacle of certain symptoms of climate change? How can we identify, tell, and create stories that are "just big enough to gather up the complexities and keep the edges open and greedy for surprising new and old connections" (Haraway 2015, p. 260)? While we don't have any easy answers to these questions, we want to suggest that this is a place in which contemporary art has a role to play.

Art, we contend, particularly works that trouble audience expectations of scale and narrative, holds out the possibility of undoing the simplistic cognitive mapping (the ways that subject position themselves in relation to the larger social totality, or in this case fail to do so) we see in the example of Fort McMurray. The inability to cognitively process a disaster at the scale of Fort McMurray without aid of representation, in this case an easily identified film genre, is where art within the Anthropocene becomes a generative lens through which to situate ourselves at both the macro and micro levels. With the Anthropocene, we must confront the scope of both the temporal and spatial scales, and of a systems complexity with which our cognitive faculties have difficulty grappling. For thousands of years, humans have employed artistic media to deal with scales that would otherwise escape our powers of representation. The cave paintings of Lascaux, for example, now 17,000 years old, span across generations of humans, mapping out relationships between environment, migration, community, and spirituality. Or, from a different cultural context, the paintings, sculpture, stone arrangements, stories, and string-figure traditions of indigenous Australian groups, used for millennia to engage with Altjira (generally translated as "Dreamtime"), to navigate and interact with the environment, and to inform all aspects of life in a holistic way. In examples such as these, art provides a way to calibrate human sensoria and faculties beyond their everyday functions.

The difficulty with art, something often seen within the apocalyptic film genre, is that it can try to represent too much, it can tell too big of a story, while, on the other hand, our individual senses can tell too small a story. Telling too big (or too small) a story leaves out many of the complexities, textures, and richness that should be the generative fodder we use to make new connections, potentials, and futures. When things become spectacular they limit their ability to become practical and usable, and, instead, become narratological and predictable. Art, then, must try and tell stories that are just big enough, as to have the possibility of breaking this narratological thinking, and open up new spaces for perception and thinking. Visual art, and particularly time-based art, has the dual ability to tell stories that are just big enough and also challenge the unthought types of narrative thinking suggested by the comparison that living in the Anthropocene is akin to viewing a Hollywood film. We want to turn, now, to two examples of contemporary art that may prove exemplary in the ways that they perform such a duality, Charles Stankievech's (2013) film *The Soniferous Æther* and Pierre Huyghe's *Untilled*, a landscape installation or biotope constructed for dOCU-MENTA 13 in 2012. The two works offer, we suggest, new ways of thinking stories, and call into question the passivity of more spectacular narratives, and the too-easy narrativization of both apocalyptic and utopian visions.

We chose these two works for a number of reasons, in terms of their content, their formal qualities, and their methods of production. First, both pieces are deeply invested in thinking through and representing landscapes; they both deploy specific environments (Canada's arctic and the compost site of a Baroque garden, respectively) in ways that call into question traditional ways of thinking about landscape, or nature more broadly. Each engages with its site as a central aspect of the conditions of its production, so that the specificity of the landscape drives the process of artistic creation. Second, both works directly challenge the too-easy closure presented by the simile discussed above, by inviting viewers to consider environments and connections that extend beyond the frame of works themselves, or beyond the scope of human perception. Third, both pieces operate as forms of antinarratives, and call into question the anthropocentrism of traditional narrative as such, by rendering intelligible human finitude, on the one hand, and the interconnectedness of the world beyond humanity. *The Soniferous Æther* foregrounds problems of spatial and temporal scale, while *Untilled* highlights the difficulty of making meaning in interconnected and complex systems, developing a logic of compost (rather than composition) to resist the power of the spectacle and undercut the anthropocentric ground from which all spectacular modes of representation ultimately spring. The two, we want to suggest, offer us new and striking images of the Anthropocene, a form of imagery missing from anything watched "like a movie."

THE SONIFEROUS ÆTHER (2013), CHARLES STANKIEVECH

First exhibited at the Ottawa Art Gallery in 2013, *The Soniferous Æther of The Land Beyond The Land Beyond* (TSA) is a 10-minute film captured in the most northern settlement on Earth—CFS Alert Signals Intelligence Station. Stankievech completed a residency there as part of his fieldwork around architectures embedded within the landscape. The film comprises a series of computer-controlled timelapse vignettes of the frozen landscape, buildings, and, later in the film, inside the facility itself. The shots are devoid of humans, populated instead by their traces: frozen machines, buildings pouring out steam and artificial light, and the interiors of bowling alleys, hallways, and workstations.

In this sense, we see the film as a kind of pseudo-post-apocalypse, in a similar vein as recent works of "science faction," a genre of texts that "depict the world after...extinction of the human race" (Bellamy and Szeman 2014: 193) by extrapolating contemporary scientific facts and projections with science-fictional interpretations of possible future outcomes. Unlike the apocalypse suggested by the Fort McMurray fire, however, or other, more didactic texts in the genre, Stankievech's science-factional film resists narrative closure, and indeed, eschews explicit narrative completely. Where the spectacle of apocalyptic film tends to limit the need for viewer engagement, providing a kind of implicit script by virtue of their generic and narrative conventions that close off the need for critical thought, TSA renders viewers bereft of such clues, obliging them to generate meaning on their own.

The film begins with a strobing light and a piercing acoustic pulse, immediately unmooring the viewer, then cutting to a slow rotation of the sky centered on a static Northstar with the rest of the cosmos being dragged around it. Slowly, a voice lifted from shortwave broadcasts begins speaking repetitive signal- and code-words with Cold War precision, joined by the slow oscillation of a high-pitched frequency, distorted sounds recorded on location, and a remixed loop of Glenn

Gould performing Bach's *Goldberg Variations*, slowly morphing to create a rich and otherworldly soundscape. The film cuts from shot to shot, not staying on any landscape for long; but the combination of the slow computer-controlled pan and time lapse makes each shot of this frozen landscape seem to escape temporal closure—this terrain, after all, has remained relatively static since the last ice age. Without any visible beings to act as our stand-in, we are left only with the slow crawl of the camera to provide footing. However, there is no agency in this position, just an incessant, pushing movement that places us in a new temporal geography, which draws attention to the stretching geological and cosmological scale of a place that seems to exist without beginning or perceivable end. This polar cap, our most northern settlement, is not just a focal point for physically witnessing climate change within our human physiological and generational time scales, but provides a slowly crawling vision of a landscape devoid of life, filled with the detritus of civilization and haunted by the specter of human habitation.

The film operates, then, in the tradition of Canadian landscape art found in the paintings of Emily Carr, Tom Thomson, and the Group of Seven, but also in the often anxiety-ridden works of Canadian landscape fictions, beginning with the works of Susanna Moodie in the 1830s and continuing through the Confederation poets to the present. Stankievech's landscapes, though, call into question the separation and conflict between the human and the environment that weighed heavily on works like Moodie's *Roughing it in the Bush*, and eschew any hint of the fantasy promulgated by the Group of Seven of the landscape as a kind of blank slate. Here, we see instead the landscape from the other side of civilization: not presented as a *terra nullius* open to colonization, settlement, and technological manipulation, but rather as both inextricably linked with the civilization that entered it, and enigmatic, even hostile to habitation.

The film also refuses to engage in another project of the Group of Seven, the Confederation poets, and other artists drawing on the Canadian landscape: the creation of a distinct national identity. TSA works against any such attempt at spatial closure. The starkly lit and bleak landscape resists integration into the narrative nationality and place, appearing as an anonymous, alien space uninterested in the concerns of human politics; indeed, some of the slow panning shots of the icy ground share a similar aesthetic to images of lunar or Martian landscapes. Coupled with the night sky stretching down to touch the horizon in all directions, this has the effect of further uncutting the stability of place: the film, it seems, might as well have been shot at an abandoned extraterrestrial outpost as on planet earth.

Stankievech talks about the landscape of TSA as an "embedded" one, which "shapes us as much as we shape it" (Wlusek and Stankievech 2013, n.p.), and it is in this reciprocal shaping that we identify a further difference between TSA and the film suggested by our opening simile. TSA draws the viewer into contemplating, and hopefully wrestling with, the complexities of meaning-making in an era where humanity and the environment have become co-constitutive to a never-before experienced degree. The more spectacular movie, in contrast, closes off such contemplation, cropping out the broader connections between event and viewer, landscape and human, film and critical engagement, and thereby reducing the problem(s) of the Anthropocene "to an apocalyptic fantasy of human finitude [and] world finitude" contained in a manageable, if terrible, event like the Fort McMurray fire (Emmelhainz 2015, n.p.).

The materiality of film is also of interest, something often not readily considered in this medium, since it too will gradually decay with each showing. The work is transferred from a series of black and white photographs on 35mm film and is exhibited on a film projector, already taking on a granular look, which will only increase each time it loops, hour on hour, day on day. A viewer on the first day of an exhibit will see a slightly clearer film than someone at the end of its exhibition. This almost imperceptible incremental destruction is very similar to our experience of climate change, a slow saturation of micro-scale changes, thickening and gaining momentum. These are changes you cannot perceive as a data point, for instance a particularly hot day or on a single viewing of the film, but, over longitudinal records and the passage of time, one might apprehend the slow but inexorable momentum of those gradual distortions. Through its use of 35mm film, and the continuous looping that gradually impinges upon its initial clarity, the material form of TSA's presentation in the gallery space draws the reader into another kind of embedded landscape, echoing that represented through the film itself.

UNTILLED (2012), PIERRE HUYGHE

Consider Pierre Huyghe's work *Untilled* from 2012, which was displayed at dOCUMENTA 13 in a composting site on the exhibition's grounds in Kassel, Germany. The outdoor installation encompasses an assemblage of disparate works: an odalisque nude made of concrete, with an active beehive as its head; a free-roaming dog named "Human" with one leg painted pink; piles of concrete rubble; a pond, full of algae and other micro-organisms; poisonous, psychotropic, and aphrodisiac plants; an uprooted oak tree from Joseph Beuys' 1982 piece *7000 Oaks*; and an on-site gardener among other objects and actors. These works rarely come within line of sight of each other as they are spread across the large expanse of the site. Instead, they create smaller sites, that, if read alone, could constitute distinct artworks; when taken together, however, they form a disparate narrative not easily read. The most noticeable connecting agents are the dog, as it freely wanders between the sites, sometimes leaving the area entirely, and the gardener who tends to the free-growing saplings, bushes, and psychotropic plants overtaking the sites. At no point can the viewer take in the totality of the work as a whole: Huyghe's piece deliberately resists the closure of the spectacular gaze.

At a residency at the Banff Centre for the Arts and Creativity that occurred in conjunction with dOCUMENTA 13, Huyghe talked about how he is unsure where his hand as an artist ends and where nature simply takes over (does it end in the puddle formed within his space, where bacteria now are multiplying and thriving? In the interactions of insects, plants, and animals?). Deep within the Anthropocene the perceived divide between nature and unnatural becomes muddled, like a bed of compost, where the unnatural (technology, (post)humans) have embedded themselves within this dying world.

In a public lecture at the Banff Centre, Huyghe describes his artistic method as deploying a kind of antinarrative, striving to "de-link, or to de-narrate, or to de-script the imaginary language of the narrative authority" that "makes a relation between things" (Banff 2013). In *Untilled* we see this method deployed in the uncontrollable nature of his materials, the sheer complexity of the various elements acting according to their natures in relation to each other, and to the environment of the site itself. Indeed, the title of the work itself suggests an unwillingness to deform what exists with one's predetermined intent: although the artist has brought the different components of the biotope together, the forms that their interaction may take are left unplanned and contingent. As he describes it, this piece in particular involves accepting "the uncertain, the unreasoned, and the unknown" in order to "intensify the speculative potential within a given body," whether of an exhibit, and institution, an artwork, and so on (Banff 2013). Such speculative potentials are precisely what spectacular representations close off: the certainty of the spectacular narrative shuts down the possibility of speculation that might allow other stories to be told, or render the existing narrative more malleable.

Huyghe's *Untilled* resists narration, and indeed, actively fights against the idea of narrative as such; he describes it stemming from "the need... to separate the sedimentation of narrative" from the works' various "elements in space," to leave the various "markers [from] history... without culture within that compost" (Banff 2013). The piece works from a logic of *compost* rather than *composition*, to become a sympoetic site in which the elements mix and change without the top-down hierarchy of planned space or narrative. There is at once too much going on—how to track the movements of each insect, the pollination of each flower, the meanderings of the dog, the shifts within the pond—and too little framing—one could stumble upon the space and not even recognize it for what it is—for *Untilled* to be transposed into a simile: it is like nothing but itself, deliberately resisting even the narrative of comparison. Huyghe's "methodology of the compostv...[his] attempt to intensify what a compost is," determines both the work's structure and its generative processes, so that the elements generate new forms as they "leak" into one another and "into the physical, chemical reality" of the space (Banff 2013).

Conclusion

Like the signs of human existence that litter the land- and soundscapes of *The Soniferous Æther*, Huyghe's compost "becomes a place where things are left without culture...indifferent to us, metabolizing, allowing the emergence of new forms" (Huyghe, cited in Godden 2012, n.p.). Also like *TSA*, the biotope refuses to conform to a logic of the spectacle, in which the object or image is put on display for the viewing subject: as Huyghe states, it is "indifferent" to the viewer, "not displayed for a public" so much as standing as "raw witness" to its own "topological operations" (cited in Godden 2012, n.p.). As Andy Weir puts it, *Untilled* offers a "continual ungrounding (or compostation) of anthropocentric experience" (2013, p. 29), much as *TSA* decentres anthropocentric spectacle in its omission of the human form, otherworldly landscapes, and more general refusal of narrative closure.

In both of these case studies, we encounter stories that attempt to represent scales (spatio-temporal, rates of change, and systems complexity) that exist outside of or challenge human sensory and cognitive faculties—they extend them beyond their everyday uses, affording us different ways of seeing and narrating, with generative opportunities for investigating and experiencing the Anthropocene. When confronted with the immense objects of the Anthropocene (climate change and so on) humanity collectively disassociates from the immediate (and distant) implications and, instead, processes them through familiar narrative structures. We cling to our familiar stories, and resist the imposition of the material world when it rises up to complicate or unmake them.

And here we return to the problematic with which we began, the issue of representing the Anthropocene beyond any single spectacular event within it. In a recent essay on visuality within the Anthropocene, Irmgard Emmelhainz phrases this problem as one of reduction:

> Instead of being conceived as speculative images of our future economic and political system, the Anthropocene has been reduced to an apocalyptic fantasy…. (2015, n.p.)

The reaction of the Fort McMurray evacuee with which we began, along with so many other representation of specific Anthropocene events (extreme weather patterns, ice-cap melt, species die-off), demonstrate precisely this form of reduction: the narrative closure of apocalypse replaces the interconnected reality and scalar shifts of the Anthropocene. Like Huyghe's *Untilled*, the Anthropocene is an assemblage of images, stories, things, and effects, not a single event or even a series of events: the stuff that happens between spectacles is just as much part of the Anthropocene as the next wildfire, flood, or heatwave. The Anthropocene needs to be confronted in all of its sheer *banality* if we are to engage with it—forestall it, limit, learn to live and die within it—in a meaningful way. As Heather Swanson points out, it is easier for those most immediately impacted by Anthropocene events to understand their own precarity within it (although that understanding can be swiftly undone through the remapping offered by familiar narratives) (2017, n.p.); how much harder will that be to those for whom such events are merely images on screens, encountered as if one were watching a movie?

The type of contemporary art exemplified by *The Soniferous Æther* and *Untilled* (one might even refer to this as Anthropocene art) performs a different way of encountering the Anthropocene and of imagining one's place within it. Such works provide something that Emmelhainz, Haraway, Nixon, and others have contended is missing: images of the Anthropocene, instead of the spectacular pictures that we have now (Emmelhainz 2015, n.p.). These works perform the Anthropocene in various ways, calling our attention to its complexity, scale, and increasing rates of change, foregrounding the interconnectedness of systems, species, and physical forces within it, and challenging the viewer to engage in a critical mode of visuality, to think critically, and to eschew the easy abstraction and closure of thinking like a movie. We need more images of the Anthropocene: microcosms of the Anthropocene; interconnected works that can exceed narrative closure, fill in the banal gaps of time from one extreme event to the next, and break with the comfortable narratives that brought us to this juncture in the first place.

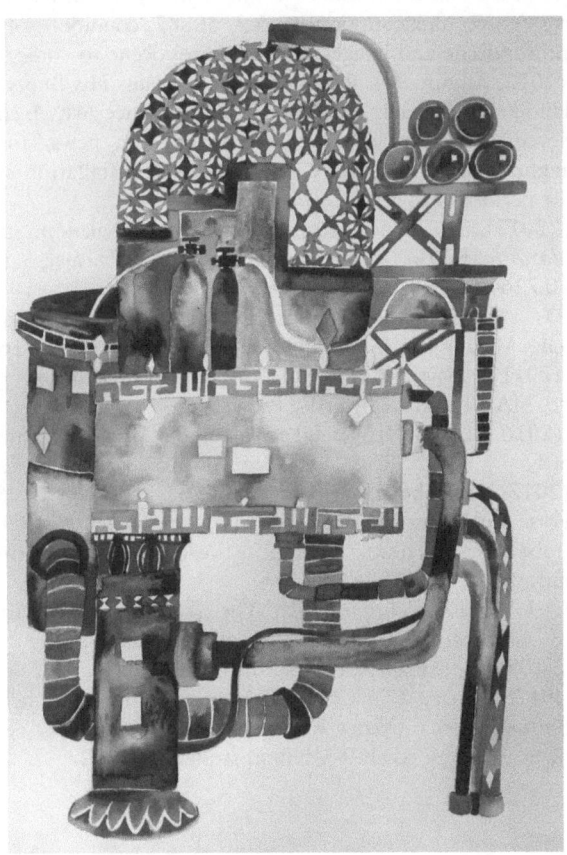

Image 9.1 Piping refinery (Mia Feuer, watercolor)

REFERENCES

Alberta Energy Regulator. (2016). ST98: 2017. Alberta's Energy Reserves and Supply/Demand Outlook: Executive Summary. Available at http://aer.ca/data-and-publications/statistical-reports/executive-summary.

Banff Centre for Art and Creativity. (2013, January 23). Pierre Huyghe—The Retreat dOCUMENTA (13). https://www.youtube.com/watch?v=aRC5iFlxfnU.

Bellamy, B., & Szeman, I. (2014). Life After People: Science Faction and Ecological Futures. In G. Cannavan & K. S. Robinson (Eds.), Green Planets: Ecology and Science Fiction (pp. 192–205). Middletown, CT: Wesleyan University Press.

Emmelhainz, I. (2015, March). Conditions of Visuality Under the Anthropocene and Images of the Anthropocene to Come. e-flux Journal, 63, n.p. Retrieved from http://www.e-flux.com/journal/63/60882/conditions-of-visuality-under-the-anthropocene-and-images-of-the-anthropocene-to-come/.

Godden, S. (2012, August 30). Pierre Huyghe Explains His Buzzy Documenta 13 Installation and Why His Work Is Not Performance Art. ArtInfo Canada, np. Retrieved from http://www.blouinartinfo.com/news/story/822127/pierre-huyghe-explains-his-buzzy-documenta-13-installation-and-why-his-work-is-not-performance-art.

Haraway, D. (2013). SF: Science Fiction, Speculative Fabulation, String Figures, So Far. Ada: A Journal of Gender, New Media, and Technology, 3, n.p. https://doi.org/10.7264/n3kh0k81.

Haraway, D. (2015). Anthropocene, Capitalocene, Plantationocene, Chthulucene: Making Kin. Environmental Humanities, 6, 159–165.

Nixon, R. (2011). Slow Violence and the Environmentalism of the Poor. Cambridge, MA: Harvard University Press.

Paterson, V. (2016, May 7). 'It Was Like Watching a Movie:' Evacuee. St. Albert Gazette, p. 1.

Quinlan, P. (2017, December 19, Updated April 5). 'The Beast' Is Still Burning East of Fort McMurray in 2017. Global News Edmonton. Retrieved from http://globalnews.ca/news/3137099/the-beast-is-still-burning-east-of-fort-mcmurray/.

Swanson, H. A. (2017, February 22). The Banality of the Anthropocene. Dispatches, Cultural Anthropology. Retrieved from https://culanth.org/fieldsights/1074-the-banality-of-the-anthropocene.

Weber, B. (2017, January 17). Costs of Alberta Wildfire Reach $9.5 Billion: Study. Business News Network. Retrieved from http://www.bnn.ca/costs-of-alberta-wildfire-reach-9-5-billion-study-1.652292.

Weir, A. (2013). Myrmecochory Occurs: Exhibiting Indifference to the Participating Subject in Pierre Huyghe's Untilled (2012) at Documenta 13. *Postgraduate Journal of Aesthetics, 10*(1), 29–40.

Wlusek, O., & Stankievech, C. (2013). The Embedded Landscape: Unveiling the Duality of Site. *Charles Stankievech: The Soniferous æther of the Land Beyond the Land Beyond.* Exhibition Catalogue. Retrieved from http://stankievech. net/dwn/Stankievech_TSA-interview.pdf.

FOAMA or … You Make Me Feel the Way Gasoline Looks on Water

Mia Feuer

Throughout *Interrogating the Anthropocene*, the reader will come across a series of watercolors that I have created in recent years. They were made in response to my time spent in the field researching concepts that relate to the origin and extraction of organic and synthetic materials, my relationship to these materials within my sculpture studio practice and the hyper vast lifespan of such materials as related to the transformation of landscapes and climate, petrochemicals, and speculative futures. These are all interconnected in the potential doom that they harbor.

Some of these sites include Suez, Fort McMurray, Svalbard, Bayou Point au Chien, and Tehran.

I visited the Suez Canal in 2011, shortly after Hosni Mubarak was ousted. I recall standing in a police station that was set ablaze by protesters a few days prior. Everything was charred black. The ceiling fans had melted as if they were right out a Salvador Dali painting. Outside, massive oil tankers continued to ooze through the tiny human-made canal. It was there that I first began questioning my relationships to petrochemicals in my studio—specifically—Styrofoam.

M. Feuer (✉)
California College of the Arts, San Francisco, CA, USA

© The Author(s) 2018
j. jagodzinski (ed.), *Interrogating the Anthropocene*, Palgrave Studies in Educational Futures, https://doi.org/10.1007/978-3-319-78747-3_10

A year later, I traveled to the Tar Sands of Fort McMurray, Alberta where I observed first-hand environmental quagmires that I could only process through a nightmarish logic. Beyond the strip mines, the upgraders and the bright yellow ziggurats made of sulfur, I toured the Reclamation Sites. These were immense areas of land formerly inhabited by boreal forests before becoming the site of desperate and unconscionable bitumen extraction. In the hopes of one day detoxifying the land and regrowing the forest, bioremediation was practiced and wheat was planted. In an effort to control the mice that this crop attracted—The energy company installed inverted dead birch trees, meant to attract the 'unkindnesses' of ravens. These ravens would perch atop the upsidedown trees and hunt the mice who were living off the toxic wheat.

In 2013, I participated in an Arts and Science Residency in the Arctic Circle. For almost a month during the season of the midnight sun, we sailed the Arctic waters on a barquentine tall ship. We explored a multitude of long-abandoned mining projects, witnessed massive glaciers calve into the sea, and were mesmerized by the turquoise glow of the melting ancient sea ice.

In 2015, I visited the devastated Gulf Coast, in Louisiana. While exploring Bayou Pointe au Chien and Isle de Jean Charles upon a local fishing boat, a fisherman explained to me that thirty years earlier a huge oak forest grew there. Now, in the fragmented slivers of remaining land that have not been swallowed by the rising sea levels, stand skeletal dead Oak corpses. These lifeless bleached trunks haunt the ever-creeping shoreline, which is lined with concrete revetment matts and tangled discarded crab traps.

During the summer of 2016, while pregnant with my son Galileo, I traveled to Iran. I recall exploring the underground vault located beneath the Federal Bank of Tehran. This repository housed a breathtaking collection of jewel-encrusted objects from Persian history and mountainous piles of sparkling rubies, diamonds, sapphires and other precious stones, and shimmering metals. While being dazzled by the actual physical material wealth glittering in front of me, former US President Barack Obama and five other super power Heads of States were meeting not too far away to discuss the future of Iran's nuclear Program ... and outside was an inferno—Iran's hottest summer on record.

Visiting these sites has been an ongoing meditation on the interconnectedness between geographical locations, geologic time, and the relationships that exist between individuals, corporations, governments, the natural + constructed world, synthetic materials, and our collective addiction to burning hydrocarbons.

In terms of my evolving relationship to Styrofoam: A wondrous material that can be manipulated into any form. Carved, sliced, rolled, cast, sanded, sculpted, and suspended. Lightweight, extremely strong, floats in water, and insulates. My attraction to suspended sculpture and being compelled to create whimsical installations where giant menacing structures can fly through the space is possible because of my experimentation with this material. Over the years, my feelings toward this petrochemical have changed. My feelings of guilt and hypocrisy for consistently utilizing such a damaging material are now beginning to transform into feelings of respect. Polystyrene is made from the stuffs of Paleolithic Sea Creatures from 50 million years ago. Polystyrene now exists as disposable coffee cups, throw away meat containers and construction materials. It lives on for an eternity in landfills, oceans, and the bodies of animals. It is heavily subsided by governments and lobbyists creating an abusive and wasteful overabundance. And yet, when I use it to create sculpture, I am starting to see it as a powerful medium that should be respected. I now demand that the material interrogate itself, and the questions and poetry of this self -examination be embedded in the formal and conceptual qualities of the pieces I create.

Image 10.1   Cargo container ship (Mia Feuer, watercolor) (*Note* Mia's watercolors appear through this book as sketches for planned projects, especially her Fort McMurray Tar Sands project)

CHAPTER 11

Catastrophism and Its Critics: On the New Genre of Environmentalist Documentary Film

Michael Truscello

Diana Liverman is a Professor of Geography, a 2014 Guggenheim Fellow, and a recipient of the Founder's Gold Medal of the Royal Geographical Society. Her online biography describes her as "an expert on the human dimensions of global environmental change."[1] In an August 20, 2014 op-ed for the *Washington Post*, Liverman described her efforts to teach university students about climate change. The problem, as she described it, was that she could not seem to "shock them out of complacency and into action."[2] Instead, many of her students left the class "feeling despondent and powerless." Liverman reminded herself of the research in her own field that "demonstrates that fearful people feel disempowered and less willing to act." And so, in her words, she decided to change her "narrative." She included "many more positive and hopeful examples and analyses" in her lectures. "Rather than lament the failures of US policy to reduce climate risks," she writes, "I point out how

M. Truscello (✉)
English and General Education, Mount Royal University in Calgary, Calgary, AB, Canada

© The Author(s) 2018
j. jagodzinski (ed.), *Interrogating the Anthropocene*, Palgrave Studies in Educational Futures, https://doi.org/10.1007/978-3-319-78747-3_11

four decades of laws helped clean up our air and waterways, saving lives, money and ecosystems." Rather than "spend so much time on the environmental injustices suffered by the poor," she adds, "I talk more about how we have halved poverty and given millions access to safe water." Liverman links to the United Nations' Millennium Development Goals and uses the UN language of people becoming "agents of change." After teaching the revised, upbeat version of her class twice, she notes how "student evaluations are different. Now, students say things like 'I'm motivated to follow a green career' and 'My roommates are fed up with me telling them all the things they can do to save the planet.'" The change of attitude in the classroom and positive response from her students also inspired Liverman to invest in solar and water harvesting for her home, and to "actively support local political candidates with strong environmental records."

Liverman's op-ed illustrates and advocates for the rhetoric encouraged by leftists who condemn what is known as *catastrophism*, the doomsaying that has characterized much of the environmentalist literature of the past few decades. Some leftists believe the emphases on catastrophic ecological realities demoralize and demobilize potentially radical communities. As Doug Henwood writes in the foreword to the 2012 collection *Catastrophism: The Apocalyptic Politics of Collapse and Rebirth*, "Wouldn't it be better to spin narratives of how humans are marvelously resourceful creatures who could do a lot better with the intellectual, social, and material resources we have?.... Dystopia is for losers."[3] Prominent liberal and progressive filmmakers, much like Professor Liverman, have taken this advice seriously, but in their attempts to produce anti-catastrophistic messages they have understated the crises we face and promoted solutions that are dramatically inadequate, maybe even counterproductive. In other words, sometimes apathy isn't the worst thing that could happen in a crisis.

This chapter examines the recent documentary films *This Changes Everything* (2015), *Racing Extinction* (2015), *Chasing Ice* (2012), and *Chasing Coral* (2017), because they exemplify both the most catastrophic ecological crises (climate change and mass extinction) and promote deeply problematic politics in response. At issue in these films and the debate over catastrophism is not the science of climate change or mass extinction, or the reality of a host of other existential ecological crises; catastrophists and their critics agree on the science. The disagreement is over what Eddie Yuen calls "the larger politics within which the science is couched."[4] Critics of catastrophism believe, based on some studies of the phenomenon, that once people are convinced of "apocalyptic scenarios" they become "more apathetic," and over time, a "catastrophe fatigue" develops.[5] There are many aspects of this discussion on which I agree with the critics of catastrophism; for example, Yuen accuses the Al Gore film *An Inconvenient Truth* of following "compelling evidence for ecological collapse with woefully inadequate injunctions to green consumption or lobbying of political representatives."[6] This is accurate, and the same could be said of *This Changes Everything*, *Racing Extinction*, and the other films under consideration below; most of these films have been made after the debate over catastrophism and appear to be deliberate responses to it, in their overall aesthetic and the solutions they promote. The past decade, especially with the emergence of YouTube, has seen an abundance of doomer documentaries, a genre Imre Szeman refers to as "eco-apocalypse discourses," in which "the deep political and economic investments in oil are assessed, the dire social-political-environmental consequences of inaction on oil are laid out, and because it becomes obvious that avoiding these results would require changing everything, apocalyptic narratives and statistics are trotted out."[7] Several peak oil films released in the wake of the US invasion of Iraq framed life under oil capitalism as too dependent on a rapidly depleting resource, with a common message that societal collapse is near and few alternatives available. In documentaries such as *The End of Suburbia* (2004), *A Crude Awakening* (2006), *Crude Impact* (2006), *What a Way to Go: Life at the End of Empire* (2007), *Sweet Crude* (2009), and *Collapse* (2009), filmmakers both amateur and professional developed narratives that often combined ecological collapse with peak oil fatalism. At the heart of this trend was an attempt to untangle the apparent paradox Gerry Canavan describes like this: "either we have peak oil and the entire world suffers a tumultuous transition to post-cheap-oil economics, or else there's plenty of oil left for us to permanently destroy the global climate through excess carbon emissions."[8] There is no utopian outcome, no "good" Anthropocene,[9] and our commitment to oil capitalism has only ensured that many potential escape routes are blocked by infrastructural overdevelopment and capitalism's rapacious colonization of the planet. Peak oil doomerism has temporarily ceded to films concerned with climate change, the other half of Canavan's paradox, largely because of a global oil glut created by a combination of market prices, geopolitical maneuvering, and the success of fracking-related technologies; by June 2017, a record amount of oil, 111.9 million barrels, was sitting in tankers at sea.[10] The abundance of oil globally will subside, and perhaps once again documentarians will produce popular films about the obstacles to change that oil capitalism has created. For now, however, it seems general interest has shifted to the ecological catastrophe generated by almost two centuries of oil capitalism.

In the context of oil abundance, a new documentary discourse is emerging, one that takes seriously the criticism of catastrophism, as well as what Yuen rightly calls "catastrophe fatigue," and attempts to address environmental catastrophe by shielding viewers from its complexities and instead emphasizing optimism based on the presumption of human ingenuity, a vague understanding of the collective capacity for change, and, most important, an almost apologetic recognition, if any recognition, of the existence of capitalism. *Racing Extinction*, for example, doesn't even mention the word capitalism, even though the mass extinction event in progress is demonstrably a by-product of global capitalism. If the eco-apocalyptic narratives of the preceding decade were too focused on the scale of ecological destruction and the complexity of system change necessary to reverse or slow that destruction, then the new leftist eco-documentaries, what I would call eco-opportunist narratives, make a series of gestures that understate the crises and promote inadequate solutions, all in the name of generating some kind of audience response, *any* kind of audience response.

What is the state of the world, to the best of our knowledge? In this world, industrial capitalism: incurred the sixth mass extinction event in the history of the planet[11]; overfished the oceans and acidified them, similar to the phenomenon that killed off 90% of ocean life 252 million years ago[12]; assaulted coral reefs to the extent that 40% are in decline worldwide[13]; cut down almost half of the trees (and cuts down another 15 billion every year)[14]; cultivated every space on the planet to the point that there is no longer such a thing as "pristine" land untouched by humans[15]; ignited catastrophic runaway global climate change that, unchecked, will produce an estimated warming of 6 degrees Celsius above baseline by the end of this century[16]; produced 250 billion tonnes of chemical substances annually, leading scientists to declare the earth a "toxic" planet saturated with 144,000 human-made chemicals[17]; produced 8.3 billion tonnes of plastic since the 1950s, and could produce an estimated accumulation of 34 billion tonnes by 2050[18]; pushed the planet into a new geological epoch termed the Anthropocene to reflect the impact of human beings being the equivalent of a geological force[19]; the scientific evidence that describes this condition is overwhelming; and yet, despite the obvious state of global ecological collapse that is clearly underway, substantial numbers of industrial capitalist societies are actually escalating the urbanization project, building engineering megastructures at an accelerated rate, instead of slowing the pace of industrialization, or even reversing it. With human extinction on the horizon, capitalists have accelerated the construction of the infrastructure that enables ecological devastation at an unprecedented pace in human history. According to a Princeton University study, to avoid catastrophic climate change and keep global warming below 2 degrees Celsius, the world must stop building new oil infrastructure by 2018[20]; a climate study from the UK determined that the world must reach zero carbon emissions many years in advance of 2040, to avoid the IPCC's target of global warming no more than 2 degrees Celsius above baseline[21]; but in 2015, the world consumed more oil and gas than at any time in human history,[22] and recently breached the carbon dioxide intensity of 410 ppm, a concentration on track to create a climate not seen in 50 million years.[23] The consumption of coal receded recently, but still remains well over nineteenth-century rates of consumption, even though the "era of coal" is supposedly long over; in fact, it is estimated that coal will remain the second-most consumed source of energy until 2030, and will average an increase in consumption of 0.6% from 2012 to 2040.[24] Concomitant with a global industrial capitalist system that has not slowed its voracious pace of destruction, climate feedback loops appear to be accelerating the ecological collapse; methane, a greenhouse gas that traps heat at approximately 30 times the rate of carbon dioxide, is being released from melting permafrost in the Arctic, which could create a catastrophic positive feedback loop.[25] Despite the comprehensive scientific consensus on these issues, expanding carbon-intensive infrastructure continues to be a priority for global capitalism. How will documentary films concerned about this situation respond?

RACING EXTINCTION

The opening scenes of *Racing Extinction* establish a clear trajectory for the film's point of view: This is a film in which the views of "humanity" are equated with the editorial choices of *The Financial Times* (one talking head expresses surprise that *The Financial Times* relegated a story about mass extinction to page six), and high technology is the principal means for saving humanity from mass extinction. The film never mentions the word capitalism. It apparently sees no correlation between the worldview of *The Financial Times* and the sixth mass extinction event in the history of the planet, even though the neoliberal phase of capitalism, its most successful global surge, corresponds precisely with what scientists call The Great Acceleration, "the most rapid transformation of the human relationship with the natural world in the history of humankind."[26] When you meet the film's alleged heroes, which include Elon Musk and a race car driver, you realize why the connection between capitalism and mass extinction is never made. Perhaps part of the reason for this particular narrative selection also has to do with the film's producers, which include the Li Ka Shing Foundation. As of 2014, Li Ka Shing was the richest person in Asia, with a net worth of over $30 billion. Li made his fortune in a variety of business ventures, which include plastics manufacturing, cement production, construction, airports, and an ownership stake in several container ports around the world. He is also the majority shareholder in Husky Energy. *Racing Extinction* is therefore co-produced by a man whose entire professional life has been devoted to the destruction of the oceans.

Racing Extinction does not avoid some of the ominous scenarios of mass extinction, but it assuredly accompanies those statistics with a confident statement by one of the film's protagonists: "It's like we're living in the age of dinosaurs, but we can do something about it." The analogy might not make sense, for several reasons: for one thing, if this is indeed a mass extinction event we are living in, then there is no way to reverse its general course. Paleontologist Doug Erwin, who does not believe the current planetary situation is a mass extinction event, contradicts *Racing Extinction*'s primary invitation to its audience to save very visible and majestic species: "if it's actually true we're in a sixth mass extinction, then there's no point in conservation biology," he says.[27] According to Erwin, an expert in past mass extinctions, a mass extinction represents a "network" failure that would eliminate almost all life on earth. At the moment, he says, we are approaching a mass extinction: "Yeah, everything's fine until it's not. And then everything goes to hell." In other words, the analogy with the dinosaurs, presented early in *Racing Extinction*, makes no sense because even a species with the ingenuity of *Homo sapiens* could not have saved the dinosaurs from the Cretaceous–Paleogene (K–Pg) extinction event, which eliminated three-quarters of the species on earth; ingenuity cannot reverse a mass extinction. *Racing Extinction* extends the analogy with the Cretaceous-Paleogene mass extinction to declare the source of the current mass extinction: "Humanity is the asteroid," says a voice from offscreen. Of course, "humanity" is not the asteroid. Overconsumptive, industrialized, capitalist humanity is the asteroid, and, even then, comparing humanity with an asteroid creates a faulty comparison of a natural disaster (the asteroid) with a definitively unnatural one (a human economic system). Most of humanity does not drive cars, produce synthetic chemicals, or test atomic bombs. As Eddie Yuen writes, "The status quo of capitalist production of unnecessary commodities and services for the global elites and 'middle classes' is the ongoing catastrophe that must be addressed."[28] *Racing Extinction* notably avoids the primary causes of mass extinction and instead condemns *all* of humanity.

The heroes of *Racing Extinction* are photographers and first world technologists. The central assumption of both is that visibility correlates with capacity for change. If you can show the problem, the problem will be solved. Of course, if this were true, the world would have turned its back on nuclear energy, polar bears would not be endangered, and war would be a thing of the past. *Racing Extinction* simplifies both the crisis and the solution, with a myopic sense of the relationship between communication and social change. The model of success for *Racing Extinction* does not involve replacing capitalism with a less ecocidal economic system; instead, the film advocates for capitalism as the answer to the mass extinction driven by capitalism itself. Shawn Heinrichs, who formerly worked in finance and international business development, is presented in *Racing Extinction* as a "marine conservationist" and one of the exemplars of the film's proposed solutions to the extinction crisis. Heinrichs is credited with transforming the shark hunting area of Isla Mujeres in the Caribbean into a shark watching location for ecotourism. After making the case for swimming with the sharks instead of hunting them, Heinrichs adds, "It's just simple economics." For *Racing Extinction*, the way to reduce an extinction crisis that sees species dying in numbers up to 1000 times the normal background rate of extinction is to promote the economic system that gave us The Great Acceleration.

In a subsequent sequence, Heinrichs asks, "Often people say, 'How can one person make a difference?'" The film then transitions to a Public Service Announcement (PSA) involving Chinese basketball superstar Yao Ming. In the example of a PSA aimed at curbing the Chinese appetite for shark fin soup is a combination of the film's favorite messages: first, consumer choices are what drive change; "Remember, when the buying stops, the killing can too," the message from the PSA, is a wonderfully ironic message for a film that does not ask its first world viewers to confront the structural violence of capitalism; second, once again the film makes the correlation between witnessing and activism, this time by demonstrating how a television commercial with an internationally known athlete can affect consumption habits (not a lesson that applies to everyone, obviously); later in the film, the fetish of witnessing reaches a crescendo of cognitive dissonance, as an "environmentalist" race car driver responds to her critics by saying if she were driving a bicycle and not a race car, she would not have "75 million" people watching her; and third, this sequence represents one of several examples in the film in which the focus is on Asia; only briefly does the film pause to remonstrate the West for its contributions to planetary mass extinctions, even though the Columbian Exchange was, prior to global warming, the largest environmental disaster in human history. For the most part, this is a film featuring white, first world men as talking heads addressing businesses in Hong Kong, mainland China, and Indonesia that traffic in endangered species. The focus on Asia is likely a consequence of where the featured photographers already work, and perhaps because of where the film gets some of its funding, but for a documentary whose primary audience is in the West on the Discovery Channel and at various film festivals, the choice to depict onscreen almost entirely Asian fishermen and traders slaughtering endangered species such as sharks and manta rays, and leaving out an extensive history of Western deforestation and plunder of the oceans, for example, produces a weirdly orientalist visual grammar. At one point, for example, there is a particularly gruesome sequence in which Indonesian fishermen hunt manta rays while Shawn Heinrichs chokes back tears as he narrates his horror at the unfolding savagery.

Toward the end of the film, famous primatologist Jane Goodall says, "Without hope, people fall into apathy." This message, as well as the message that says incremental changes are better than no change at all, is repeated throughout the film and is clearly the guiding ethos of the film, a message very much consonant with the critics of catastrophism. *Racing Extinction* oscillates between dramatic statistics of mass extinction and the "green" capitalists who are retrofitting skyscrapers or turning shark hunting into eco-tourism. The entrepreneur who retrofitted the Empire

State Building says, "The best way to move the needle, when people are talking about the environment, is the bottom line." This sentiment echoes the "simple economics" statement of Shawn Heinrichs and reaffirms the film's commitment to capitalist metrics. When viewers combine the two central messages of *Racing Extinction*—incremental change and economic (capitalist) metrics are the solution to the extinction crisis— they receive a profoundly misguided set of instructions for responding to an extinction crisis that threatens to erase at least half of all species by the end of this century; however, clearly, this message lowers the bar to entry for those who want to be perceived as environmentalists and therefore gives the appearance of solving the dilemma of catastrophism.

Toward the end of *Racing Extinction*, Heinrichs appears on screen in a moment of triumph and declares, "Most people say you can't beat money... with just hope and inspiration." His statement is followed by one of the celebrated photographers in the film demonstrating how the photographs they have taken of manta rays have produced a change because "people get it" when they see the problem represented visually. It is this scene that appears to fold the argument of *Racing Extinction* in on itself. The audience has been told repeatedly that the way to address environmental issues is "the bottom line" and "simple economics." The audience has also been told that witnessing is powerful, and witnessing with high technology is perhaps the most powerful form of advocacy. Money loses, apparently, because images of manta rays have been shared with government regulators who declare the manta ray off limits to the market. Images trump money, at least in a limited way that the filmmakers use to illustrate the efficacy of their chosen form of advocacy. The photographers at the center of the film are champions for an endangered species, as a result, but the incantations about market metrics and environmentalism are undermined. In order to prove that "hope" *can* beat money—that is, in order to maintain the critique of catastrophism—the filmmakers must contradict themselves and their free market preferences.

A common feature of eco-opportunist documentaries is their desire to appeal to audiences beyond the assumed progressive vanguard of North American film festivals. One could, I suppose, argue that the inclusion of an environmentalist race car driver is such an attempt, a gesture beyond the choir. Later in the film, the environmentalist race car driver is behind the wheel of a car designed by Wall Street's favorite techno-entrepreneur, Elon Musk, as they tour New York City using a car-mounted overhead projector to transmit images of endangered species onto the surfaces of various buildings, including the United Nations head-quarters. The rousing finale of the film affirms the value of the "small choices we make each day," once again encouraging the apathetic North American viewer to accept the idea that individual consumer choices, and not collective action, are all that is required to combat what the film itself has portrayed as an existential threat to humanity that could very well cause our extinction in a matter of a few decades. Instead of the eco-apocalyptic narrative in which the crises created by industrial capitalism are global, complex, and intractable, the eco-opportunist nar-rative is ebullient and optimistic because the mass extinction crisis only requires that we as individuals eat less meat, install solar panels, and drive one of Elon Musk's Tesla vehicles into a latte-filled future. The prob-lem with *Racing Extinction* is not that all its recommendations are bad ideas; the problem is that it sets the bar so low that success in its terms would not come close to solving the extinction crisis; in fact, there is a strong likelihood that following only the film's solutions would simply extend the life of capitalist ecocide, deferring the mass extinction event for a few years or decades. The muddled message of the film ultimately seems irrelevant within the context of its stunning imagery and exhila-rating musical crescendo that implores viewers to "start with one thing." This is not pragmatism. This is techno-libertarian fantasy.

This Changes Everything

Compared with *Racing Extinction*, *This Changes Everything* is in many ways successful at combining the message of hope with at least a sem-blance of adequate solutions. Avi Lewis' film is of course based on the Naomi Klein bestseller of the same title. From the opening scene of the film, it is clear that the critique of catastrophism is very much on the mind of the filmmakers. "Can I be honest with you?" begins Klein's narration. "I've always kinda hated films about climate change.... Is it really possible to be bored by the end of the world?" The first gesture in the film is to identify the jaded first world viewer and attempt to thwart that person's cynicism. Klein says we are told that the cause of climate change is the innate greed of human beings, and "if that's true, then there is no hope." Immediately, before any facts about climate change have been dispensed, *This Changes Everything* sets out to counter the effects of catastrophism. "What if the real problem is a story, one we've been telling ourselves for 400 years?" Klein asks rhetorically, in a form of cultural critique that sounds similar to 1990s-era post-structuralism and remnants of the linguistic turn in the humanities. This massive and complex problem of climate change can be solved, according to Klein, if we just change the narrative. She reached a comparable conclusion in her popular 2007 book *The Shock Doctrine*, in which she argued that learning the history of disaster capitalism was the best way to shield a populace against its effects. With the documentary *This Changes Everything*, the opening gambit is once again about changing narratives; however, given how substantially the infrastructure of oil capitalism has expanded in the past century, even if one looks at only the time that has passed since the publication of *No Logo*, Klein's emphasis on changing the narrative—rather than, say, disrupting or destroying the infrastructure of capitalism's supply chains—seems somewhat quaint. In particular, in that opening scene Klein's voice-over draws our attention to the incredible duration of the colonial narrative of mastery over nature. What makes the film think we, or even many of *us*, can change such a durable and materially embedded narrative in a matter of decades or less?

In the same way that *Racing Extinction* condemns *all* human beings for mass extinction, *This Changes Everything* uses the royal *we* to describe the origins of climate change, instead of more accurately historicizing climate change primarily as a consequence of European colonialism and industrial capitalism. In her book, Klein uses similar language when assessing an aggregate human capacity for change: "It seems to me that if humans are capable of sacrificing this much collective benefit in the name of stabilizing an economic system that makes daily life so much more expensive and precarious [the austerity agenda], then surely humans should be capable of making some important lifestyle changes in the interest of stabilizing the physical systems upon which all of life depends."[29] In this passage, Klein is trying to use the rhetoric of the capitalist austerity agenda against capitalism, as if to say, "Calls for collective sacrifice on behalf of the austerity agenda presume we are capable of such sacrifice, so why not apply that capacity to saving the planet?" The intended target of such rhetoric, much like the opening gambit of the film, is not the dedicated leftist but the skeptics in the audience, especially the conservatives. Of course, the question is not whether "humans" are capable of changing economic systems, but whether the overconsumptive and exploitive capitalist class can be brought to heal by a socialist revolution. That is, Klein's liberal use of "we" and "humans" in the film obscures significant social divisions based on economics, gender, race, colonialism, and other oppressive structures.

By referencing the "leaders" of "most" nations as sharing a consensus that the world must not pass the 2 degrees Celsius above baseline warming cautioned by the IPCC, *This Changes Everything* strangely avoids using the scientific consensus to make the same point. More troubling is the emphasis the film places on nation-states to be the solution for climate change, such as the section in the film that deals with Hurricane Sandy, when the state form has accompanied capitalism at every stage of this catastrophe. How many decades of failed climate talks will it take for Klein, Bill McKibben, and 350.org to realize that the state form collaborated with capitalism in the conquests of the Columbian Exchange and every petrodollar produced since the beginning of oil capitalism? The only viable option to slow catastrophic climate change involves rapid decolonization and deindustrialization, and neither of these projects can be accomplished through the state form; the state form is actively complicit in the twin disasters of the colonial project and industrial capitalism.

The scene in which a Greek comrade in Halkidiki questions whether identifying capitalism as the primary problem will help "the struggle" condenses all the anxieties of the left and its critique of catastrophism into a single, reluctant admission. "What is the core problem?" Klein asks the Greek comrade. "Do you want me to state it on camera?" she responds, giggling nervously. "Yeah, I would say it's the economic system," she says, then stumbles. "Capitalism, I guess." "You're not sure if you should say that on camera?" asks Klein. "No. I don't know if it helps the struggle," the comrade laughs in response. While Klein and Lewis deserve credit for actually naming capitalism as the source of the climate crisis, in her book and in their film, the admission in the film doesn't happen until about halfway through, and the filmmakers allow its identification to be framed sheepishly by a speaker who isn't sure whether saying the word capitalism will help defeat it. In her book, Klein uses the phrase "unregulated capitalism" to describe what she sees as the enemy of the environmentalist movement. In public appearances, Klein echoes Doug Henwood's belief, articulated in the foreword to the *Catastrophism* anthology, that "it doesn't seem fruitful to argue that there's no way to save the earth without ending capitalism."[30] Klein wrote an online column in response to early critics of her book, in which she makes the same point: we can fight climate change before ending capitalism.[31] Her defense of her book, when writers on the left and the right accused it of calling for the end of capitalism, essentially denies that she is anti-capitalist, even though her book makes the case about capitalism and the environment that anarchists and Marxists made decades earlier:

I argue forcefully that capitalist logic and neoliberal policies are sabotaging the actions required to avert catastrophe. But I have never said that we need to "slay," "ditch" or "dismantle" capitalism in order to fight climate change. And I most certainly didn't say we need to do so *first*. Indeed I say the opposite, very early on in the book (page 25), precisely because it would be so dangerous to make such a purist claim.

The sentence "And I most certainly didn't say we need to do so *first*" conveys a specific meaning to the sentence that precedes it: Klein is clearly saying that she does not believe capitalism must be abolished. The idea that humanity could fight climate change without dismantling the ecocidal system that created climate change is the kind of logic eco-opportunists engage when trying to lower the bar for environmental activism, and when attempting to make their books and films accessible to a liberal audience in the Global North. Slowing climate change is clearly a greater priority than ending capitalism. But can you do the first thing without doing the second? And exactly how much capitalism does Klein believe we can live with and still fight climate change? And how many people will interpret this strategy as liberal reformism that makes excuses for the continued existence of capitalism? The film *This Changes Everything* does little to convince viewers that Klein is advocating for revolutionary change, rather than liberal reform. At one point, Klein describes the role of right-wing think tanks in the extension of the neoliberal agenda. The outcome is described in decidedly liberal terms: "Governments whither, the middle class dissolves, and profits soar." Klein mourns the dissolution of the state form and the so-called middle class, a liberal invention to mask the class war by which capitalism is defined.

This Changes Everything does contain some coverage of necessary and exciting forms of resistance, from the "environmentalism of the poor" in India to the resistance of indigenous people in Canada. By focusing on problematic elements in the film, I am not denying that it has value; I am simply trying to condense a critique of the film into a small space. When the film encounters what might be the biggest obstacle in the fight against climate change, China, Klein acknowledges that China is not a "model" country; in fact, it's "more like a battlefield." The section on China is where *This Changes Everything* pulls punches, reserves some of the most demoralizing statistics and tendencies for the sake of preserving a viewer with hope. It is also in the section on China, very late in the film, that Klein finally frames the crisis of climate change in the terms it deserves, but the terms the film has been reluctant to offer: "capitalism versus the climate." Is withholding the principal culprit in this story a canny strategy for inviting the skeptical liberal or conservative to watch over an hour of a documentary about climate change? The earlier mention of capitalism was from a Greek comrade who sheepishly offered the word, and wasn't sure if it would help to name the problem. This kind of dodging of the real issue is a staple of the eco-opportunist documentary, almost a parallel of the very easy way in which catastrophism documentaries discuss the intellectual cul-de-sac one finds when trying to solve the problem of peak oil. The eco-opportunist ethos returns at the end of *This Changes Everything*, to frame global warming as an opportunity: "So here's the big question: What if global warming isn't *only* a crisis?" Klein asks provocatively. "What if it's the best chance we're every going to get to build a better world?" The eco-opportunist strays so far from the catastrophist that global warming becomes an opportunity, not simply a crisis.

Chasing Ice and Chasing Coral

Chasing Ice (2012) and *Chasing Coral* (2017) are both directed by Jeff Orlowski, and so it is not surprising that they share the same aesthetic and political valences. Both films contain some of the most incredible images of natural phenomena ever recorded. A substantial portion of these films is devoted to the technical challenges of obtaining photographs of glacial melt and coral bleaching. The goal of each film is a form of witnessing that the filmmakers believe will activate radical eco-activism necessary to reverse the respective crises. The techno-utopian narrative trajectory emphasizes the passion and technical skill of the first world technicians and scientists engaged in capturing rare images. The narrative structure matches that of *Racing Extinction*, and in *Chasing Coral*, an important subplot involves a former ad man, similar to Shawn Heinrichs' character arc in *Racing Extinction*. Characteristic of the eco-opportunist film, then, is the redemption of the capitalist, but the word capitalism must not be uttered in the film, and the solution to the ecological crises created by capitalism must also involve capitalism. In *Chasing Ice*, former Shell Oil employee Richard Ward declares that the images of the melting glaciers were a revelation, never mind the fact that Exxon knew about the effects of global warming almost 40 years ago and chose to invest millions in forms of climate denial,[32] and Shell Oil produced a film in 1991 that warned of the dangers of climate change but continued to conduct its business without concern for its role in greenhouse gas emissions.[33] Eco-opportunist films substitute ineffectual individual defections of capitalists for systemic change that might have a chance of reducing significantly the ecological impact of capitalism.

Chasing Ice follows environmental photographer James Balog in his quest to capture images of the effects of global warming. This is a noble goal, and those images will likely contribute to various forms of public education. Toward the end of the film, Balog is seen presenting to the 2009 United Nations Copenhagen Climate Change Conference; presumably, the filmmakers believe this type of political arena is where significant emissions cuts can be accomplished. The conference in 2009 ended with "all references to 1.5 degrees Celsius [warming as a target for emissions cuts] in past drafts...removed at the last minute," and "the earlier 2050 goal of reducing global CO_2 emissions by 80% was also dropped."[34] Future conferences marked similar failures, with no legally binding emissions targets that could prevent warming of under 2 degrees Celsius. As suggested above, the IPCC target of 2 degrees Celsius no longer seems tenable, and more likely scenarios predict anywhere from 2 degrees Celsius to 10 degrees Celsius warming by the end of the century. *Chasing Coral* presents a similarly catastrophic future for coral reefs, with the probable outcome that most reefs will be dead in 30 years. Like *Chasing Ice* and *Racing Extinction*, there is no mention of capitalism in *Chasing Coral*, and the solutions proposed at the end of the film direct the viewer to the film's Web site. The downloadable pamphlet of recommended actions includes many sensible activities: petition political leaders to support "clean energy and clean jobs"; commit to reducing your carbon footprint; join local NGOs such as 350.org and the Sierra Club; and participate in a number of groups dedicated to supporting coral reefs. None of these ideas is necessarily inappropriate; however, each by itself is potentially problematic and certainly does not address the overlapping systems of oppression within capitalism. For example, an appeal to a political leader is easily ignored; the state is structurally a hostage of capitalist influence. Individual commitments to carbon reduction do not address massive institutional abuses (the military, oil companies, etc.). NGOs routinely greenwash corporate exploitation of the environment and buttress capitalist oppression by providing a recuperative form of charity for what might otherwise be revolutionary struggle.

By not identifying the structural source of the environmental crises, and by providing piecemeal, reform-oriented solutions, eco-opportunist documentaries such as *Racing Extinction, This Changes Everything, Chasing Ice,* and *Chasing Coral* might be more inviting to a broader audience and might even overcome some of the problems with films that preached catastrophism, but ultimately these eco-opportunist documentaries understate the crises and provide insufficient solutions to such an extent that they can be something worse than unproductive; they can be counterproductive.

NOTES

1. "Diana Liverman Gold Medal (RGS) Awarded In 2010," *Edubilla.com,* 2010. Available online: http://www.edubilla.com/award/gold-medal-rgs-/diana-liverman/.
2. Diana Liverman, "How to Teach Your Students About Climate Without Making Them Hopeless," *Washington Post,* August 20, 2014. Available online: https://www.washingtonpost.com/posteverything/wp/2014/08/20/how-to-teach-about-climate-without-making-your-students-feel-hopeless/.
3. Doug Henwood, "Dystopia Is for Losers," in *Catastrophism: The Apocalyptic Politics of Collapse and Rebirth* (Oakland: PM Press, 2012), xv.
4. Eddie Yuen, "The Politics of Failure Have Failed: The Environmental Movement and Catastrophism," in *Catastrophism: The Apocalyptic Politics of Collapse and Rebirth* (Oakland: PM Press, 2012), 19.
5. Yuen, 20.
6. Yuen, 19.
7. Imre Szeman, "Oil, Futurity, and the Anticipation of Disaster," *South Atlantic Quarterly* 106.4 (Fall 2007): 815.
8. Gerry Canavan, "Retrofutures and Petrofutures: Oil, Scarcity, Limit," in *Oil Culture,* Ross Barrett and Daniel Worden, eds. (Minneapolis: University of Minnesota Press, 2014), 333.
9. Clive Hamilton, "The New Environmentalism Will Lead Us to Disaster," *Scientific American,* June 19, 2014. Available online: https://www.scientificamerican.com/article/the-new-environmentalism-will-lead-us-to-disaster/.
10. Brian Wingfield, "Oil Tanker Storage Hits 2017 Record Despite OPEC's Cuts," *Bloomberg,* June 19, 2017. Available online: https://www.bloomberg.com/news/articles/2017-06-19/oil-tankers-store-most-oil-this-year-as-glut-proves-hard-to-kill.
11. Elizabeth Colbert, *The Sixth Extinction: An Unnatural History.* (New York: Henry Holt & Co., 2014).

12. Steve Connor, "Ocean Acidification Killed off More Than 90% of Marine Life 252 Million Years Ago, scientists believe," *The Independent*, April 9, 2015. Available online http://www.independent.co.uk/news/science/ocean-acidification-killed-off-more-than-90-per-cent-of-marine-life-252-million-years-ago-scientists-10165989.html.

13. Carl Zimmer, "Ocean Life Faces Mass Extinction, Broad Study Says," *The New York Times*, January 15, 2015. Available online: https://www.nytimes.com/2015/01/16/science/earth/study-raises-alarm-for-health-of-ocean-life.html.

14. T. W. Crowther, H. B. Glick, K. R. Covey, C. Bettigole, D. S. Maynard, S. M. Thomas, J. R. Smith, G. Hintler, M. C. Duguid, G. Amatulli, M.-N. Tuanmu, W. Jetz, C. Salas, C. Stam, D. Piotto, R. Tavani, S. Green, G. Bruce, S. J. Williams, S. K. Wiser, M. O. Huber, G. M. Hengeveld, G.-J. Nabuurs, E. Tikhonova, P. Borchardt, C.-F. Li, L. W. Powrie, M. Fischer, A. Hemp, J. Homeier, P. Cho, A. C. Vibrans, P. M. Umunay, S. L. Piao, C. W. Rowe, M. S. Ashton, P. R. Crane & M. A. Bradford, "Mapping Tree Density at a Global Scale," *Nature* 525 (10 September 2015): 201–205.

15. Ada Carr, "'Pristine' Landscapes No Longer Exist and Haven't For Thousands of Years, Study Says," *Weather.com*, June 7, 2016. Available online: https://weather.com/science/environment/news/pristine-nature-humans-global-biodiversity-landscape-ecosystems-earth#/! "These findings suggest that we need to move away from a conservation paradigm of protecting the earth from change to a design paradigm of positively and proactively shaping the types of changes that are taking place," lead author Nicole Boivin told The Post. "The reality is that there are 7 billion people living on an already heavily altered planet. It is a pipe dream to think that we can go back to some sort of pristine past."

16. Joby Warrick and Chris Mooney, "Effects of Climate Change 'Irreversible,' U.N. Panel Warns in Report," *The Washington Post*, November 2, 2014. Available online: https://www.washingtonpost.com/national/health-science/effects-of-climate-change-irreversible-un-panel-warns-in-report/2014/11/01/2d49aeec-6142-11e4-8b9e-2ccdac31a031_story.html; Jeremy Lovell, "Clean Energy Lags Put World on Pace for 6 Degrees Celsius of Global Warming," *Scientific American*, April 26, 2012. Available online https://www.scientificamerican.com/article/clean-energy-lags-put-world-on-pace-for-6-degrees-celsius-of-global-warming/.

17. "Scientists Categorize Earth as a 'Toxic Planet'," *Phys.Org*, February 7, 2017. Available online: https://phys.org/news/2017-02-scientists-categorize-earth-toxic-planet.html.

18. Matthew Taylor, "Plastic Pollution Risks 'Near Permanent Contamination of Natural Environment'," *The Guardian*, July 19, 2017. Available online: https://www.theguardian.com/environment/2017/jul/19/plastic-pollution-risks-near-permanent-contamination-of-natural-environment.

19. Damian Carrington, "The Anthropocene Epoch: Scientists Declare Dawn of Human-Influenced Age," *The Guardian*, August 26, 2016. Available online: https://www.theguardian.com/environment/2016/aug/29/declare-anthropocene-epoch-experts-urge-geological-congress-human-impact-earth.
20. Stephen J. Davis and Robert H. Socolow, "Commitment Accounting of CO_2 Emissions," *Environmental Research Letters* 9.8 (26 August 2014): 1–9.
21. Ian Johnston, "World Must Hit Zero Carbon Emissions 'Well Before 2040,' Scientists Warn," *The Independent*, April 13, 2017. Available online: http://www.independent.co.uk/environment/world-zero-carbon-emissions-before-2040-two-decades-climate-change-global-warming-greenhouse-gases-a7682001.html.
22. Robert Rapier, "World Sets Record for Fossil Fuel Consumption," *Forbes*, June 8, 2016. Available online: http://www.forbes.com/sites/rrapier/2016/06/08/world-sets-record-for-fossil-fuel-consumption/.
23. Brian Kahn, "We Just Breached the 410 PPM Threshold for CO_2," *Scientific American*, April 21, 2017. Available online: https://www.scientificamerican.com/article/we-just-breached-the-410-ppm-threshold-for-co2/.
24. U.S. Energy Information Administration, "International Energy Outlook 2016," May 11, 2016. Available online: https://www.eia.gov/outlooks/ieo/coal.cfm.
25. John Abraham, "Methane Release from Melting Permafrost Could Trigger Dangerous Global Warming," *The Guardian*, October 15, 2015. Available online: https://www.theguardian.com/environment/climate-consensus-97-per-cent/2015/oct/13/methane-release-from-melting-permafrost-could-trigger-dangerous-global-warming.
26. Quoted in Will Steffen, Wendy Broadgate, Lisa Deutsch, Owen Gaffney, and Cornelia Ludwig, "The Trajectory of the Anthropocene: The Great Acceleration," *The Anthropocene Review* 2.1 (2015): 82.
27. Peter Brannen, "Earth is Not in the Midst of a Mass Extinction," *Atlantic Monthly*, June 13, 2017. Available online: https://www.theatlantic.com/science/archive/2017/06/the-ends-of-the-world/529545/.
28. Yuen, 29.
29. Naomi Klein, *This Changes Everything* (New York: Vintage Canada, 2015), 17.
30. Henwood, xv.
31. Naomi Klein, "No, We Don't Need to Ditch/Slay/Kill Capitalism Before We Can Fight Climate Change. But We Sure as Hell Need to Challenge It," *The Leap*, September 27, 2014. Available online: https://theleapblog.org/no-we-dont-need-to-ditchslaykill-capitalism-before-we-can-fight-climate-change-but-we-sure-as-hell-need-to-challenge-it/.

32. Shannon Hall, "Exxon Knew About Climate Change Almost 40 Years Ago," *Scientific American*, October 26, 2015. Available online: https://www.scientificamerican.com/article/exxon-knew-about-climate-change-almost-40-years-ago/.

33. Damian Carrington and Jelmer Mommers, "'Shell knew': Oil Giant's 1991 Film Warned of Climate Change Danger," *The Guardian*, February 28, 2017. Available online: https://www.theguardian.com/environment/2017/feb/28/shell-knew-oil-giants-1991-film-warned-climate-change-danger.

34. John Vidal, Allegra Stratton, and Suzanne Goldenberg, "Low Targets, Goals Dropped: Copenhagen Ends in Failure," *The Guardian*, December 19, 2009. Available online: https://www.theguardian.com/environment/2009/dec/18/copenhagen-deal.

Image 11.1 Boat (Mia Feuer, watercolor)

Slow Motion Electric Chiaroscuro: An Experiment in Glitch-Anthropo-Scenic Landscape Art

Patti Pente

INTRODUCTION

For almost twenty years the Anthropocene has been discussed in academic and popular cultural circles. For a similar period, artists have been practicing forms of glitch art, interrupting normative systems and digital structures; in many cases, in order to critique selected aspects of online society. In light of these parallel developments, I consider, as an artist and educator, ways to position my artistic practice within these two cultural developments. In this chapter, I compare aspects of glitch art and the late-Anthropocene through the theoretical cleaving of the bounded relationship of *fail and fix*. I draw attention to ways that glitch art and the Anthropocene exceed their intended uses and are unpredictable: how divergent failures can open thought towards multiple potentials for becoming posthuman. This could confound prosumers'[1] expectations in surprising ways so that alternative ideas about what it might mean to live relationally in these contemporary times are highlighted. I find

P. Pente (✉)
University of Alberta, Edmonton, Canada

© The Author(s) 2018
j. jagodzinski (ed.), *Interrogating the Anthropocene*, Palgrave Studies in Educational Futures, https://doi.org/10.1007/978-3-319-78747-3_12

purchase in this mutability, and close this chapter with an experiment in glitch landscape art, filtered through the Deleuzo-Guattarian concept of disjunctive synthesis.

I connect the Anthropocene to glitch, calling it glitch-Anthropocene, to indicate the failing of a popular understanding of the Anthropocene as a form of effective public pedagogy. Specifically, the version of Anthropocene as warning indication of irreversible planetary destruction, taken up on various websites,[2] does not go far enough to help people shift from seeing themselves as humans who use/save the planet to entities who are symbiotic parts of the planet, sharing this existence with many equally integrated non/inhuman entities. In other words, in much information that acts as public pedagogy, the Anthropocene still remains at the level of the human who manages the planet, often with the intention to sustain natural resources for longevity of this late-capitalist life. Glitch, in this case, is not a noticeable interruption in the normative system of affluent, capitalist society or a disruption that makes us pay greater attention. On the contrary, it fails to glitch in that way. We are not noticing Earth's glitches as opportunities for deeply onto-logical change—what I would suggest is necessary for lasting social action. Instead, if noticed at all, we, as wealthy citizens, embrace the more desirable and perhaps easier idea of humanism aimed at sustainability and better stewardship. In this scenario, we do not really have to let go of prosperous lifestyles, just tweak our consumerism somewhat. It may also reflect William Connolly's take on Nietzsche's passive nihilism whereby, "many (people) refute climate denialism but slide away from stronger action" (2017, p. 9). Perhaps, within the tensions of public/private relations, the daily routines of life leave little time, desire, and/or opportunity to take action. On a global scale, change seems out of our individual control. Ironically, we could think then, that there is no effective glitch in the Anthropocene at all, despite superficial attempts to act like one within the multiple failures found in climate change, pollutants, and rampant land use: there are, indeed, many glitches and/or planetary malfunctions. In this case, the paradox is that the glitch-Anthropocene *fails* as a glitch. It misses the educational opportunity to develop a lasting means by which change might happen.

Furthermore, the articulation of the Anthropocene as a warning supports a unification of all humans through the potential of extinction, or where wealthier nations are called upon as major polluters to unite in fixing the planet, thereby protecting and supporting the poor

(Chakrabarty 2009; Luke 2015). While there is no denying this important reality of global disparity, the problem with this unification is that it smooths all difference under this perspective of *the* human. As Claire Colebrook (2017, p. 4, italics in original) explains,

> The more "we" reflect upon "our" mark on the planet the more we appear to be a single polluting species, while also being more and more divided by the causes and consequences of what has come to be known as the Anthropocene. In short, the more "we" appear to be unified as a species, differentiated from other species, and the more we become defined by the claim of *the* Anthropocene, the more mindful we should become of all the forces and tendencies too minimal to appear as differences.

Forms of human impact—such as pollution or the destruction of ecosystems—have long been acknowledged, but with the claim made for the Anthropocene as an irreversible event, its particular difference is deemed to be dramatically different. Colebrook (2016a) notes that this unity (in this case, based upon species survival) eliminates so many other possible ways of sharing this planet/universe. We cannot see this important limitation of the Anthropocene from our location so deeply within it. However, there may be disruptive and alternative ways to understand our relationship with the Earth available within artistic practices that also depend upon the power of the glitch.

Glitch art, and its relative, glitch aesthetics, for the purposes of this paper, are interchangeable. Glitch aesthetics can be described as a cultural practice whereby an interruption of a system focusses attention politically, aesthetically and/or socially towards a characteristic flaw of that system (Betancourt 2016; Cloninger 2010; Nunes 2011; Roy 2014). Thus, glitch art is also involved in forms of failure: it typically operates by creating an intended failure within its system in order to draw attention to the materiality of that system and its inherent manipulation of the user (Cloninger 2010; McCormack 2010). In some cases, glitch art acts as a form of pedagogy. For example, the glitch research of French artist, Benjamin Gaulon, centres on issues of technology, specifically around the ubiquity of electronic waste. His initiative, the Recyclism Hacklab, now situated in Paris, opens to the public opportunities to learn more about hacking and/or glitching through various workshops. The public is encouraged to bring in obsolete devices, including phones, toys, computers, etc., so that they can repurpose them through new-found skills.

However, glitch art, like the glitch-Anthropocene, also *fails* in degree as a critique. Firstly, it can be misrecognized and ignored by the user who interprets it as a mere moment of dysfunctionality of the system, and thus the perceived transparency of the system is maintained (Betancourt 2016). This is similar to the ways that the glitch-Anthropocene fails to effectively draw awareness to the limitations within Anthropos unity, as previously noted. Secondly, glitch art can be cooped by the very system it critiques, ending up as another form of consumable, immaterial "art" (ibid., 2016). It becomes absorbed into the system of cultural production, which can quickly depoliticize a work of art.[3] This is indicative of the history of the avant-garde throughout Western art history. Glitch artist and theorist, Rosa Menkman (2011, p. 55), notes that the fate of glitch has, in some cases, morphed into such a path. She explains,

> When the glitch becomes domesticated into a desired process, controlled by a tool, or technology – essentially cultivated – it has lost the radical basis of its enchantment and becomes predictable. It is no longer a break from a flow within a technology, but instead a form of craft. For many critical artists, it is considered no longer a glitch, but a filter that consists of a preset and/or a default: what was once a glitch is now a new commodity.

Thus, glitch art, in popular culture, can become severely limited. Significantly, glitch-Anthropocene and glitch art can *fail to fix* as public pedagogies for change towards more ecologically sensitive relations. More needs to be explored within this impasse, however, because a focus on these concepts as predictable pedagogical strategies presupposes an instrumental approach. To stop there would miss the potential to theoretically wander and drift from within these concepts' shortcomings, because the glitch can emerge as an unintended event, revealing an unruly, inhuman agency. In other words, the glitch can instigate new occurrences if it is left to interrupt in truly unanticipated ways. I explore slippages within glitch as art through the Deleuzo-Guattarian concept of disjunctive synthesis, which signals an opportunity for continual morphing and multiplying of concepts within experimentation. In doing so, I look towards the challenge of what Colebrook calls an "intensifying of the tendency of the Anthropocene… to a higher deterritorialization" (2016b, p. 442). In other words, using the disjunctions that emerge from considering glitch art and the glitch-Anthropocene, I take up such a challenge through an imaginative exploration of Anthropo-scenic, digital/analogue art, where these limitations and impossibilities of fail and fix can highlight the porous nature of the posthuman.

ANTHROPOCENE

When atmospheric chemist, Paul Crutzen, introduced the popular term "Anthropocene" along with colleague, Eugene Stoermer, to describe what they interpreted as the permanent sign of human destruction and abuse upon the Earth, it triggered the popular spread of the term within culture (Crutzen 2002; Crutzen and Stoermer 2000). With the current concept evolving into a *post* or *late* stage, the Anthropocene is now firmly entrenched as a signifier for a new geological epoch of the Earth's deep geologic time that anticipates a lasting residue of human effects upon geo/bio/ecosystems. As the popularization of the term has grown, interpretations of it through disaster cinema, TV, and other forms of culture continue to thrive (Colebrook 2016a; jagodzinski 2015). For example, there are artists who focus on the Anthropocene (Anderson 2015); there are websites dedicated to it as a form of online pedagogy; there are scholarly journals about the Anthropocene; and there are advertisements that exploit the term to promote "greener" consumerism. The guilt and/or fear regarding the state of the planet is palpable in the West but there is a strong level of protectionism and nationalism emerging in the United States at the time of this writing that thwarts positive action on these concerns.[4] Whether or not the term affirms a legitimate layer of change within geological time based on the paradigms of geological science remains to be seen, and, as Ellis et al. (2016) urge, the decision regarding such classification should include input from the humanities and social sciences. Then again, officially registering the Anthropocene into geological time remains moot because the cultural acceptance of the term has already occurred and as such, it holds a certain sway. It is not news that extensive overuse of fossil fuels, species extinction, ozone depletion and many, many more destructive changes, some irreversible, have been linked to human activity: typically, activity that maintains or drives the lifestyle of affluence. The critique that prosperous Western countries are responsible for the majority of the ecological damage due to their continued reliance on fossil fuels has been presented as the *Capitalocene* by Thomas Moore (2016), and the *Chthulucene* by Donna Haraway. They warn that the late-capitalist lifestyle that continues to grind away at the planet is inequitable and short-sighted. In this view, to blame all of humanity for such levels of pollution when a relatively small portion of the global population is responsible is unethical and blinds us to the systemic inequities of late-capitalism (Bignall et al. 2016; Chakrabarty 2009; Luke 2015). Ethically, this suggests, of course, that the wealthy should lead in the clean-up, for it is easy to recognize the suffering of many peoples in various sites due to climate change, despite the fact that they were not complicit in major polluting activities. While such identifications and critiques have served to alert the public of the complicity between affluence and the health of the planet, William Connolly (2017) warns against a simplistic understanding of planetary systems as smooth and uneventful before human intervention. This is not the case, as evident in the violent ruptures within non-human events throughout the life of the planet (volcanic action, meteors, extinction events, etc.). These kinds of planetary events continue in concert with increased human activities, resulting in today's complexities of climate change. As Connolly concedes, multiple names for the degraded planetary situation that has resulted from the human footprint are acceptable, as long as concern and action are the results (2017). Of greater issue than the exact naming of our situation, however, is the way that the identification of these complex systems and human destructive forces as the Anthropocene have had unintended negative effects as forms of public pedagogy. Politically, there are important considerations regarding this rise of the Anthropocene.

Firstly, the development of the global "human" united in its potential destruction oversimplifies a complex situation. Thus, questions occur about how this popularity of the term, Anthropocene, has created a version of unified humanism, whereby,

> ...the very notion of "the human" (however varied, diverse, or multicultural) is the false commonality from which other violent captures have emerged, including the notion of "the planet" the "we" must save. A change in critical climates would require situating the human as one of the effects of a broader milieu of multiple, conflicting, and diverging forces, where "the human" both occupies a position of seeming universality and appears as an odd parochial exception. (Cohen and Colebrook 2017, p. 133, quotation marks in original)

The misguided perception of the Anthropocene identified here is an unhelpful simplification that further separates the "human" from the rest of the complexities of the planet. Based upon species survival, such as Chakrabarty (2009) has outlined, theorists warn that this perspective opens the door to institutional action that impedes the voicing of multiple, contradictory positions that is the hallmark of democratic process (Colebrook 2016a; Dillon 2007). The tone of urgency about the destruction of the planet gives a popular acceptance for governments to act unilaterally, or as Colebrook warns (2016a, p. 87), "states of emergency seem to call for a suspension of the free reign of opinion along with the resurgence of authority". Secondly, the identification of the human enacting a form of stewardship for survival of the species leaves a problem whereby the Anthropocene is "insufficiently capable of capturing all the differences among humans, while also creating too much difference between 'the human' and its others", (Colebrook 2016b, p. 446). Thus, the Anthropocene, in this simplified understanding that seems to frequently be taken up in the public imaginary, flattens out all possible difference, both politically and morally, into the narrative of fail and fix, whereby either rich humans have failed and must fix the planet for sustainability, or we all, as humans, should band together to survive. Either way, humanity is placed in a privileged position with respect to the rest of the planet and our collective blindness to the complexities of contemporary life sustains a glitch-Anthropocene in the making. It erodes the posthuman efforts in recent past to dethrone the human and to build new, more symbiotic relations with living and non-living entities.

THE POSTHUMAN

The posthuman is politically and socially significant because it questions the tendencies of humanism which give licence to this managerial model of utility instead of one of mutual synergy. In using a trope of porous boundaries, a Braidotti (2013) theory of the posthuman focuses attention on the ways that the human, politically and historically coded as white and male, misunderstands or ignores the position that the subject is ontologically and physically entwined with the machine, the animal and the Earth. The posthuman offers a critique of Western, unified notions of individuality, rationality, and neoliberalism. In this, Braidotti is similar to other posthuman scholars such as Hayles (1999), Herbrechter (2013), Nayar (2014), and Wolfe (2010). Posthuman perspectives challenge

Anthropocentric dominance by contesting the unifying trend and pressures in the ways that the Anthropocene has been taken up in society.

Challenging such humanism, Noah Gough (2015) urges for an awareness that we are, at any given moment, co-constituted by the non-human materials within our milieu. Specifically, as posthumans, we are as much influenced and dynamically formed by the materials of our environment as we influence them. Thus, posthuman theory suggests *becoming with* the world. This echoes ethical, aesthetic, and political considerations as to how humans relate to non-humans as equals as if, "technological and natural materialities were themselves actors along side and with us" (Bennett 2010, p. 47). Consequently, the move from human to posthuman within discussion of the Anthropocene has an ethical importance. Braidotti (2013, 2017) calls for an ethics formulated on the subject as polyvalent, connected to multiple animal, inhuman and human connections. Her use of the term *zoe* references this ontological and physical link among multiple entities of life/non-life. She notes, "*Zoe* . . . is the transversal force that cuts across and reconnects previously segregated species, categories and domains. Zoe-centered egalitarianism is, for me, the core of the post-Anthropocentric turn: it is a materialist, secular, grounded and unsentimental response to the opportunistic trans-species commodification of Life that is the logic of advanced capitalism" (2017, p. 32). Rather than a rational, planned, structural fix based on the limited humanist version that I have described as the glitch-Anthropocene, the posthuman preference for chance, unknowability, and emergence can result in multiple, unanticipated connections and relations among entities.

This awareness has been sought in various discussions throughout feminist, new materialist thought with an acknowledgement of the legitimacy of performativity as a means by which we live materially with the world (Barad 2007; Bennett 2010). The concept of the posthuman could lead to greater ecological change than the idea of the Anthropocene as it has generally been understood so far. This is so because of the porous nature of the posthuman theorized as ontologically connected to cyber-animal-vegetal-mineral life instead of maintenance of the Anthropocene as human manager; recalcitrant and sorry, but in charge of the planet—in this case, in order to fix things. This Anthropocene gives little credence to the fact that we actually share this planet with multiple human and non-human entities (jagodzinski 2017). This blindness is why I consider the Anthropocene glitched: broken. A shift in awareness and understanding is necessary to instil changes in approaches that mirror the relationality of the posthuman. The posthuman is one alternative path towards how Colebrook suggests "one might superimpose a thousand tiny Anthropocenes, or all the lived and unlived potentialities of the Earth" (Colebrook 2016b, p. 449). With respect to the machine and the human, Katheryn Hayles' (1999) focus on the dissolution of corporeal and cybernetic boundaries is relevant. As a nod to the cyborg that echoes back to Donna Haraway's (1985) seminal essay on the topic, subjectivity takes on a kind of multiplicity in its relationality by focusing upon this porous nature of boundaries among species and materials. Hayles writes, "When system boundaries are defined by information flows and feedback loops rather than epidermal surfaces, the subject becomes a system to be assembled and disassembled rather than an entity whose organic wholeness can be assumed" (1999, p. 160). By creating an effective glitch within the human/machine, these flows can be made prominent for consideration.

GLITCH AESTHETICS

One of the first uses of the term is by musician, Kim Cascone (2000), in his exploration of digital music where glitches in output were intentionally created. Glitch art is now firmly part of the cultural scene, with artists working on various online platforms. Within the glitch community, the targets and methods of glitching vary, including the manipulation of software, and/or hardware, often with the insertion of information into existing code.[5] With respect to networks and their glitch failures, art educator, Robert Sweeney (2015) describes three common strategies of glitch instigation: overloading a system by adding to it; creating noise through additions in the network; or a break of some sort in the network. Artist and blogger, Mallika Roy (2014) also lists a number of particular strategies for manipulating software and hardware, including the manipulation of non-computerized devices such as cameras. Whatever the strategy used, the significance of glitch art is not necessarily on the methods utilized but the effects of such glitches on audiences/users. Glitch art is gaining popularity through the work of artists such as Rosa Menkman. In her online work and manifesto, *Moment(um)*, she envisions the glitch as a form of powerful inhuman force:

A glitch represents a loss of control. The 'world' or the interface does the unexpected. It goes beyond the borders of its known and programmed territories, changing viewers' assumptions about technology and its assumed functions...and it comes to seem profoundly irrational in its 'behaviour'. The glitch makes the computer itself suddenly appear unconventionally deep, in contrast to the more banal, predictable surface-level behaviours of 'normal' machines and systems. In this way, glitches announce a crazy and dangerous kind of moment(um) instantiated and dictated by the machine itself. (2011, p. 41)

While Menkman anticipates that these failures in the computational systems will bring a critical awareness to the user, it is in the audience interpretation that the success of such intentions might be measured. There is never a conclusive result to be born from an intervention such as glitch art (Betancourt 2016). On the one hand, glitch is a way to pause and take note of our human/non-human condition, but on the other hand, glitch feeds into the very system it proposes to denounce, through its inclusion in the late-capitalist art machine. Although glitch is hot and heralds the contemporary postdigital moment (Berry and Dieter 2015), I remain unconvinced that glitch aesthetics could instigate the kind of social changes that might posthumanize the Anthropocene— perhaps because of glitch art's trendiness, or perhaps because my perspective is somewhat jaded after years of observing how the art world gobbles up every resistant strain into its gaping cultural, late-capitalist maw. Menkman voices a similar concern when she writes, "I wanted to move glitch artists beyond these burgeoning conservative impulses into rethinking and expanding out from the standardizing and only aesthetically engaging forms of glitch." (2011, p. 50).

It is worth the effort, then, to explore whatever capabilities may be hidden in the unexpected nature of glitch that might yet be effectual. Glitch, as inhuman agency, continually escapes from control of the human touch. A way forward might be found in Mark Nunes' (2011) tracing of the genealogy of error, where he asks, "Would we not find error its double sense: not only the predefined misstep, but also a going astray that poses a challenge to purposive intent? In doing so, we would be establishing the place of error – within and exterior to the programmatic - as a kind of real, yet excluded possibility..." (p. 9). Significantly, Nunes takes the glitch beyond the notion of a critique, and, following Deleuze, traces the ways that a glitch can be an opportunity for avoiding the apolitical formalism that much glitch art reproduces. As jan jagodzinski notes,

The glitch is the ontological response of technology in the sense that it is not controlled by programmers, and (theoretically) presents the case of technology as a sentient agency in-itself – that is, when such agency cannot be equated as "human" in any remote way. This is the problematic of glitch aesthetic: raising this ontological difficulty. This inhuman "agency" does not disappear simply because such agency is always "beyond" – whether in terms of machinic language or speed and/or glitch. (personal conversation, 2017)

The notion of inhuman agency is amplified by the glitch before actualization through art and/or artistic intention. The glitch must be understood artistically as posthumanly working alongside the artist, rather than as material/process controlled by the artist. In this way, it may be possible to generate conditions that simply act as creative outlet from within which the glitch would expand and generate. This inhuman promise of the glitch to unhinge rather than become yet another depoliticized art style is articulated by Menkman when she states,

Just as the understanding of a glitch changes once it is named, so does the notion of transparency or systemic equilibrium supposedly damaged by the glitch itself. The "original" experience of rupture is moved beyond its sublime moment(um) and vanishes into a realm of new conditions. The glitch has become a new mode; and its previous uncanny encounter has come to register as an ephemeral, personal experience of a machine. (2011, p. 31)

It is this inhuman agency within the glitch that can only be supported through a kind of art that remains open to being led by the glitch itself, rather than the artist procedurally "using" the glitch for a set intention. So too, the Anthropocene, if lead by the glitch, might first instigate our recognition of a restricted, global humanism, and second, instill our submission to the planet through a coexistence among multiple entities. Eivind Rossaak (2016) echoes this aptitude of the glitch to shape or change normative systems when he wonders, "Perhaps error is the excluded possibility between the virtual and actual" (2016, p. 230). As glitch art continues to morph, it is this elusive quality that may affect a different culture and life in late-Anthropocene times. For example, rethinking and experimenting around the issue of electronic waste, as Benjamin Gaulon,[6] has done. The virtual potential in such inhuman agency is unpredictably actualized in multiple forms through various methods of hacking.

Glitch art has expanded from strictly online environments to gallery spaces or amalgams of online and physical spaces that are under threat from human activity (Ellsworth and Kruse 2012). This aligns with a move from the digital to postdigital. While glitch art is associated with networks, computers, and everything "digital", the terms digital and analogue have been put under scrutiny. Using the term *postdigital,* David Berry and Michael Dieter (2015) make the argument that we have moved beyond the *new* in media art and are in a phase of expansion of the many varied instances where the digital is entwined with the analogue. The use of the prefix, *post,* reflects this expansion, rather than merely representing a sense of the chronological. Along these lines, in our everyday understandings, digital and analogue are typically viewed as distinct from one another. However, Florian Cramer (2015) argues that although popular culture still divides digital from analogue by the criterion of computerization, in fact, this division is too simplistic. Cramer identifies any whole unit that is divided into discrete, smaller parts as digital. Analogue, on the other hand, is that which cannot be divided or counted. He gives the example of musical notation being digital and the directions as to how to play the piece as analogue (e.g., crescendo). Robert Sweeney (2015) concurs with this description, and from the perspective of cybernetics, he describes anything that is computational, but not necessarily computerized, as digital. Importantly, Cramer (2015) emphasizes that the digital and analogue exist in enmeshed and interdependent states. This entanglement can be pushed further, as jagodzinski suggests, through the joining of computation and the organic in much creative exploration of wetware (biologically based) and dryware (computer-driven actions and/or concepts): in other words, the postdigital may well reside in the "blurring of dry and wet ...where... analogue and digital occurs in its most profound sense regarding life and the question between zoe and bios" (jagodzinski, personal conversation, April 10, 2017). Such explorations include the work of many artists, such as Eduardo Kac, who use organic micro/macro materials within bodies, plant systems or the planet itself in combination with a computational component.[7] With an opening of a postdigital glitch to embrace the computational (whether computerized or not) and analogue in approaching implications of the Anthropocene, glitch art can locate where, "the gaps and fissures in systems of control provide for possibilities of creative, unintended consequences" (Nunes 2011, p. 14). Within the organic/inorganic tangles in current conditions of the world, the work of Deleuze and Guattari (1983) theorizes these kinds of disjuncture that can extend towards and support such virtuality.

DISJUNCTIVE SYNTHESIS

In their opus, *Anti-Oedipus*, Deleuze and Guattari (1983) conceptualize disjunctive synthesis as one of three syntheses that constitute the unconscious. While well beyond the scope of this paper to detail the force of the theoretical and practical implications of all three areas, I touch on the second one: disjunctive synthesis. Through its abilities to disrupt an additive notion of the psyche, the disjunctive synthesis breaks tendencies towards repetition and instead, difference and the variable are introduced; thus, the result is the possible creation of an open and flexible subject (Deleuze and Guattari 1983; Holland 2002; Nakib 2013). Instead of remaining in a relationship between two choices, (one or the other), Deleuze and Guattari conceive difference through an,

> unknown force of the disjunctive synthesis, an immanent use that would no longer be exclusive or restrictive, but fully affirmative, nonrestrictive, inclusive. A disjunction that remains disjunctive, and that still affirms the disjoined terms, that affirms them throughout their entire distance, *without restricting one by the other or excluding the other from the one*, is perhaps the greatest paradox. 'Either ...or...or,' instead of either/or. (1983, p. 76, italics in original)

In the disjunctive, the glitch breaks from its capture as a piece of art within systems of digital communication, and the glitch-Anthropocene breaks from its narrowing humanism as one-who-fixes-the-planet. Thrown in together, these potential series can mix in multiple combinations, and I pull from this tendency to temporarily synthesize aspects of both into my art. As multiple series of differences within and between, in this case, the Anthropocene and glitch art, there can occur an opening up of lines of flight about what it means to *fail or fix or... form.... or free....* or.... This is not a call to replace one way that these two terms relate with another, constrained relationship; additionally, tapping into such multiple differences does not erase the singularity of each term (Nakib 2013). I suggest that each such series can be considered simultaneously in both singular and multiple ways so that focus expands thoughts and elements about both that result in space and time of delay:

a pause that thwarts the tendency towards signification and identification. This space is critical in the creation of that which is newly thought. Within this delayed time, previously unimagined articulations can unfold into the ecological, artistic, political, cultural, and so on. This does raise the question of multiple scales of time; the human time, the geologic time, the nanosecond of electron flight; all of which add further complexity to this disjunctive synthesis.

GLITCH-ANTHROPO-SCENIC LANDSCAPE ART: A PROPOSAL

Title: Slow Motion Electric Chiaroscuro

Inspiration: Scales of deep time, or nano-time or human time or glitch or Anthropocene or Landscape painting or the electron or contradiction or disjunction or environmental degradation, or...

Description: In this project, the glitch is not sudden but occurs slowly within the structural system of the electrical grid. This landscape art proposes that the supply of electricity browns out a chosen location very slowly—as slowly as the setting of the summer sun in northern climates. This browning and then final blackout of power is triggered, not by the rotation of the Earth (as with the setting sun), but by the frequency of algorithmically-identified words collected from social media. In this example, accumulations of the phrase "carbon emissions" is targeted, along with the word "electricity", using a trolling algorithm.[8]

As the collected data increases, an inversion occurs so that the electrical current decreases. A full blackout happens when the number of occurrences of the targeted words on Twitter reaches an algorithmic equivalency of the amount of carbon dioxide gas that has been emitted in the previous 24 hours by the fossil-fueled turbine that runs the electrical generating plant for that particular locale. As a loosely designed feedback loop, the CO_2 levels dictate the point at which the electricity shuts off completely, and thus creates a full blackout. This cycling continues to monitor the CO_2 levels and when there is a change of .5 ppm measured in the location of the turbine, it triggers the key pairing of recovery words, *carbon/tax,* that then generates the reverse effects and gradually the power returns. Announcements regarding the CO_2 emissions and their connection to the flow of electricity would coincide with the slow glitch. As soon as the blackout occurs, people would be free to load their Twitter accounts with comments using the recovery words of *carbon* and *tax.*

The CO_2 is then monitored for another 24 hours and this cycle continues, indefinitely, or until intervention from within the system reboots the programming.

A digital chiaroscuro is the result, as the electronic "shadows" change with the changing flow of electricity. Shadow is not theorized here as a visual phenomenon. Similar to the light and dark areas characterizing chiaroscuro painting, which is known for dramatic lighting contrasts, I consider the continual movement of electronic shadow as a fluctuation of slow depletion and regeneration of electrons. Inspired by Michelangelo Caravaggio's chiaroscuro effects,[9] the electronic shadows produce an increasingly dramatic duration of disjunction in which multiple unexpected events/thoughts could happen. This may first be noticed in questions regarding the stability of the electrical grid, followed by comments made on Twitter about carbon tax.

I acknowledge that this project holds a metaphorical undercurrent of colour and light, and therefore, it references the sense of sight, as does all landscape painting. Furthermore, in artistic media practices, applications of data mining are typically visually and/or aurally synthesized.[10] In this glitch landscape art, I leave the senses behind as much as possible. The flow of electricity is beyond human perception until it is transferred into a physical manifestation and sensed as light or heat. My interest in this chiaroscuro glitch landscape art is not primarily in this manifestation, but in the ways that this synthesis of multiple components from within the Anthropocene and glitch art assemblage link to/with the world. It is not about visualizing the land. Instead, it actualizes the unseen force of the electron that inhabits the land, alongside myriad other forces/materialities (including (in)human/animal/cyber/bodies).

The change of glitch from its typically quick nature into a slow emergence that gradually affects our affluence is similar to the slow boiling of a frog that does not jump out of the pot. Allegorically, this refers to our chosen Western lifestyles in this current time of the polluted world. Soon it may be impossible to get out of the hot water. This stasis (and in some cases, ignorance) parallels the limitations of human abilities to conceive geologic and electronic scales of time, "where human perceptions of time are challenged" (Tyszczuk 2016). This slow electronic fading and then darkening from the waning/waxing of power takes place on a scale of the Earth's time with respect to the levels of CO_2, but also uses the very quick speed of electron movement. I focus on this convergence of these two speeds: the nanosecond and geological time that emphasize our limited human awareness of the world we share and, as Colebrook (2016c) points out, always have shared. Both scales are beyond human discernment and may also be beyond each other's understanding (one can only presume this is so). Of interest is the way that this art taps into imaginative *"relation of non-relation"*, that Levi Bryant (2011, italics in original) identifies between two series within Deleuzo-Guattarian disjunctive synthesis, where the two "do not converge yet somehow manage to communicate by virtue of a *difference* that passes between them like a spark". Both timescales expose our human limitations and dependencies. We do not notice geological time because it is too slow, nor nano seconds because that scale is too fast. However, when they mash together through the attrition of power from the grid, we may begin to sense the limitations of human Anthropocentric time. We may begin to notice other times of non-living and living entities. This may be the affordance of this slow glitch, in the spirit of Davis and Turpin's description of ideas on art and the Anthropocene as "operating as a conceptual centrifuge for further speculation and future action" (2015, p. 3).

It synthesizes multiple scales of time in localized physical places that are linked globally through a social media platform. This slow glitch reaches into alternative images of thought as these timescales converge. The limitations of our humanity can be pedagogically humbling.

CONCLUSION

I began this chapter with a challenge to deterritorialize the Anthropocene further than it has typically been understood within the public sphere, following Colebrook (2016b). I mobilize the concept of disjunctive synthesis from Deleuze and Guattari (1983) as a means by which to consider glitch aesthetics and the late-Anthropocene. From the disjunctions that emerge, I take the Anthropocene and "post" it through "glitch", sieving it through concepts of the postdigital and posthuman. The "post" does not refer to linearity but to an infinite expansion of permutations of terms. Using these two areas as inspiration, I rive the theoretical rigidity within the binding causality of *fail and fix* and respond with my artwork, *Slow Motion Electric Chiaroscuro*. In a cycle of repetitive failure and its immaterial "labour" in the form of Tweeting towards an (almost) fix, an unpredictable potential within the convergence of glitch art and the Anthropocene is actualized in local places, and decodes the normative relationship of utility between electricity and the human.

As a form of public pedagogy, it points towards a posthuman life lived in accordance with electricity—and without it—given that about a billion of the global, human population is always in blackout because they have no access to electricity in the first place (World Bank 2017). Human and non-human relate in the artwork in ways that cannot be anticipated; but always with a nod to issues of inequity. As Donna Haraway (2016, p. 4) poetically puts it, "we require each other in unexpected collaborations and combinations, in hot compost piles". As a thought experiment in glitch landscape art, the posthuman is linked to the Anthropo-scenic, and while it does not offer any definite solutions to the overuse of power within the West, through a consideration of timescales and limitations, it can open capacity for rethinking our complicity in this assemblage. Such a focus may have credence for alternative modes of becoming, and might ultimately generate social change with respect to climate change.

Notes

1. This term is a combination of producer and consumer. In digital environments, one uses and produces at the same time. For more detailed explanations of this immaterial labour, see, https://en.wikipedia.org/wiki/Prosumer.
2. As a form of public pedagogy, see, for example, this richly visual example that captures an affect in an almost antithetical way of expressing the destructive realities that it proposes. http://www.anthropocene.info/.
3. There are now outlets whereby one can purchase a reproduction of a screenshot glitch. As glitch artists become better known, the market for their images will continue to grow. Also related to the purchase of reproductions (posters) is the rise of DIY glitch applications that one might purchase (see for example Joseph Gordan's app Glitch Art—Glitch Effect and Trippy Effects Editor https://itunes.apple.com/us/app/glitch-art-glitch-effect-trippy-effects-editor/id1024492593?mt=8 or social media sites where one posts DIY creations). In many of these cases, the aestheticizing of glitch art has depoliticized its potential (https://www.reddit.com/r/glitch_art/).
4. At the time of this writing, I am listening to Donald Trump withdraw from the Paris Accord on climate change, citing that it is an economically bad deal for America. While this superpower and second-highest polluter at this time steps away from the global table of leaders, we enter into an unanticipated future where ideas about the Anthropocene may shift dramatically (Liptak and Acosta 2017).

5. For a general introduction to the most common methods of creating glitches see Michael Betancourt's (2016) glossary in his book, *Glitch Art in Theory and Practice*. Also review artist and blogger, Mallika Roy's descriptions of glitch art, viewed from a position firmly entrenched in art and its institutions, http://www.theperipherymag.com/on-the-arts-glitch-it-good/.
6. Benjamin Gaulon (2017) is a French artist who investigates and creates opportunities for the public to question normative associations and activities with respect to technology. See his work where he initiated the Recyclism Hacklab, a collaborative workspace that provides mentoring for people who are interested in aspects of electronic life, including gaming, hacking, programming, etc. http://hacklab.recyclism.com/. Other research called *RandomMe CCTV Randomizer* where Gaulon investigates the ubiquity of surveillance cameras in our lives, http://www.recyclism.com/randomme.php.
7. Wetware has expanded into all areas of bio/digital exploration and while it is beyond the scope of this paper to give detailed accounts, a general introduction into these artistic practices, including a number of artists whose work combines these postdigital phenomena, can be found here: http://www.arts.uci.edu/press-room/beall-center-art-technology-presents-wetware-art-agency-animation. https://uniavisen.dk/en/a-new-alchemy-biotechnology-as-art/. Also see the work of Eduardo Kac who has been instrumental in art that combines complex systems from technology to biology. Here is a recent interview with Jens Andermann and Gabriel Giorgi (2017), in the Journal of Latin American Cultural Studies (online) where Kac considers bio art from its inception and its relationship to philosophy and science. Available from http://www.tandfonline.com/doi/full/10.1080/13569325.2016.1274646?scroll=top&needAccess=true.
8. Analysing big data from social media is a common use in research and artistic circles and a number of software applications such as MAXQDA, http://www.maxqda.com will accomplish this task. Other data collection/manipulation applications such as open frameworks is useful for creative applications, http://openframeworks.cc/about/.
9. Caravaggio was a seventeenth-century Italian painter who was famous for his painting of dramatic Biblical scenes using chiaroscuro, which is a strategy of contrasting dark and light areas in a painting from dramatic effect introduced by Leonardo da Vinci. For more information, see Encyclopaedia Britannica, https://www.britannica.com/art/chiaroscuro.
10. I include the work of Manovitch (2016) and his area of Software Studies. See http://manovich.net/.

Image 12.1   Gas cans (Mia Feuer, watercolor)

REFERENCES

Andermann, J., & Giorgi, G. (2017). We Are Never Alone: A Conversation on Bio Art with Eduardo Kac. *Journal of Latin American Cultural Studies, 26*(2), 1–19.

Anderson, K. (2015). Ethics, Ecology, and the Future: Art and Design Face the Anthropocene. *Leonardo, 48*(4), 338–347.

Barad, K. (2007). *Meeting the Universe Halfway: Quantum Physics and the Entanglement of Matter and Meaning.* Durham, NC: Duke University Press.

Bennett, J. (2010). A Vitalist Stopover on the Way to a New Materialism. In D. Coole & S. Frost (Eds.), *New Materialisms: Ontology, Agency and Politics* (pp. 47–69). Durham, NC: Duke University Press.

Berry, D., & Dieter, M. (2015). Thinking Postdigital Aesthetics: Art, Computation and Design. In D. Berry & M. Dieter (Eds.), *Postdigital Aesthetics: Art, Computation and Design* (pp. 1–11). New York, NY: Palgrave Macmillan.

Betancourt, M. (2016). *Glitch Art in Theory and Practice: Critical Failures and Post-digital Aesthetics*. New York, NY: Routledge.

Bignall, S., Hemming, S., & Rigney, D. (2016). Three Ecosophies for the Anthropocene: Environmental Governance, Continental Posthumanism and Indigenous Expressivism. *Deleuze Studies, 10*(4), 455–478.

Braidotti, R. (2013). *The Posthuman*. Cambridge, UK: Polity Press.

Braidotti, R. (2017). Four Theses on Posthuman Feminism. In R. Grusin (Ed.), *Anthropocene Feminism* (pp. 21–48). Minneapolis: University of Minnesota Press.

Bryant, L. (2011). *Love*. Retrieved from https://larvalsubjects.wordpress.com/2011/05/19/love/.

Cascone, K. (2000). The Aesthetics of Failure: "Post-digital" Tendencies in Contemporary Computer Music. *Computer Music Journal, 24*(4), 12–18.

Chakrabarty, D. (2009). The Climate of History: Four Theses. *Critical Inquiry, 35*(2), 197–222.

Clark, N. (2014). Geo-Politics and the Disaster of the Anthropocene. *The Sociological Review, 62*, 19–37. https://doi.org/10.1111/1467-954X.12122.

Cloninger, C. (2010, October). *GltchLnguistx: The Machine in the Ghoat/Static Trapped in Mouths*. Retrieved from http://lab404.com/glitch/.

Cohen, T., & Colebrook, C. (2017). Vortices: On "Critical Climate Change" as a Project. *The South Atlantic Quarterly, 116*(1), 129–143.

Colebrook, C. (2016a). What is the Anthropo-Political? In T. Cohen, C. Colebrook & J. Hillis Miller (Eds.), *Twilight of the Anthropocene Idols* (pp. 81–125). London: Open Humanities Press.

Colebrook, C. (2016b). 'A Grandiose Time of Coexistence': Stratigraphy of the Anthropocene. *Deleuze Studies, 10*(4), 440–454.

Colebrook, C. (2016c). Time That Is Intolerant. In S. Groes (Ed.), *Memory in the Twenty-First Century: New Critical Perspectives from the Arts, Humanities, and Sciences* (pp. 147–158). London, UK: Palgrave Macmillan. https://doi.org/10.1057/9781137520586.

Colebrook, C. (2017). We Have Always Been Post-Anthropocene: The Anthropocene Counter-Factual. In R. Grusin (Ed.), *Anthropocene Feminism* (pp. 1–20). Minneapolis: University of Minnesota Press.

Connolly, W. (2017). *Facing the Planetary: Entangled Humanism and the Politics of Swarming*. Durham, NC: Duke University Press.

Cramer, F. (2015). What is "Postdigital"? In D. Berry & M. Dieter (Eds.), *Postdigital Aesthetics: Art, Computation and Design* (pp. 12–26). New York: Palgrave Macmillan.

Crutzen, P. (2002). Geology of Mankind. *Nature, 415,* 23.

Crutzen, P., & Stoermer, E. (2000). The Anthropocene. *International Geosphere-Biosphere Programme Newsletter, 41,* 17–18.

Davis, H., & Turpin, E. (Eds.). (2015). *Art in the Anthropocene: Encounters Among Aesthetics, Politics, Environments and Epistemologies.* London, UK: Open Humanities Press.

Deleuze, G., & Guattari, F. (1983). *Anti-oedipus: Capitalism and Schizophrenia* (R. Hurley, M. Seem, & H. R. Lane, Trans.). Minneapolis: University of Minnesota Press. (Original Work Published 1972).

Dillon, M. (2007). Governing Terror: The State of Emergency of Biopolitical Emergence. *International Political Sociology, 1,* 7–28.

Ellis, E., Maslin, M., Bolvin, N., & Bauer, A. (2016). Involve Social Scientists in Defining the Anthropocene. *Nature News, 540,* 7632. Retrieved from http://www.nature.com/news/involve-social-scientists-in-defining-the-anthropocene-1.21090.

Ellsworth, E., & Kruse, J. (2012). *Making the Geologic Now: Responses to Material Conditions of Contemporary Life.* Brooklyn, NY: Punctum Books. Retrieved from http://www.geologicnow.com/.

Gaulon, B. (2017). *Philosophy: Critical Making.* Retrieved from http://hacklab. recyclism.com/.

Gough, N. (2015). Undoing Anthropocentrism in Educational Inquiry: A Phildickian Space Odyssey? In N. Snaza & J. Weaver (Eds.), *Posthumanism and Educational Research* (pp. 151–166). New York: Routledge.

Haraway, D. (1985). A Manifesto for Cyborgs: Science, Technology and Social Feminist in the 1980s. *Socialist Review, 15*(2), 65–107.

Haraway, D. (2016). *Staying with the Trouble: Making Kin in the Chthulucene.* Durham, NC: Duke University Press.

Hayles, N. K. (1999). *How We Became Posthuman: Virtual Bodies in Cybernetics, Literature and Informatics.* Chicago, IL: University of Chicago Press.

Herbrechter, S. (2013). *Posthumanism: A Critical Analysis.* London, UK: Bloomsbury.

Holland, E. W. (2002). *Deleuze and Guattari's Anti-oedipus: Introduction to Schizoanalysis.* London, UK: Routledge.

jagodzinski, j. (2015). Environment or Sustainability? *Psychoanalysis, Culture & Society, 20,* 84–85.

jagodzinski, j. (2017). A Critical Introduction to What Is Art Education? In j. jagodzinski (Ed.), *What Is Art Education for? After Deleuze and Guattari* (pp. 1–61), New York: Palgrave Macmillan.

Liptak, K., & Acosta, J. (2017, June 2). Trump on Paris Accord: 'We're Getting Out'. *CNN*. Retrieved from http://www.cnn.com/2017/06/01/politics/trump-paris-climate-decision/index.html.

Luke, T. (2015). Introduction: Political Critiques of the Anthropocene. *Telos, 172,* 3–14.

Manovitch, L. (2016). *Cultural Analytics Lab.* Retrieved from http://manovich.net/ and http://lab.culturalanalytics.info/.

McCormack, T. (2010, October 13). Code Eroded: At GLI.TC/H. *Rhizome*. Retrieved from http://rhizome.org/editorial/2010/oct/13/code-eroded-at-glitch/.

Menkman, R. (2011). *The Glitch Moment(um).* Retrieved from http://network-cultures.org/_uploads/NN%234_RosaMenkman.pdf.

Moore, T. (2016). Introduction: Anthropocene or Capitalocene? Nature, History and the Crisis of Capitalism. In J. Moore (Ed.), *Anthropocene or Capitalocene? Nature, History and the Crisis of Capitalism* (pp. 1–11). Chicago, IL: PM Press.

Nakib, M. A. (2013). Disjunctive Synthesis and Arab Feminism. *Signs: Journal of Women in Culture and Society, 38*(2), 459–482. https://doi.org/10.1086/667220.

Nayar, P. (2014). *Posthumanism.* Malden, MA: Polity Press.

Nunes, M. (2011). *Error: Glitch, Noise, and Jam in New Media Cultures.* New York: Continuum.

Rossaak, E. (2016). Who Generates the Image Error? From Hitchcock to Glitch. In B. Cohen & A. Streitberger (Eds.), *The Photofilmic* (pp. 217–232). Leuven and Belgium: Leuven University Press.

Roy, M. (2014). Understanding the Glitch Art Movement. *The Periphery.* Retrieved from http://www.theperipherymag.com/on-the-arts-glitch-it-good/.

Sweeney, R. (2015). *Disfunction and Decentralization in New Media Art and Education.* Chicago: Intellect.

Tyszczuk, R. (2016). Anthropocene Unconformities: On the Aporias of Geological Space and Time. *Space and Culture, 19*(4), 435–447.

Weinstein, J., & Colebrook, C. (2017). Introduction: Critical Life Studies and the Problems of Inhuman Rites and Posthumous Life. *Posthumous Life: Theorizing Beyond the Posthuman* (pp. 1–14). New York: Columbia University Press.

Wolfe, C. (2010). *What Is Posthumanism?* Minneapolis: University of Minnesota Press.

World Bank. (2017). *State of Electricity Access Report 2017* (Vol. 2). Full Report (English). Washington, DC: World Bank Group. http://documents.world-bank.org/curated/en/364571494517675149/full-report.

CHAPTER 13

Situations for Empathic Movement

Leslie Sharpe

TRACKING

Most of us sit at a distance, away from Arctic sea ice, away from the unmitigated harshness of it, away from the harsh reality of its shifting state from solid to liquid in growing amounts as ocean temperatures rise due to global warming. It has long been the home and habitat for northern people, a site of colonial exploit, and more recently is documented in numerous books and online sites as the major indicator of climate change.[1] In addition to the expanding literature on this subject, publicly accessible government and science websites such as the Canadian Ice Service or the American National Snow and Ice Data Center[2] contribute to the growing public awareness and concern about the impact of global warming on our planet. This awareness and narrative often play on continued romanticism about the north as a site for mapping colonial desire and as a source for non-indigenous land- and water-exploitation; as passages open up with less multi-year sea ice, non-indigenous traffic increases through its waters.

So, we may even find ourselves there as tourists, near the sea ice, overwhelmed as we pass it in the zodiac boats, wondering what is happening beneath the surface of the water where we see it. Will it pierce the

L. Sharpe (✉)
MacEwan University, Edmonton, Canada

© The Author(s) 2018   299
j. jagodzinski (ed.), *Interrogating the Anthropocene*, Palgrave Studies in
Educational Futures, https://doi.org/10.1007/978-3-319-78747-3_13

zodiac so we sink into the icy waters? Are we seeing the last remaining sea ice? We clamber off the boats onto the ice and feel a connection that goes beyond our presence at that time and place to a pre-colonial past and its continuation into the present situation that forces us to wonder about traversing the Northwest Passage and other locations without danger.[3]

TRANSMITTING PRESENCE

Meanwhile, some of us sit at a distance, not able to travel there, but still maintaining a connection of sorts with the North, on screen via telemetry. In addition to following the ice, one can follow tracked animals online through a wide range of publicly available sites, from World Wildlife Foundation's (WWF) tracking of endangered animals (polar bear, narwhal and bowhead whale in the arctic, or marine turtles, jaguars, and tuna in the southern hemisphere), to sites specializing in the migration of birds.[4]

While the WWF site relies on information gathered by scientists and transmitted to non-specialists to solicit political and economic support of their cause, sites such as *eBird.org* and *Audubon.org* present data and information on bird presence and migration gathered by citizen scientists. This latter site relies on specialist information and analysis as birds are temporarily tracked and given a round of tests to monitor health, and to see if they are on course. So, while specialists are necessary to observe, hear, catch, drug, tag, photograph, test, and/or release animals, these sites also rely on a brigade of citizen scientists who provide information on animal movement and health, by participating in counts of birds spotted at various locations on a specific date, or by sending in imagery taken at bird feeders or in the wild that present evidence of states and rates of various health conditions.

For very large parts of time and space traversed by the tracked animal, humans are left behind, traveling in a vast unknown as far as human direct contact goes, but open to the imagination as they are followed online remotely. As we follow the animals on screen and through transmitted data, we may grow fond of the animals, choosing the various 'characters' who interest us among these tracked beings (e.g., Iola, Lucky, Larry). We check back daily on their progress, watching their movement to consider what they might encounter on their routes, and why they might stray off-course or disappear from the screen, and perhaps to make sense of what we might be doing to them, why it matters to follow them (Fig. 13.1).

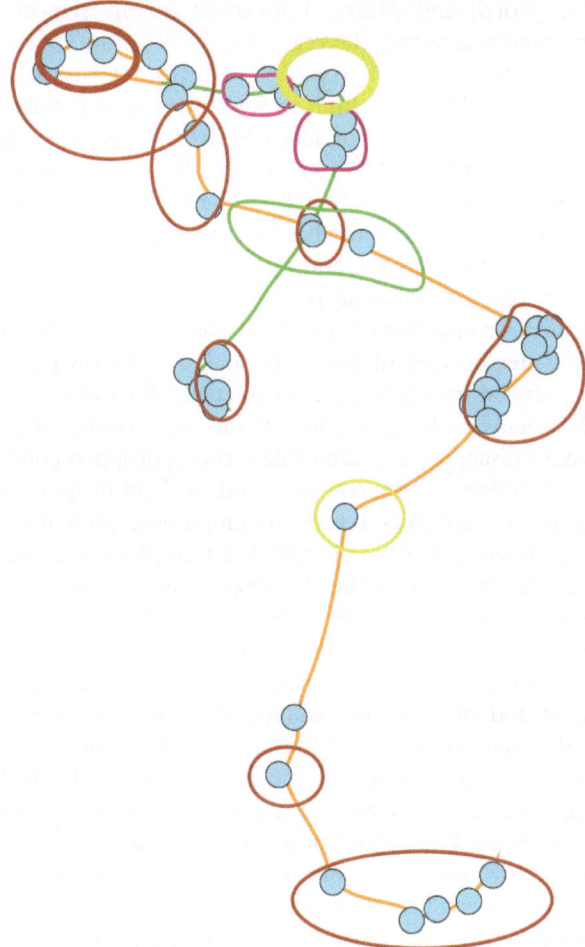

Fig. 13.1   "Iola Lines and Stoppages," digital image, 2010

My series "Situations for Empathic Movement" begins with *follow-ing*. Like so many others—scientists, citizen scientists, 'interested' pub-lics, I *follow the animals online*. Around 2008, I began to actively look for tracked animals online and eventually found myself exploring the WWF site as well as the Porcupine Caribou Management Board's site following caribou herds traveling through the arctic and subarctic of the Canadian North and Alaska. I followed up my remote following with brief moments of seeing the real thing in situ—during a visit to the Northwest Passage from Nunavut to Northwest Territories, observing narwhal being hunted by local Inuit in Resolute Bay, watching polar bear and marine animals from a boat just off Beechey Island, walking near caribou on Victoria Island, NWT, and years later tracking animal presence through prints and hair in Ivvavik in the Canadian Yukon. However most of my connection to animals has been through online following.

I connect to even more sites, and *follow*, more recently expanding this following with reading scientific papers on migration, or changes and threats to animal or bird habitat and health as a result of climate change, looking at a wider portion of the globe as I start following bird migrations. I get seduced into following 'characters,' for instance the cuckoo "Larry" who seems to be going far off his usual route, or the caribou "Lucky," whose disappearance indicated either a dropped collar or death. The following allows me to connect, and as I anthropomorphize the creatures in these narratives I tend to empathize even more. What I must remind myself as I follow is that I am employing a system that is every part of the creation of the Anthropocene as it is of the tracking of anthropogenic interruption; the telemetry used to track these animals and to convert their existences into a spread of data and information is a creation of satellite and information gathering that has been used as an aid to map and plunder the habitats that these animals live within. To observe the changed routes of animals is both symptom and evidence of the effects of their lives within the Anthropocene. Each disruption, each interruption, each aberration is a shift in what was a fairly predictable route, where we knew more or less certain risks at various points of the migration. Now, the not-knowing is predictable, and the following is necessary to discover and understand what is happening.

So, some of us sit at a distance, at a screen and *follow*. We see the animal's route as a tracked line on a flat map-plane, and with Google Earth, we might be able to discern the kind of terrain that animal follows. Is it traveling over ice, water, mountains, desert? If we do some armchair research, we might be able to see where the ice was last year when the WWF-tracked polar bear Noortje (N26241) traveled there, if it was multi-year pack ice or too thin to traverse, or if Larry the cuckoo flew over the desert or didn't bother. With luck, the tracking sites give us further information, for instance new things the animal is exposed to: human interventions (roads, pipelines, and building sites that have removed, interrupted, or displaced habitat, gmo- or monoculture-crops that have eradicated bugs and plants that are part of the animal's diet), changes due to extreme weather (solid ice sheets over snow that make it impossible to dig for food, flooding that destroys habitat with essential diet and protections, warming that invites predators, pests, and the diseases they carry farther north where non-migrating animals have no protection, or plant food that grows and disappears in the warmer weather before the animal reaches it). This latter knowledge helps us see that Larry and Iola and Nootje's lives are in peril, and the narratives that emerge from those stories are opportunities for empathic understanding of their plight.

As an artist following these animals, I wondered how I can contribute to sharing this knowledge set with my audience, so they too can follow, empathize, perhaps know for a moment, perhaps know enough to take a moment of action. So, several years ago, I went back online and imagined a scenario for a project.

First, Google for images of caribou (in this case, Google Images for "caribou movement in the north," or "barren-ground caribou," which is a species of caribou I have been following). A set of images of caribou moving will display onscreen, and if I look at them one by one, I am then:

>imagining the body
>imagining the body moving over terrain
>imagining the passage
>imagining lines of movement
>imagining shapes for protection
>imagining pests
>imagining predators
>imagining responses to objects in the land

In this way, I began to match the lines and stops of Iola's tracked migration route with particular movements as she encountered interruptions caused by humans, geography, weather, or animals, as she was:

>crossing over land, over snow, over ice
>crossing borders
>digging for lichen
>dodging black flies and mosquitos
>heading to calving grounds
>giving birth
>avoiding grizzlies
>crossing pipelines

>steering clear of hunters
>eating mushrooms
>rutting
>heading south
>running from wolves
>crossing rivers

I then began to create a series of walks and movements that one would do away from Iola—not only remotely from her geographically, but distant from the screen and the disembodied empathy it created. I imagined these walks taking place far away from Iola, in a rural landscape she was not built to survive in, or in an urban environment that was designed solely for human passage. In 2012, I traveled to Brazil and met choreographer and dancer Amy Burrell, a former grad student of mine who I had worked with before in USA.[5] We met in Brasilia and drove out to Pirenópolis, Goiás, where we pored over and discussed my plans along with photos and videos I brought of caribou movement, so she could understand how they might move in various situations.

I gave her the sets of 'imaginations' and 'scenarios' as above and we went out to the hills outside Pirenópolis, where I photographed her enacting some of the movements, to which she brought her own sensibility as a dancer and choreographer who understands the body in space. We then went back to Brasilia, where again we enacted the movements along the planned central area and buildings designed by architect Oscar Niemeyer. I photographed Burrell's movements for a series of photos and instruction cards to use back in Canada with an audience, for anyone to do anywhere, remote from Iola, in order to recall an experience of the caribou (Fig. 13.2).

In the main square of Brasilia, I directed Burrell to enact Iola's migration route as a walk based on a set of instructions and a gps-route track I created to help plan the walk. On my return to Canada, I showed several of the photographs at *Latitude 53 Gallery*, an artist's run gallery in Edmonton, and installed one of Iola's walk patterns on the wall. I also took advantage of their 'patio party' to engage the audience to follow the instruction cards which I printed on coasters for them to take with them, calling these "Situations for Empathic Movement."

Some of the cards explicitly referenced climate change, while others referenced the usual encounters a caribou might meet, like nose bot flies that get up your nose and lay eggs. One instruction card 'Dig' reads:

Situation: you are hungry but the extreme weather has made digging through heavy snow difficult to get at the lichen.

Movement: Dig and Claw. Keep on digging even if it tires you out, you have to get food or you will die.

Situation:
you've encountered a
human-made obstacle.

Movement: Choose one of
these obstacles
and choose how you will navigate
without encountering a person:
road, fence, pipeline

Situations for Empathic Movement
Leslie Sharpe
Latitude 53, August 2012

Situation:
waist high snow makes
lichen on ground hard to get at

Movement: dig, dig, dig.

Situations for Empathic Movement
Leslie Sharpe
Latitude 53, August 2012

Fig. 13.2   "Situations for Empathic Movement (detail)," digital images, 2012

The instruction cards call on the audience to imagine, and perform an action. One needs to imagine a self that is not us, yet we may have had a hand in its situation. By enacting the action in response to the situation, we might not only empathize with that animal's situation and imagine scenarios of relief and remediation. We might also place ourselves in those moments, fast-forwarding in time to the moment when we must respond to an interruption along our own paths—to a flood, to a heat-wave, to a shortage of water, to a flood of refugees, to an oil spill, to a toxic plume in the ocean, to an acidic sea, to a shortage of food, etc. Whether or not we bring it back to our*selves* at that moment—as specula-tive, responsible, or guilt-ridden—we can begin to wonder: what can be done, and decide on an action in response to a proposed scenario, espe-cially as we know that the imagined scenario is already upon us.

That scenario can be also imagined spatially if we look at the entire habitat range of an animal that is being affected by climate change. In my series (*Endangered Animal*) *Spaces as Nation*, I have created sev-eral prints derived from the tracked routes of animals that I have fol-lowed, including polar bears, caribou, narwhal, and various birds. In each image, the space of an animal's tracked movement covering their migration and habitat range from several days to several months or years, is conceived of as though it were a *nation*, to be mapped out with borders, sometimes placed near a mapped political nation for compari-son. For instance, in "Polar Bear Space as Nation," the shape of a sin-gle tracked polar bear's habitat range over several months reveals that it traveled over vast spaces of frozen ice in the Beaufort Sea between Alaska and Russia. Mapped out as a 'nation' in a map-like image, the habitat of the polar bear positions itself as a geographical place to be recognized and reckoned with. In previous talks about this work I have jokingly imagined Sarah Palin looking out of the governor's mansion in Alaska at Polar Bear nation, worrying that she might need to defend her state from invasion. Hilarity aside, we must understand that the border of the habitat range depicted is in a flux posing dangers for the polar bear's continued survival, for as warming waters melt Arctic sea ice sooner and in greater amounts, the polar bear is faced with enormous spaces of water to cross to reach food.[6]

Other prints in this series include my own imagery of following ani-mals or photographing their habitat, including my own tracking of animals and myself in Ivvavik, and gathering changing water habitat imagery from Google Earth, as I have done with "Narwhal Space as Nation," an image that shows a briefly tracked route of a single narwhal, floating above Google Earth water as an island that looks like it has been hacked and flooded at the edges (Fig. 13.3).

Fig. 13.3 "Narwhal Space as Nation," digital image, 2017

ART, CITIZEN SCIENCE, SCIENCE: SCREEN TRACKING
AND BIRD SIGHTINGS

In 2013, I was invited to create an on-site installation project as part of "Ramble in the Bramble," a transitory outdoor exhibition curated by Kristy Trinier at Whitemud Creek park in Edmonton, Alberta for the Edmonton Arts Council's public art program. I found a desire path with red markings on the trees that meandered into the woods from the main walking path. This path had several offshoots, including two stops at Whitemud Creek, a grove of trees, and a final exit at a culvert that flowed into the creek below a high cliff. I took a gps route of the track and created several images that were posted along the path showing one's location, with information from *eBird.org* on what birds were passing through that location over the season. If one followed the path, you could stop at two blinds I created to either view my signage of bird flight patterns across the creek, or to view a sculptural installation of cultural 'birds' that were 'nesting' in the culvert: stuffed birds I created from bird imagery on recycled t-shirts. An additional stop in the grove of trees held a sound installation I created of sounds of birds not normally found in the area. This project aimed to draw attention to site, to changes in the environment that were both created by humans and to observe occurrences in nature. The light pattern signs were of birds that flew through the area, so if one didn't see a bird, they could imagine their flight. This walking path, titled "Redpath," was up for the duration of the show, so that visitors could traverse the path, and through each element consider the birds that live or migrate through this part of their habitat, along with other aspects of the site (Fig. 13.4).

While walking in a park, we may alternately be in a state of reverie, companionship, or walking with agendas (exercise, relief of stress, solitude, hiding). Some walk in the woods with the intention of discovery, of themselves or something other.[7] The particular path I chose was likely carved out by city employees, now found in faded red spray marks along the trees leading across fallen logs that would disappear under new plant growth and then return, ending across from the cliff and culvert. This path was then appropriated by me and others for an aberrant and at times unpredictable, walk; at one time I found someone camping along the path, at another the path was flooded and regrew through the muck. I encountered the dregs of a party, an older couple having a tea break near my sound piece, and on the final weekend stones were thrown at the ceramic nests in the culvert, breaking several.

Fig. 13.4 "Redpath," outdoor installation (detail), 2013

When exploring off the main path, one's senses are acute. No longer guided with official signage and safe passage through plants or underfoot, you learn to look off to the side, above and below—in case of danger as much as in search of something new. You part branches while making sure they are not toxic or prickly, you lean under and step over and leap. You prepare to get messy and acknowledge that you are part of an environment. You listen and smell and see things that wanted to stay hidden. We seldom apply this same awareness to walking on a main park path, where we give into being led through selections of nature. This aware approach is that of a citizen scientist, such as a birdwatcher. In this way, they can discover and contribute to a well of information about species that help us know more about animal presence and changes to their habitat.

It was only through walking the path in this way that I realized I had to add a real bird-watching component, even if it was so late that I would gather a sample rather than compile a lot of useful 'data.' For the final days of the exhibition, I set up one bird blind as a bird-watching station, with a notebook at a blind to record what you see. I painted stripes on the blind to match species that were spotted, and created an online twitter feed so people could tweet their records of birds that they spotted in the region (Fig. 13.5).

Fig. 13.5 "Ramblebirds," Website and Twitter feed, 2013

The data I gathered was only a sample, but provided a valuable insight into what kind of participation I wanted in a 'participatory' artwork that reflected on environment. While the movement instructions for the caribou empathic walk provided a situation and a call for an 'action,' I realized that a work that provoked discovery and hinted at data-gathering held potential for others to imagine possible social engagement toward climate change, and am exploring that for future iterations of this work.

WATCHING THE BIRDS

While doing online research on songs of migrating and local birds, I found an image of a bird with a twisted and elongated beak. Normally, the little chickadee in the image would have looked like so many I see in my yard, flitting from branch to feeder to branch. But this online chickadee was compelling; it looked exotic and oddly beautiful, with its curving upper beak that turned under the lower one. Dangerous notes about its existence started to infiltrate the beauty; I wondered how it could catch any seed in that beak, how it could preen its feathers or feed its young. I clicked on the image. It turned out that all of those normal behaviors were difficult, if not impossible: as I read the scientific article detailing the symptoms and problems for birds with 'avian keratin disorder,' a still misunderstood illness that causes not only beak deformation, but other symptoms of declining health such as skin lesions, and loss of feathers.[8]

Further reading revealed the role that a concerned public could play in contributing to understanding the spread of the disease and symptoms among various bird species over a wide geographical range. Through citizen science, government scientists in Alaska were able to draw on photographs and reported data so that they could track the spread of the disease, and locate diseased birds. This mix of hard science and citizen science signaled to me that I could mix specialized high-tech approaches with non-specialized, DIY approaches in a new work that would imagine a woods full of birds who had contracted AKD, using 3D printed 'beaks' combined with recycled local wood and a sound installation.

I designed the 3D beaks in the software Rhino based upon an actual chickadee beak as well the various rates and examples of distortion in the studies of AKD, 3D printing and showing 20 of these at the 2015 sound art exhibition SONAR in Edmonton, Canada, and then in 2016 created 50 new beaks for the exhibition *BalanceUnBalance: Data Science and Eco Action* at Museo de Arte de Caldes in Manizales, Columbia. In each case, I used local wood that was gathered by me or regional assistants, and included a sound mix of chickadees and crows that I had recorded at a bird sanctuary near Edmonton. The sound mix began with chickadees singing but grew increasingly distorted and in crisis as the crows joined them, then rapidly falling to silence. For Manizales, these sounds were also accessible online through wall images of QR codes that the audience could access either in the gallery or download and play later on their devices (Fig. 13.6).

I have described the process and effect of this work as follows for the book *Exhibiting Sound*: "I decided the beaks had to be removed from 'bird-like' bodies and function as signifiers for missing birds—whether those were birds that had already contracted the disorder, or for those that would develop it in the future. So rather than phantom limbs, the beaks would be nagging reminders of phantom birds, and they would be situated within a 'woods' with a soundscape befitting of this tragedy."[9]

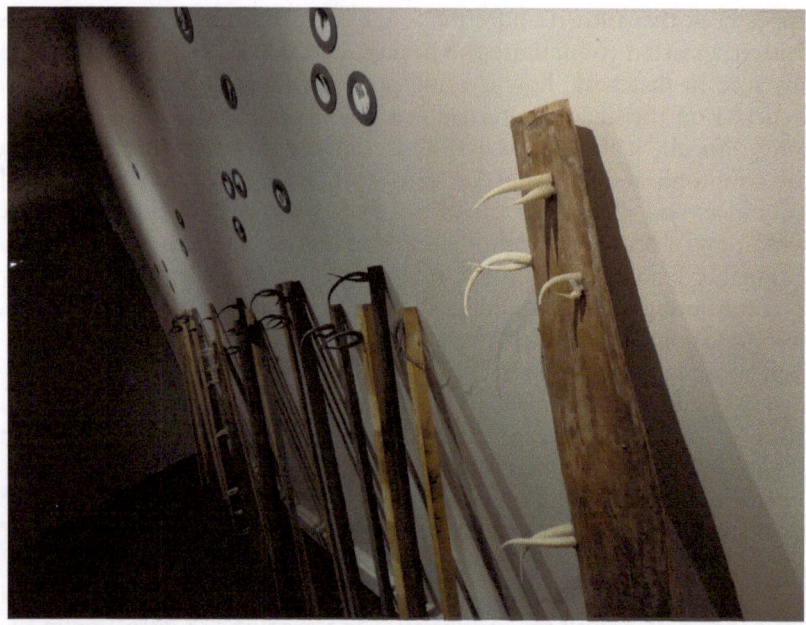

Fig. 13.6 "Beak Disorder," 3D printed beaks, found/recycled wood, and 2-channel soundwork (detail), 2016

In researching for "Beak Disorder" and my other works on migration and transmission, I am increasingly encountering anthropogenic habitat changes that will radically change the lives and existence of many animal and bird species[10] (as well as for indigenous people who have lived in and used these lands for centuries).[11] Researching for "Beak Disorder" brought me back again to the stories of animals who at times relay the crisis of their changed habitat to us not just as closely-watched neighbors, but also through their emergence as remotely tracked data, as imagined online characters, and also as the *disappeared*, inhabiting telematic, cultural, scientific, and memory spaces to remind us that within these real and telematic connections lies an urgency and a call for action and empathy, before the disappearance is irreversible. While a participatory or socially-engaged approach in the development or presentation of work critical of the Anthropocene offers the chance to engage an audience in a call to action (as documented by the critic T.J. Demos), the poetics or actions of the call need to be such that they embed reminders of what is at stake and at least hint at the urgency of the moment.

NOTES

1. For examples of recent writing on ice and climate change, see Shelley Wright, *Our Ice Is Vanishing/Sikuvut Nunguliqtug: A History of Inuit, Newcomers, and Climate Change* (McGill-Queens University Press, 2014), and Mariana Gosnell, *Ice: The Nature, The History, and The Uses of an Astonishing Substance* (New York: Alfred A. Knopf, 2005).

2. See *Canadian Ice Service* maps at https://www.ec.gc.ca/glaces-ice/ and *American National Snow and Ice Data Center* at http://nsidc.org/.

3. In 2008, I traveled through the Northwest Passage to photograph and record sounds for an art project, traveling by zodiac boat through sea ice.

4. I have followed tracked animals online since around 2008, starting with following the porcupine caribou herd being tracked in Canada and Alaska, WWF's tracking site: http://wwfgap.org/tracker/; and other sites including https://www.ebird.org, and https://www.audubon.org.

5. Burrell played a waking and walking 'polar bear' transformed to human in my 2010 performance project "Nostalgia and Myths of North."

6. I later created this work as a rya 'polar bear rug' with electronics for lights indicating two Arctic Dew-line stations, with portions of the white 'ice' melting into blue in the rug, exhibited at *Future Station*, the 2015 *Alberta Biennale of Art* along with photographs taken along the Northwest Passage of sites of human intervention during British colonial expansion, the cold war or with contemporary tourism.

7. Two recommended books on this subject are Herb Belcourt, *Walking in the Woods: A Métis Journey* (Canada: Brindle & Glass, 2006), and Sara Maitland, *Gossip from the Forest: The Tangled Roots of Our Forests and Fairytales* (London: Granta, 2012).

8. Caroline Van Hemert, Colleen M. Handel, and Todd M. O'Hara, "Evidence of Accelerated Beak Growth Associated with Avian Keratin Disorder in Black-Capped Chickadees (Poecile atricapillus)," *Journal of Wildlife Diseases* 48.3 (2012): 686–694.

9. Leslie Sharpe, 2016, "Boundary Crossings in Exhibiting Sound'" in *Exhibiting Sound* (Canada: McGill-Queen's University Press, Publication Forthcoming).

10. Habitats of birds that migrate to Central America and Africa are changing due to climate change and human activity, reducing the range of these spaces for a huge number of the worlds' migrating birds, as evidenced through scientific articles cited at *eBird.org*, *Audubon.org*, and published in numerous journals, e.g., *The Auk*.

11. In my contribution to Jane Marsching and Andrea Polli, eds., *Far Field: Digital Culture, Climate Change and the Poles* (UK: Intellect Press, 2012), I reference indigenous mapping of Nunavut's Lancaster Sound based on their land use, Inuit environmental knowledge which aided in creating a marine conservation area in that area.

PART IV

Pedagogical Responses

Image IV.1   Rotunda (Mia Feuer, watercolor)

CHAPTER 14

Against Climate Stoicism: Learning to Fight in the Anthropocene

Ted Stolze

In the first century of the Common Era, the manumitted Roman slave philosopher Epictetus famously distinguished between what we can control and what we cannot. In the succinct compilation of his teachings known as the *Enchiridion* or "Handbook," Epictetus argued that:

> On the one hand, there are things that are in our power, whereas other things are not in our power. In our power are opinion, impulse, desire, aversion, and, in a word, whatever is our own doing. Things not in our power include our body, our possessions, our reputations, our status, and, in a word, whatever is not our own doing. Now, things that are in our power are by nature free, unhindered, unimpeded; but things not in our power are weak, slavish, hindered, and belong to others. Remember, therefore, that whenever you suppose those things that are by nature slavish to be free, or those things that belong to others to be your own, you will be hindered, miserable and distressed, and you will find fault with both gods and men. If, however, you suppose to be yours only what is yours, and what belongs to another to belong to another (as indeed it does), no one will ever compel you, no one will hinder you; you will find fault with no one, reproach no one, nor act against your own will; you will have no enemies and no one will harm you, for no harm can touch you.[1]

T. Stolze (✉)
Cerritos College, Norwalk, CA, USA

© The Author(s) 2018
j. jagodzinski (ed.), *Interrogating the Anthropocene*, Palgrave Studies in Educational Futures, https://doi.org/10.1007/978-3-319-78747-3_14

Recently, Jules Evans has expanded on and refined Epictetus's list,[2] which we can visualize by using two lopsided columns:

Not in our control	In our control
Our body	Our beliefs
Our property	
Our reputation	
Our job	
Our parents	
Our friends	
Our co-workers	
Our boss	
The weather	
The economy	
The past	
The future	
The fact we're going to die	

Although this is a debatable interpretation,[3] Evans's point is clear: For Epictetus—and, by extension, for other Stoic philosophers as Musonius Rufus, Seneca, and Marcus Aurelius—there is precious little that human beings can do to alter the external world; and so we should get on with adapting to, especially by reevaluating, forces, obstacles, and opportunities coming from outside of themselves.

But why must we draw such a sharp contrast between "things that are in our power" and "things that are not in our power"? It would appear that many things are better described as *partially* or *largely but not exclusively* in our power. Moreover, many things are not in our *individual* power but to some extent may be in our *collective* power.[4]

For example, in *The Eighteenth Brumaire of Louis Bonaparte* Karl Marx addressed the blasted hopes of the revolutionaries of 1848 by pointedly observing that human beings cannot make history "just as they please in circumstances they choose for themselves"[5]; but this hardly entailed for him that they cannot make history at all. Indeed, Marxists have standardly argued that human acting in the world cannot be reduced to an all or nothing prospect: Either full-fledged causal determinism or else an "out-of-gear freedom" that would disengage "our choices from causal interaction with the world" and so "ward off the threat that the nature of that world might limit or determine them."[6] Perhaps the liveliest illustration of the classical Marxist position on the "relative autonomy"[7] afforded human beings in their individual and collective pursuits has been provided by Norman Geras, who devised the following thought experiment:

> A length of chain secures me by the ankle to a stout post. This limits what I can do but also leaves me a certain freedom. I can stand or sit, read or sing. I cannot play a decent game of table tennis, however, and cannot attend social functions or political meetings at all. The chain not only limits me, negatively; it also compels me to certain actions. The way it is fixed to my leg, I must keep adjusting how it lies, otherwise it begins to hurt me. I must apply medicaments periodically to sores which develop around my ankle. And so on. Understanding my situation more or less, I enjoy a relative autonomy: the chain and post are fundamental determinants of my lifestyle but they do still leave me scope for independent decisions.[8]

As I shall argue below, Epictetus's searching question of what lies is in our human power and what does not has profound bearing on the philosophical challenges posed by the new geological epoch which humanity has forced upon the Earth—the *Anthropocene*.[9] However, as I shall equally argue, in the Anthropocene the key question for philosophy is not learning to die but learning to *fight*.

* * * *

We have entered a period in which extreme skepticism about climate change—usually termed *climate denialism*[10]—has been increasingly been replaced by what we could call *climate stoicism*, namely, the attitude that dangerous climate change must be accepted as an external force beyond human control.[11] There is supposedly nothing that can be done except to prepare for, and cope with, the inevitable deterioration of life on Earth. For example, in his existentially riveting but unremittingly bleak book *Learning to Die in the Anthropocene*, Roy Scranton has written that "we have entered humanity's most philosophical age, for this is precisely the problem of the Anthropocene. ... The rub is that we have to learn to die not as individuals, but as a civilization."[12] Although Scranton's philosophical frame of reference is not narrowly Stoic (he draws on a wide range of authors and texts[13]), his perspective remains one that falls back onto the classical distinction between what lies (a) within our power (not much other than beliefs) and (b) what lies outside of our power (virtually everything else):

> We have failed to prevent unmanageable global warming and that global capitalist civilization as we know it is already over, but ... humanity can survive and adapt to the new world of the Anthropocene if we accept

human limits and transience as fundamental truths, and work to nurture the variety and richness of our collective cultural heritage. Learning to die as an individual means letting go of our predispositions and fear. Learning to die as a civilization means letting go of this particular way of life and its ideas of identity, freedom, success, and progress. These two ways of learning to die come together in the role of the humanist thinker: the one who is willing to stop and ask troublesome questions, the one who is willing to interrupt, the one who resonates on other channels and with slower, deeper rhythms.[14]

In the rest of his book Scranton sets forth and develops several key points that may be enumerated as follows:

1. An empirical claim that extremely dangerous climate change cannot be mitigated;
2. An empirical claim that climate change is the ultimate outcome of a human nature that is deeply flawed;
3. A conceptual claim (borrowed from Peter Sloterdijk) that philosophy serves as a kind of "interruption" of social background assumptions and daily interactions.[15]

Allow me to consider these claims in order. The first two chapters of Scranton's book offer a genealogy of "human ecologies" in order to highlight the extent to which "carbon-fueled capitalism" has become a "zombie system, voracious but sterile," and ecologically unsustainable.[16] Although Scranton offers ample evidence to convey both the *seriousness* of human-caused climate change and the *urgent* need for action, he maintains that it is likely too late for *effective* action. This is because climate change poses an especially "wicked problem."[17]

What is at issue, though, is not the difficulty of addressing climate change—and other human pressures on the Earth System—but soberly trying to conceptualize feasible solutions and figuring out how best to motivate and sustain political action on behalf of such solutions. For example, Johan Rockström and other climate scientists have drawn up an inspiring "roadmap for rapid carbonization" by 2050.[18] Their guiding idea is to frame the "decarbonization challenge" in terms of a "carbon law" (analogous to "Moore's Law," which projects that the number of components per integrated circuit will double approximately every two years[19]) that will lead to cutting in half "gross anthropogenic carbon dioxide (CO_2) emissions every decade" and ultimately result in "net-zero emissions" around mid-century, a path necessary to limit warming to well below 2°C."[20] The key question remains, though, how to build and sustain the global political movement required to mount sufficient pressure on the world's governments to implement such a roadmap.

Scranton, however, derides concerted efforts at sociopolitical and technological change as arising from a primal but ultimately futile drive:

> You've heard the call: We have to do something. We need to fight. We need to identify the enemy and go after them. Some respond, march, and chant. Some look away, deny what's happening, and search out escape routes into imaginary tomorrows: a life off the grid, space colonies, immortality in paradise, explicit denial, or consumer satiety in a wireless, robot-staffed, 3D-printed techno-utopia. Meanwhile, the rich take shelter in their fortresses, trusting to their air conditioning, private schools, and well-paid guards. Fight. Flight. Flight. Fight. The threat of death activates our deepest animal drives.[21]

Let us neither fight nor flee, he suggests, but resolutely stand our moral ground and die with dignity.

****

If indeed the rise of the Anthropocene had imposed a collective death sentence on humanity, then Scranton's principled position would in its own way be admirable. In the last instance, however, his grim vision of climate destabilization derives from his equally grim view of human nature as fundamentally greedy, violent, and self-deluded; an original sin without prospect for redemption. He writes that:

> for most of human history, violence has been a central element of social conflict. The first clear evidence of mass human violence is as old as civilization; the first evidence of its end has yet to be seen. ... The long record of human brutality seems to offer conclusive evidence that both individual and socially organized violence are as biologically a part of human life as are sex, language, and eating, that aggression and the drive for dominance are neither vestigial atavisms nor social maladaptations but rather species traits, and that we have little reason to hope that war and murder might someday disappear.[22]

This is a tendentious passage, to say the least! A wide range of social scientists have offered compelling evidence that human violence is largely socially constructed and has to do with background conditions of social inequality.[23]

Likewise, primatologist Frans de Waal's research has led him to conclude that nonhuman animals exemplify a range of moral behavior, from fairness and reciprocity to altruism.[24]

In this passage and throughout his book, Scranton expresses a perspective that human beings are innately disposed to violence and war. Following Douglas Fry, we could call such a view "man the warrior," according to which,

> humans (especially males) are warlike by nature. Advocates of this perspective forge a tight evolutionary link between chimpanzee and human violence, emphasize sex differences in aggression, and recite a litany of barbarity, atrocity, and brutality to support this portrait of humanity. The validity of this 'man the warrior' view may seem rather obvious; after all, we all know that humans make war and that wars always seem to be raging somewhere. However, a different – but *not* polar opposite – perspective will be suggested in this book. According to this new view, clearly humans are capable of creating great mayhem, but they also have a remarkable capacity for working out conflicts without resorting to violence. Specifically, a careful reexamination of the actual evidence will lead us to the conclusion that humans are not warlike by nature.[25]

By contrast, then, Fry provides substantial "macroscopic"[26] anthropological evidence that "warfare is not inevitable and that humans have a substantial capacity for dealing with conflicts nonviolently."[27] He worries that:

> widespread beliefs that war is natural and acceptable hinder the search for alternatives—and thus the inevitability of war becomes a self-fulfilling prophecy. Such beliefs may be detrimental both to preventing particular wars and to abolishing the institution of war. Perhaps this insight can help us to overcome the problem.[28]

Yet it remains possible to "create alternative ways of dealing with international conflicts." Fry continues:

> Humans have a solid repertory of conflict management skills to draw upon. Across societies, people are apt preventers and avoiders of violence. Over a vast array of societal circumstances, humans deal with most conflicts without any physical aggression at all. Regularly, the language-using primate 'talks it out,' airs grievances verbally in the court of public opinion, negotiates compensation, focuses on restoring relationships bruised by a dispute, convenes conflict resolution assembles, and listens to the wisdom

of elders or other third parties who, acting as peacemakers, strive to end the tension within the group and among disputants. As we have considered, humans also routinely show a great deal of self-restraint against acting aggressively. Such restraint makes evolutionary sense and has numerous parallels in other animal species.[29]

Consequently, what is required for the resolution of major and conflicts[30] is a new form of identification with others:

> Anthropology shows clearly that through millennia and across continents humans experience tremendous variation in ways of life and social organization. In foraging bands, individuals identify with their relatives and friends in their own and neighboring bands, in nation states, as Darwin noted, the level of identification generally rises to the country as a whole. This shows that both the social organization and the unit of identification (the 'us' compared to the 'them') are extremely malleable. A global identification, 'all of us,' in addition to lower-level 'us' identifications, seems well within the realm of human capacities, especially when our common survival depends on at least enough common identification to put a halt to war and to cooperate to solve global problems that threaten all of us.[31]

From the standpoint of a more explicitly *philosophical* anthropology than Fry's, let us try to situate Scranton's conception of human nature in terms of a fourfold classification advanced by Norman Geras.[32] Geras distinguishes the following possible views:

a. Human nature is intrinsically evil.
b. Human nature is intrinsically good.
c. Human nature is intrinsically blank.
d. Human nature is intrinsically mixed.

Throughout his book Scranton not only rejects (b) and (c); he effectively endorses (a). However, he overlooks option (d), according to which members of our species are not saints, brutes, or empty vessels; rather they are flawed—but improvable—*human beings*. As Geras writes,

> base or egregious human impulses ... are not so all-consuming as to make pervasive and enormous evil forever inevitable, but nor are they so weak or insignificant that they might be conceived as entirely eliminable, as one day gone, as even now 'really' something else than they appear,

not human impulses after all, but alienated, capitalist, class-oppressive or class-oppressed, patriarchal, corrupted ones. Conversely, benign and admirable tendencies … are not so dominant as to make the possibility of serious human evil only a temporary, albeit long, historical phase which may one day pass, nor so feeble or so sparsely distributed as to make attempts to limit and counteract that baleful possibility a pointless quest. Both sorts of impulses or tendency are conceived … as being permanent features of our nature, realities to be negotiated, lived with, if possible understood – and if possible tilted toward the more benign and admirable, and tilted as far that way as possible.[33]

Yet it is precisely option (d) that sustains the "hope of socialism." As Geras proposes,

the goal of a much better and a more just society is to be fought for not just because human beings are by nature overwhelmingly or essentially good, nor because they do not have an intrinsic nature; but because and in spite of the bad combination in their nature of bad impulses with good ones. Because of the bad impulses, this struggle is necessary. In spite of them, it is to be hoped, a socialist society may yet be possible.[34]

In sum, the survival of complex human societies in the Anthropocene hinges not only whether or not (d) is a plausible conception of human nature but also on whether or not we can struggle to fashion more egalitarian societies that will both reinforce our good impulses and restrain our bad impulses. This is not just a question of ethical self-fashioning; it remains a properly political project. The only way out of the Anthropocene—and a return to a "Holocene-like" state of the climate—is by moving forward both individually and collectively, indeed, *transindividually*.[35]

****

In an unsettling passage Scranton wonders if some of our descendants will "build new cities on the shores of the Arctic Sea, when the rest of the Earth is scorching deserts and steaming jungles."[36] One could play this survival game and speculate that the fortunate few will relocate to New Zealand, mountainous regions, or abandoned missile silos converted into condo bunkers to serve as a "doomsday prep for the super-rich."[37] Unfortunately, the elite who could afford to relocate would be the "worst and the dullest" among humanity—those most responsible for the devastation of the Earth System and least likely to have any interest in the new humanism sought by Scranton. The reclamation of humanism can occur only under conditions of mass survival. Moreover, in this twisted variation on "lifeboat ethics" thought experiment,[38] one could insist that it would be better that *no one* survive than that the affluent minority responsible for the disaster survive while the impoverished majority drown. The world's non-culpable majority would be justified in capsizing such a lifeboat.

Christian Parenti has evoked precisely this specter of a "politics of the armed lifeboat: responding to climate change by arming, excluding, forgetting, repressing, policing, and killing."[39] But such a politics must at all cost be avoided: "a world in climatological collapse—marked by hunger, disease, criminality, fanaticism, and violent social breakdown—will overwhelm the armed lifeboat. Eventually, all will sink into the same morass."[40] Parenti holds out the hope, then, that there is "another path." Indeed, there is another path: It is both morally preferable and feasible to avoid *in advance* the brutalizing politics of the armed lifeboat by compelling those in charge of the global economy to remain within the *planetary boundaries* identified by such Earth System scientists as Will Steffen and Johan Rockström.[41]

A symptomatic omission in Scranton's book is not the ecological threats posed by the Anthropocene but instead any reference to these nine boundaries, which are interlinked Earth System processes and biogeophysical constraints: climate change, changes in biospheric integrity, biogeochemical flows, stratospheric ozone depletion, ocean acidification, freshwater use, land-system change, atmospheric aerosol loading, and the introduction of novel entities. Crossing even one of these boundaries would risk triggering abrupt or irreversible environmental changes that would be very damaging or even catastrophic for society. Furthermore, if any of these boundaries were crossed, then there would be a serious risk of crossing the others. However, as long as these boundaries are not crossed, "humanity has the freedom to pursue long-term social and economic development."[42]

Unfortunately, four boundaries have already been crossed: climate change, changes in biospheric integrity, interference with biogeochemical flows, and land-system change.[43] Humanity soon faces a planetary "tipping point" by which "what we like to think are gradual environmental changes in fact turn into sudden ones that we don't expect."[44]

However, Ian Angus has provided a vivid analogy to distinguish planetary boundaries from tipping points: The former can "be compared to guardrails on mountain roads, which are positioned to prevent drivers from reaching the edge, not on the edge itself."[45] As a result, Angus argues, provided that we remain within these boundaries, humanity retains the power to avoid ecological "tipping points" and successfully to resist the Anthropocene even if we cannot reverse it: "The question is not whether the Earth System is changing, but how much it will change, and how we will live on a changed planet."[46]

Jeremy Davies has sharply criticized the concept of "fixed" planetary boundaries as presuming a stable, unchanging Earth.[47] But by analogy one can allow that a building is subject to earthquakes that could devastate it without thereby rejecting the need for build "to code," which will enable it to survive likely quakes in the future. There is no guarantee, of course, but neither is there any need to reject the use of building codes. Moreover, just because some foolish persons will ignore warning signs near dangerous terrain doesn't mean that we shouldn't install them. A road or trail can certainly even collapse where there is no warning sign; but again, this doesn't mean that such signs are useless. The Anthropocene reminds us that we have remade the Earth in our image and disrupted what came before us. However, we can repair some—if not all—of the damage in ways that protect humanity and other species within a "safe operating space."

Following Geras, it remains possible to defend to a "limited notion of progress and of socialist utopia." As he insists,

> To advise resigned acceptance of the world as it is – life-and-death inequalities, universal exploitation, widespread political oppression, festering communal hatreds, genocide, recurring war – … is to eschew a naïve, optimistic teleology, only to speak the script of another, grimmer one. It is to risk making oneself, in a certain manner, the willing voice of ugly moral forces.[48]

In this light of this "modest or minimal" conception of utopia, ecosocialists need not claim that human civilization will endure forever or that extinction is ruled out in advance; the question is how to respond *now* to this climate emergency. The implication of climate stoicism is that human beings are—and should remain—largely passive in the face of a rapidly changing world. But this is morally unacceptable. As Angus writes,

We know that disaster is possible, but we refuse to despair. If we fight, we *may* lose; if we don't fight, we *will* lose. Good or bad luck may play a role, but a conscious and collective struggle to stop capitalism's hell-bound train is our only hope for a better world.[49]

We can and must act *collectively*—and urgently—without guarantee of success. What then is to be done?

****

Finally, let us consider the role of the philosopher in the midst of climate destabilization and social upheaval. For Scranton, philosopher best serve as "interrupters." Following Peter Sloterdijk, Scranton asserts that:

> The enemy isn't out there somewhere – the enemy is ourselves. Not as individuals, but as a collective. A system. A hive. How do we stop ourselves from fulfilling our fates as suicidally productive drones in a carbon-addicted hive, destroying ourselves in some kind of psychopathic colony collapse disorder? How do we interrupt the perpetual circuits of fear, aggression, crisis, and reaction that continually prod us to ever more intense levels of manic despair? One way we might begin to answer these questions is by considering the problem of global warming in terms of Peter Sloterdijk's idea of the philosopher as an interrupter.... What Sloterdijk helps us see is that responding autonomously to social excitation means not reacting to it, not passing it on, but interrupting it, then either letting the excitation die or transforming it completely. Responding freely to constant images of fear and violence, responding freely to the perpetual media circuits of pleasure and terror, responding freely to the ongoing alarms of war, environmental catastrophe, and global destruction demands a reorientation of feeling so that every new impulse is held at a distance until it fades or can be changed. While life beats its red rhythms and human swarms dance to the compulsion of strife, the interrupter practices dying.[50]

Doubtless, there is much to be said for this variation on the Socratic style of doing philosophy as a "gadfly" who seek to arouse the sleeping democratic beast.[51] However, as admirable as the philosophical practice of dying may be in the face of insurmountable external threats,[52] it changes nothing and only leaves such threats in place.[53]

Much more useful for learning to fight in the Anthropocene is the model of the philosopher as a *militant*. A militant is not someone who is especially angry or impatient but instead someone who pursues a course similar to the one identified by Paul of Tarsus, who movingly wrote in his letter to the assembly of Jesus loyalists in the Roman colony of Philippi in northern Greece that "this one thing I do: forgetting what lies behind and straining forward to what lies ahead, I press on toward the goal for the prize."[54] Not surprisingly, the Black Freedom Struggle of the 1950s and 1960s in the United States reclaimed this image of keeping your "eyes on the prize" of social justice.[55] It remains an apt image and slogan for the climate justice movement in the Anthropocene.[56]

****

But militant philosophers need a concrete way to join with others, by helping to formulate what we might call a transitional program for the Anthropocene as one of the central planks in what Rutger Bregman has called "a utopia for realists"—a utopia that calls upon contemporary "underdog socialists" to abandon their tales of doom and gloom and to reclaim and retell "a narrative of hope and progress ... that speaks to millions of ordinary people."[57] Bregman argues that his own utopian proposal for a radical reduction in the workweek to fifteen hours would not only reduce stress and reduce inequality, it would also "cut CO_2 emitted this century by half."[58] The rise of the Anthropocene demands of humanity not serene acceptance of the end of carbon-fueled capitalist civilization but a bold demand that *Another Anthropocene is Possible!*

Consider in this light also the eminent biologist Edward O. Wilson's recent intriguing proposal—one that fulfills Dipesh Chakrabarty's requirement for a *zoecentric* approach to addressing the moral challenges posed by the Anthropocene.[59] In his book *Half-Earth: Our Planet's Fight for Life*,[60] Wilson compellingly documents the dire threat to biodiversity posed by the Anthropocene and advances a bold project to dedicate half of the Earth's surface to natural preservation. Unfortunately, toward the end of the book, he implausibly claims that the "evolution of the free market system and the way it is increasingly shaped by high technology" will tend to favor "both shrinkage of the ecological footprint and the resulting improvement of biodiversity conservation ... because of the acceleration of the replacement of extensive economic growth by intensive economic growth."[61] Moreover, he timidly concludes his book by appealing to economic "evolution" that minimizes the global sociopolitical upheaval required to achieve what he calls a "shift in worldview from wealth based on quantity to wealth based on quality, with the latter made permanent through ecological realism."[62] Yet if Wilson's "central idea is to view the entire planet as an ecosystem, to see Earth as it is and not as we wish it to be,"[63] then the basic structure of capitalist degradation of the Earth System—whether extensive or intensive—remains the chief obstacle that must be resisted, dismantled, and surpassed.

"Intensive economic growth" scarcely provides a framework for human emancipation within the safe operating space afforded by planetary boundaries. As Félix Guattari once presciently observed,

> Automatized and computerized production no longer draws its consistency from a basic human factor, but from a machinic phylum that traverse, bypasses, disperses, miniaturizes, and co-opts all human activities.
>
> These transformations do not imply that the new capitalism completely takes the place of the old one. There is rather coexistence, stratification, and hierarchalization of capitalisms, at different levels ·....[64]

Capitalism, Guattari continued, increasingly "seizes individuals from the inside" as it "bears down on the basic functioning of the perceptive, sensorial, affective, cognitive, linguistic, etc. behaviors grafted to capitalist machinery." In short, technological miniaturization under capitalism generates new forms of "machinic enslavement."[65] How shall we resist this assault on mental ecology as well as on social and environmental ecologies?[66] What is the way forward for humanity in our effort to achieve a sustainable Earth System through what Timothy Lenton and his colleagues have called a "solar-powered material-recycling revolution"? Such a revolution in sustainability would have to be grounded in

> a level of social organization that can implement the changes in energy source and material cycling without preventing present and future generations from attaining similar achievements in standard of living and individual liberation associated with industrial societies.[67]

Where do we begin?

****

Jeremy Brecher has written a refreshingly detailed "manual" for how to build a "climate insurgency" that could contest and delegitimize the ideological "pillars of support" for fossil capitalism and prepare the way for an ecosocialist future.[68] For example, participants in this movement could argue that "government actions are illegal and unconstitutional" and that they are performing a vital duty to reclaim, renew, and defend the "public trust."[69] As Brecher writes,

Constitutional and public trust principles make it possible for the climate insurgency to turn the tables on the governments that purport to represent the world's people and to have the authority to rule the world. They stand for the proposition that governments do not have the right to destroy the climate – and that the people have the right to stop them when they do so. Governments have no more right to authorize the emission of greenhouse gases that destroy the climate than the trust officers of a bank have to loot the assets placed under their care. The people of the world have a right to our common natural resources. And we have a right, if necessary, to protect our common assets against those who would destroy them.[70]

An insurgent movement could even set up "climate justice tribunals" to make the case that:

> the governments and corporations of the world are systematically violating human rights, international law, and their duty to protect the public trust by allowing greenhouse gas emissions that are destroying the earth's climate. Future climate tribunals could examine the evidence in greater detail. They could issue declaratory judgments and injunctions. They could also make findings on the rights and responsibilities of global citizens to enforce the law and their legal rights vis-à-vis governments that try to subdue them when they do so. Tribunals can be convened as part of the legitimation and public education activities of specific campaigns. Although they may be initiated by the insurgency, the validity of their judgments can be based on the fairness of their conclusions and the evidence and argument on which they are based. Some tribunals could become permanent institutions. In specific instances, people could apply to such tribunals for 'advisory opinions' on questions like the need to halt new fossil fuel infrastructure or the adequacy of Climate Action Plans. Tribunals could weigh the evidence and issue judgments. They could then negotiate consent decrees or issue advisory orders. The people could then attempt to impose or implement those orders by mass action.[71]

Such tribunals would operate as a form of "dual power" aiming not just at ideological disruption but at the creation of a new world: an ecologically sustainable planet, a planet whose boundaries still allow for the flourishing of human beings and other species, a planet fit for our children and theirs. Yet the climate justice movement might founder or fail long before such advanced forms of struggle as dual power ever arise. As Angus soberly reminds us, there are no guarantees. Marxism is not deterministic. An ecosocialist revolution is not inevitable. It will only happen if people consciously decide it is necessary, and take the steps needed to bring it about.[72]

What if people do not take these necessary steps? In the last instance, it would seem, we must be prepared not only for the collapse of carbon-fueled capitalism but also for the demise of complex human societies. In this sense, Scranton remains correct: We have to learn "to let go."[73] Even as we fight.

NOTES

1. *Manual* 1 (Epictetus 2014, p. 287). This text is paralleled by *Discourses* 1.1 (Epictetus 2014, pp. 4–6).
2. Evans (2012, pp. 29–30).
3. Jérôme De Sousa, for instance, argues that classical Stoic philosophy allowed for "disobedience" whenever politics deviated from the "divine natural order" (De Sousa 2010).
4. Along these lines, Massimo Pigliucci (2017) has persuasively argued that the Stoic dichotomy between what is and what is not under one's individual control should be reformulated to allow for degrees of control over one's mind and body. Moreover, he insists, one should not succumb to fatalism or despair, "for resignation goes against not just what the Stoics themselves said but also, more importantly, what they practiced" (p. 38). Yet Pigliucci is concerned with *individual* decisions and deeds—whether by citizens, subjects, or political leaders; he does not address the problem of *collective* deliberation and action. As a result, he effectively reduces politics to ethics, instead of conceiving each as dialectically imbricated with the other.
5. Marx (1996, p. 32).
6. Collier (1994, p. 98).
7. What Collier calls "in-gear freedom," namely, "interacting causally with the world in order to realize our intentions" (p. 98).
8. Geras (1990, p. 74).
9. Jeremy Davies (2016) not only provides a superb introduction to the panoply of "versions of the Anthropocene," he also carefully defends a specifically stratigraphic interpretation.
10. For an overview of the varieties and causes of climate change denial—and strategies for how best to challenge it—see Washington and Cook (2011), Norgaard (2011), and Marshall (2014).
11. Clive Hamilton has similarly called this attitude a "new kind of existential defeatism" and offered a forceful philosophical rejoinder to it; see Hamilton (2017), especially pp. 112–135.
12. Scranton (2015, p. 24).

13. Indeed, his only explicit mention of a classical Stoic philosopher is Marcus Aurelius (p. 92).
14. Scranton (2015, p. 24).
15. My thanks to Cerritos College philosophy department colleagues—especially Ana Torres-Bower, Bob Sliff, Tim Chatman, and J.P. Pereira—for our seminar discussion of Roy Scranton's book. I am especially indebted to J.P. for the general distinction between Scranton's empirical and conceptual claims. My specific reference to a conceptual claim about philosophy and philosophers, though, derives from Justin Smith's six-fold classification of the "curiosa," or inquirer in the natural world; the "sage," the "gadfly," the "ascetic," the "mandarin," and the "courtier" (Smith 2016). Sloterdijk and Scranton's conception of the philosopher as an "interrupter" would presumably fall under the second category of "gadfly." As I shall argue later, we should add to this typology what could be called the *militant* philosopher.
16. Scranton (2015, p. 23).
17. Ibid., pp. 39–53.
18. Rockström et al. (2017).
19. For a more technical discussion of Moore's Law, see Mody (2015). On the utopian implications of Moore's Law, see Bregman (2017, pp. 179–181). Edward O. Wilson has also invoked Moore's Law in order to envision an accelerated means to shrink humanity's ecological footprint; see Wilson (2016, pp. 192–193).
20. Rockström et al. (2017, p. 1269).
21. Scranton (2015, pp. 76–77).
22. Ibid., p. 75.
23. For an overview of this literature, see Wilkinson and Pickett (2011, pp. 129–144).
24. De Waal (2010).
25. Fry (2007, p. 2).
26. Ibid., p. xiii.
27. Ibid., pp. 1–2.
28. Ibid., pp. 201–202.
29. Ibid., p. 205.
30. Fry allows that lesser-scale acts of violence such as "fights, murders, executions" may not be eliminable: "abolishing war will not mean an end to conflict. It will mean that conflicts are handled in less destructive ways" (pp. 8–9).
31. Fry (2007, p. 232).
32. Geras (1998).
33. Geras (1998, pp. 88–89).
34. Geras (1998, p. 89).
35. Read (2015). See also Read (2017) for a different approach to Anthropocene political struggles and a more positive appreciation of Scranton's book.

36. Scranton (2015, p. 109).
37. Osnos (2017).
38. Classically devised by Garrett Hardin to illustrate the so-called "tragedy of the commons." See Hardin (1974).
39. Parenti (2012, p. 11).
40. Ibid., p. 20.
41. Rockström et al. (2009), Rockström and Klum (2015), and Steffen et al. (2015).
42. Rockström et al. (2009).
43. The planetary boundaries associated with stratospheric ozone depletion, ocean acidification, freshwater use, and land-system change have not yet been crossed; and those boundaries associated with atmospheric aerosol loading and the introduction of novel entities have yet to be quantified scientifically.
44. Barnosky and Hadly (2016, p. 6).
45. Angus (2016, p. 74).
46. Ibid., p. 213.
47. Davies (2016, p. 198).
48. Geras (1998, p. 118).
49. Angus (2016, p. 223).
50. Scranton (2015, pp. 85–88).
51. See Socrates's famous defense of the practice of philosophy at his trial (Plato 2012, p. 35); see also Martin Luther King Jr.'s reference to Socrates and the need for "nonviolent gadflies" for social justice in his 1963 "Letter from Birmingham Jail" (King 2015, p. 130).
52. See especially Brandatan (2015).
53. This marks a fundamental difference between Socrates and Martin Luther King, Jr., whose gadflies did not simply criticize—and ultimately submit to—unjust laws but actively *disobeyed* them.
54. Phil 3.13-14. By "prize" and "goal" Paul alludes to Greek popular sports competitions (especially in Philippi), namely, foot races, their finishing posts, and their awarded prizes. For all of its shortcomings (some of which I discuss in Stolze 2016), this is the merit of Alain Badiou's book on Paul, namely, to reclaim Paul as a model for what he calls a "new militant figure" (Badiou 2003, p. 2).
55. Indeed, "Keep Your Eyes on the Prize" was the title of a popular "movement song" of that era.
56. Compare Wen Stephenson's invocation of the Abolitionist movement as a way to orient the next steps of the climate justice movement; see Stephenson (2015), especially pp. 23–45.
57. Bregman (2017, p. 258).
58. Ibid., p. 142.
59. On the distinction between a homocentric and a zoecentric perspective, see Chakrabarty (2015, pp. 153–156).

334   T. STOLZE

60. Wilson (2016).
61. Ibid., pp. 191–192.
62. Ibid., p. 193.
63. Ibid.
64. Guattari (2009, pp. 249–250).
65. Ibid., pp. 261–262.
66. On the entanglement of these three ecologies, see Guattari (2000).
67. Lenton et al. (2016), p. 363. See also Lenton (2016, pp. 107–123).
68. Brecher (2017).
69. Ibid., p. 85.
70. Ibid., p. 86.
71. Ibid., p. 91.
72. Angus (2016, p. 222).
73. Scranton (2015, p. 92).

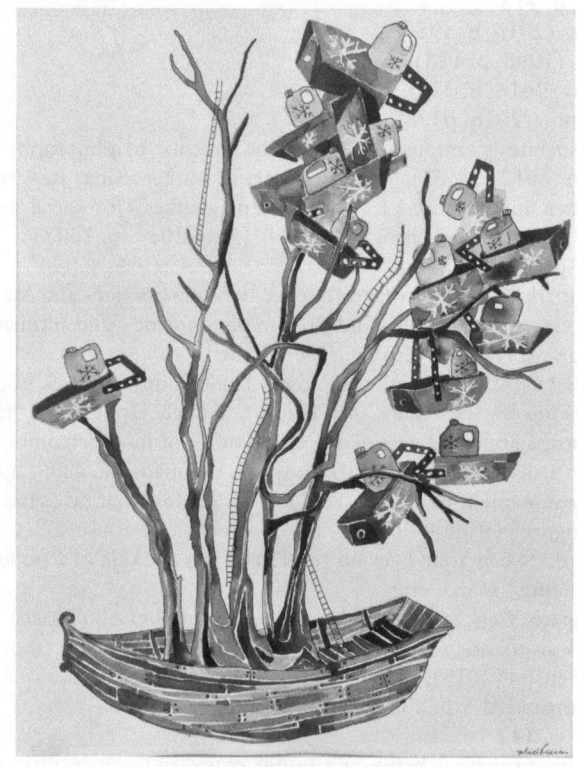

Image 14.1 Machine study (Mia Feuer, watercolor)

REFERENCES

Angus, I. (2016). *Facing the Anthropocene: Fossil Capitalism and the Crisis of the Earth System*. New York, NY: Monthly Review Press.

Badiou, A. (2003). *Saint Paul: The Foundation of Universalism* (R. Brassier, Trans.). Palo Alto, CA: Stanford University Press.

Barnosky, A. D., & Hadly, E. A. (2016). *Tipping Point for Planet Earth: How Close Are We to the Edge?* New York, NY: Thomas Dunne Books.

Brandatan, C. (2015). *Dying for Ideas: The Dangerous Lives of the Philosophers*. New York, NY: Bloomsbury.

Brecher, J. (2017). *Against Doom: A Climate Insurgency Manual*. Oakland, CA: PM Press.

Bregman, R. (2017). *Utopia for Realists: How We Can Build the Ideal World* (E. Manton, Trans.). New York, NY: Little, Brown.

Charkrabarty, D. (2015). *The Human Condition in the Anthropocene*. http://www.tannerlectures.utah.edu/Chakrabarty%20manuscript.pdf/. Last accessed March 25, 2017.

Collier, A. (1994). *Critical Realism: An Introduction to Roy Bhaskar's Philosophy*. New York, NY: Verso.

Davies, J. (2016). *The Birth of the Anthropocene*. Berkeley: University of California Press.

De Sousa, J. (2010). *Stoïcisme et politique: essai sur la désobéissance philosophique*. Paris: Editions Grammata.

De Waal, F. (2010). *The Age of Empathy: Nature's Lessons for a Kinder Society*. New York, NY: Broadway Books.

Epictetus. (2014). *Discourses, Fragments, Handbook* (R. Hard, Trans.). New York, NY: Oxford University Press.

Evans, J. (2012). *Philosophy for Life and Other Dangerous Situations: Ancient Philosophy for Modern Problems*. Novato, CA: New World Library.

Fry, D. P. (2007). *Beyond War: The Human Potential for Peace*. New York, NY: Oxford University Press.

Geras, N. (1990). *Discourse of Extremity: Radical Ethics and Post-Marxist Extravagances*. New York, NY: Verso.

Geras, N. (1998). *The Contract of Mutual Indifference: Political Philosophy After the Holocaust*. New York, NY: Verso.

Guattari, F. (2000). *The Three Ecologies* (I. Pindar & P. Patton, Trans.). New Brunswick, NJ: The Athlone Press.

Guattari, F. (2009). Capital as the Integral of Power Formations. In S. Lotringer (Ed.), C. Wiener & E. Wittman (Trans.), *Soft Subversions: Texts and interviews 1977–1985* (pp. 244–264). Los Angeles, CA: Semiotext(e).

Hamilton, C. (2017). *Defiant Earth: The Fate of Humans in the Anthropocene*. Malden, MA: Polity Press.

Hardin, G. (1974). Lifeboat Ethics: The Case Against Helping the Poor. *Psychology Today*. www.garretthardinsociety.org/articles/art_lifeboat_ethics_case_against_helping_poor.html/. Last accessed March 31, 2017.

King, M. L., Jr. (2015). *The Radical King* (C. West, Ed.). Boston, MA: Beacon Press.

Lenton, T. (2016). *Earth System Science: A Very Short Introduction*. New York, NY: Oxford University Press.

Lenton, T., Pichler, P. P., & Weisz, H. (2016). Revolutions in Energy Input and Material Cycling in Earth History and Human History. *Earth System Dynamics, 7*, 353–370. www.earth-syst-dynam.net/7/353/2016/. Last accessed May 23, 2017.

Marshall, G. (2014). *Don't Even Think About It: Why Our Brains Are Wired to Ignore Climate Change*. New York, NY: Bloomsbury Press.

Marx, K. (1996). The Eighteenth Brumaire of Louis Bonaparte. In T. Carver (Ed., & Trans.), *Later Political Writings* (pp. 31–127). New York, NY: Cambridge University Press.

Mody, C. (2015). What Kind of Thing Is Moore's Law? *IEEE Spectrum*. http://spectrum.ieee.org/semiconductors/devices/what-kind-of-thing-is-moores-law/. Last accessed March 31, 2017.

Norgaard, K. M. (2011). *Climate Change, Emotions, and Everyday Life*. Cambridge, MA: The MIT Press.

Osnos, E. (2017). Doomsday Prep for the Super-Rich. *The New Yorker*. http://www.newyorker.com/magazine/2017/01/30/doomsday-prep-for-the-super-rich. Last accessed March 24, 2017.

Parenti, C. (2012). *Tropic of Chaos: Climate Change and the New Geography of Violence*. New York, NY: Nation Books.

Pigliucci, M. (2017). *How to Be a Stoic: Using Ancient Philosophy to Live a Modern Life*. New York, NY: Basic Books.

Plato. (2012). *A Plato Reader: Eight Essential Dialogues* (C. D. C. Reeve, Ed.). Indianapolis, IN: Hackett.

Read, J. (2015). *The Politics of Transindividuality*. Boston, MA: Brill.

Read, J. (2017). Anthropocene and Anthropogenesis: Philosophical Anthropology and the Ends of Man. *The South Atlantic Quarterly, 116*(2), 257–273.

Rockström, J. et al. (2009). A Safe Operating Space for Humanity. *Nature, 461*, 472–475.

Rockström, J. et al. (2017). A Roadmap for Rapid Decarbonization. *Science, 355*(6331), 1269–1271.

Rockström, J., & Klum, M. (2015). *Big World, Small Planet: Abundance within Planetary Boundaries*. New Haven, CT: Yale University Press.

Scranton, R. (2015). *Learning to Die in the Anthropocene: Reflections on the End of a Civilization*. San Francisco, CA: City Light Books.

Smith, J. (2016). *The Philosopher: A History in Six Types*. Princeton, NJ: Princeton University Press.

Steffen, W. et al. (2015). Planetary Boundaries: Guiding Human Development on a Changing Planet. *Science, 347*(6223), 1–10.

Stolze, T. (2016). Paul of Tarsus, Thinker of the Conjuncture. In A. Hamza (Ed.), *Althusser and Theology* (pp. 129–151). Leiden: Brill.

Stephenson, W. (2015). *What We're Fighting for Now Is Each Other: Dispatches from the Front Lines of Climate Justice*. Boston, MA: Beacon Press.

Washington, H., & Cook, J. (2011). *Climate Change Denial: Heads in the Sand*. Washington, DC: Earthscan.

Wilkinson, R., & Pickett, K. (2011). *The Spirit Level: Why Greater Equality Makes Societies Stronger*. New York, NY: Bloomsbury.

Wilson, E. O. (2016). *Half-Earth: Our Planet's Fight for Life*. New York, NY: Liveright.

CHAPTER 15

The Earth Is Not "Ours" to Save

Nathan Snaza

Allow me to begin by stating, very directly, my working hypothesis. The Anthropocene is defined as an epoch of Earth's history, the one in which we find ourselves today, in which the human has become a geological agent. Attempts to register what the Anthropocene means for politics and education end up—in uneven ways, to be sure—re-affirming the Anthropocene as they seek to mobilize human action *in response to* climate change or global warming. That is, in response to the human becoming a geological actor, the vast majority of ecological thought ends up calling for humans to take this agency upon themselves and direct it better than it has thus far been directed. The consequence, of course, is that while we can imagine—not that this is easy, as I'm sure you all know—a radical redirecting of human actions and institutional changes that would avert some of the scariest anticipated outcomes of what I can only call everyday life in a highly petroliumized world, we don't seem to be able to imagine *not* being the most important geological agent. The Anthropocene thus restages, at the geological level, a humanism that always sought to install Man as the measure of all things (by violence when necessary... and it is always necessary).

N. Snaza (✉)
University of Richmond, Richmond, VA, USA

© The Author(s) 2018   339
j. jagodzinski (ed.), *Interrogating the Anthropocene*, Palgrave Studies in
Educational Futures, https://doi.org/10.1007/978-3-319-78747-3_15

The immediate corollary, for me, is that this *givenness* of the human's agency at a geological level is structured into educational institutions and practices as well. At best, ecological education seems to be able to imagine teaching young humans (and I can, here, no more than mark how it *always* takes it for granted that only humans are caught up in the *dispositifs* of education) how to behave better, more responsibly, more sustainably. And, in some of its registers, it comes to function as an update on original sin: no human alive today was there to be "responsible" for the events in the eighteenth century (metonymically linked to James Watt's invention of the steam engine in the middle of that century) that initiated the Anthropocene. We are always already, then, guilty and asked to atone. Unsurprisingly, the educational praxes imagined from this guilty knot—so full of bad affect—are not particularly exciting. We get calls for responsible stewardship (which is, in the most direct way possible, an affirmation of human dominion), or for "green" consumerism. We get a dizzying shuttling back and forth between hellfire and brimstone visions of a total ecological collapse destroying humanity (which have, to be sure, their own delicious powers of attraction), and lists of maxims that we are expected to embrace with a Can Do! attitude ("Reduce Reuse Recycle"! "Think Globally, Act Locally!"). I'd like to offer up what I call "bewildering education" as an alternative, one that neither seeks to embrace the human dominion over the world nor one that would pretend there isn't a ethical choice to be made here at all (out of denial or cynical nihilism). Instead, I think we can get lost, we can loose ourselves from the cartographies drawn up through humanism, and we can embrace *failing* to be geological agents. And this may end up teaching us that the human, despite its geological hegemony in the moment, is surrounded on all sides by a host of material agencies that can and will affect the Earth's future, that can and do affect the ways "the human" is practiced.

What I'd like to do in this chapter is draw out what the Anthropocene means for thinking about agency, especially for nonhuman agency, before dwelling with three versions of ecological politics that all attempt to grapple with the shift in the human's relations to itself and every other thing that adheres within the Anthropocene. Although I will begin with one specifically tied to theories of pedagogy, I will take it as axiomatic that the connection between politics and education is insoluble. That is, every theory of political action that requires a shift away from how things operate in the present is also, and must need be, at least implicitly a pedagogy. And so, I will end by offering bewildering education as something like an event horizon emitting political pedagogies driven by a desire that humans *fail* to be geological agents[1] by failing, in fact, to be "human."

What I'm calling bewildering education precludes me from offering any practical guidance for how we teach. Indeed, I think in our present moment we have all too many people—politicians, corporations, an army of researchers with their best practices, and even critical pedagogues who think a little more hermeneutical suspicion will awaken us to the oppressions we have to learn to overthrow—telling what to do and how to do it. Bewildering education is not a plan or program. It has no method. The march toward proscription ends up factoring out too many differences for my taste. Bewildering education will be based on the simple axiom that the Earth is not "ours," and my semi-audible quotation marks here are meant to signal at least two things. First, a radical refusal of the idea that the Earth is the property or is under the care of human beings. And second, a suspicion about how quickly a rush to the first person plural forgets differential experience of being "human." The "our," even or especially when it seeks to encompass humanity as such, stinks to me of fascism and of imperialism. Moreover, I am fairly convinced that we don't really know who we are even in the singular, and that humanism has caused us to radically misunderstand ourselves as subjects, as biological beings, as material agents (or bundles of agentic matter). I have no idea what "I" am, or "you" are, or "we" are, and I think the contemporary political and educational institutions have not only limited our abilities to imagine and experiment with possibilitites, but that they've tricked us into thinking that there is no imagining and experimenting to do.

The "Anthropocene" is a phase of the Earth's history—I come back to this phrase in a moment—that comes after the Holocene, a relatively warm phase of Earth's history that was the material, ecological, climatological condition of possibility for most of the technologies, institutions, and practices that have come to mark this strange thing called "humanity." As Dipesh Chakrabarty noted in "The Climate of History," this is the period of "the beginnings of agriculture, the founding of cities, the rise of the religions we know, the invention of writing" (2009, p. 208). In displacing the Holocene, the Anthropocene suggests that the particular genre of "human" made possible in the Holocene has shifted from being a merely "biological agent" (a particular kind of animal or species) to being a "geological agent." In Chakrabarty's gloss, "There was no point in human history when humans were not biological agents. But we become geological agents only historically and collectively, that is, when we have reached the numbers and invented technologies that are on a scale large enough to have an impact on the planet itself" (pp. 206–207). The shift in epochs, then, requires both a quantitative shift (the explosion of human populations) and a qualitative one (the invention of particular technologies, specifically those enabling the petroliumization of societies via extraction and processing of fossil fuels).

Chakrabarty's concern, here, is specifically anchored in the discipline of history, a discipline that is in the process of being forced—perhaps like all the others—to rethink the very "human" around which it has been axiomatically constructed. In being forced to contend with a possibility that the Anthropocene has engendered "climate change" or "global warming" on a scale that will make human biological—and therefore cultural—life untenable, what we get is "a sense of the present that disconnects the future from the past by putting such a future beyond the grasp of historical sensibility" (p. 197). This happens in two distinct ways. First, it undercuts the virtually universal practice of imagining the future as ontologically and epistemologically continuous with the past: the future will be like the past, but different in its particulars. We can no longer take the past as the ground of imagining futures in any (materially) stable way. Second, it throws into crisis the humanist dogma that human history and natural history—or, perhaps more strongly human History and a nature that knows no history properly speaking—are separate and separable things. Time here mutates in ways that are inextricable from the mutation of a particular subject called "the human." Before getting into how Chakrabarty's project can be constellated with a set of concerns clustering around formations like "posthumanism," "new materialism," and nonanthropocentric ecologies, I want to stay for a few minutes on why Chakrabarty's formulations seem particularly compelling to me.

Chakrabarty anchors his engagement with the Anthropocene not just in history, but in postcolonial history and what he calls "our postcolonial suspicion of the universal" (p. 220). The study of modern, imperial forms of conquest and colonization quickly disabuses one of the shaky proposition that there is or ever has been a single, monolithic thing called "humanity." Chakrabarty works quickly through some formulations by Michael Geyer and Charles Bright to say that, "humanity... is not one" (p. 214). We can note, without having to go into too much detail here, that the genre of the human whose practices (material, political, ideological, economic—where these are hardly distinct things) have produced what we now call "climate change" is the genre of a Western, imperialist, masculinist humanity that has used its military, scientific, and legal machineries to force as many other "humans" (re-coded as less than humans) as possible into its orbit.

A quick example will help. In an interview with Katherine McKittrick (2015), Sylvia Wynter discusses a 2007 *Time* magazine article on the Anthropocene in relation to an unanticipated number of wildebeest deaths in the Masai Mara reserve in Kenya. The article attempted to teach the public both that "*human* activities" produced global warming, and that this geological epoch began in (approximately) 1750 and accelerated in the 1950s. Wynter diagnoses "the terrifying thing with the *Time* report": "It thinks the causes of global worming are *human* activities, but they are not! The Masai who were (and are) being displaced have nothing to do with global warming! It's all of us – the Western and mimetically Westernized middle classes – after we fell into the trap of modeling ourselves on the mimetic model of the Western bourgeoisie's liberal monohumanist Man..." (pp. 21–22). Although Wynter's terminology here can be a bit offputting, she insists on a crucial difference between the human—as a biological and sociological being—and "Man," a specific genre of Western, colonial, masculinist, *doing* humanity that becomes "overrepresented" as the whole of the human. Her reading turns on the fact that the Masai are not represented *as* human, they are Man's dialectical constitutive outside: the less-than-human. What we see in *Time* magazine then is a kind of synecdochic logic that appears, without much critical engagement, in a lot of contemporary ecological, posthumanist, and speculative realistic thought: a particular human is made to stand in for the whole. In the process, the ways that Man outsources its most pernicious ecological effects onto those represented as less than human, and their also fragile ecologies is rendered illegible. In fact, the universalist logic here ends up making the Masai as responsible for the catastrophe as those in the West who, on Wynter's reading, *really* drive the Anthropocene.[2]

Let me dwell on why this matters. To the extent that a specific "genre of human as verb," to stay with Wynter's vocabulary, is equated with the whole of humanity, the question of what is to be done now that we recognize the geo-historical fact of the Anthropocene is always asked of a radically limited tradition of concepts and institutions (Western metaphysics and its politics), and it flattens out all actual plurality in the lived experiences of being human as verb. I am going to suggest that not staying with the irreducible plural that constitutes humanness as praxis—a plurality that has been rendered "inhuman" or nonhuman by what Wynter calls Man's colonialist project—leaves us in an imaginatively and empirically poor place from which to think politics and education in and after the Anthropocene. I want to trace this now by looking to three attempts to imagine a political—and hence educational—response to the Anthropocene.

I want to begin with Tina Lynn Evans's *Occupy Education: Living and Learning Sustainability* (2012). Evans's title evokes the Occupy Wall Street movement that, in rhizomatic relation with the Arab Spring, Idle No More, and the more recent Black Lives Matter movement, seemed to signal a sea change in human political action at a crucial moment of despair (a despair that is, perhaps, a result of the ways that economic and ecological crisis are not so secret sharers). There was, for a long moment that has not yet passed, a possibility of feeling that a challenge to state, corporate, and NGO stranglehold on politics could be loosened, allowing some as yet undetermined and unnamable potential to affect change (Hardt and Negri 2012; Butler 2011). The writing of Evans's book, as the introduction states, mostly precedes the Occupy movement, but she recognized a crucial resonance between their projects before publication.

What *Occupy Education* does is outline how theories of ecology based on indigenous knowledge systems, systems theory, and ecological science can be linked to Critical Social Theory (CST), a formation that is more or less Frankfurt School critical theory plus Gramscian pragmatics. Although the book doesn't pitch itself in relation to what we can call the nonhuman or in-human or posthumanist turn in theory, it gestures, sometimes quite gracefully, in those directions: "It takes a transdisciplinary approach to integrating the academic disciplines and seeks to heal the dichotomous and destructive fractures within the modern worldview such as the separation of humans from nature" (p. 5). In this sentence, and elsewhere, Evans approaches the insight that particular forms of knowing (epistemologies crucially linked to ontologies) are directly implicated in "fractures" such as that between humans (and we can add here metonymically, "culture") and nature. A holism is proposed, one that Evans grounds in "indigenous spirituality [which] offers many examples of ontologies that recognize the holistic circle of life in which every thing and every being is related to every other thing and being" (p. 153). These passages would seem to suggest that Evans is articulating an ontology that does not differentiate humans from nonhumans. And, at times, she comes very close to doing this: "I argue for an explicitly ecological framework for CST of sustainability, one that conceptualizes humans and nature as an inseparable human/nature complex and one that addresses how the concept of human separability from nature informs Western cultural notions of domination of both nature and other people" (p. 56). This carefully disentangles an ontological "complex" from "cultural" concepts that describe and distort this complex in the service of some humans (those Wynter calls Man) who dominate non-Man genres of human being as verb, and all nonhuman beings and things.

The issue here is that this ontology slips into what she calls an "ecological framework" for CST. That is, an ontology that could almost rhyme with what Haraway (2008) calls "naturecultures" or what theorists of Object Orientated Ontologies (OOO) might call a "flat ontology" (Bogost 2011) is pressed into the service of doing CST differently without, strangely, calling into question its drives and methods.[3] Ecology here enables a redirection of CST, but not a substantial revision of it. This might not matter were it not for the fact that CST is almost wholly Anthropocentric and based in ontologies that are resolutely humanist. Thus, the second paragraph of Evans's book can say, "Addressing th[e sustainability] crisis in human institutions and systems of power is therefore an ethical and survival imperative" (p. 3). Even as she sometimes positions humans as ontologically inseparable from everything else, when it comes to politics Evans takes for granted that existing human and humanist institutions (including the settler colonial state) are the theater of political action. And, lest we forget, nonhumans have no role at all in these institutions save as *objects* of human action. Subjectivity here—in the grammatico-political sense—is reserved for humans.

I want to dwell for just a moment on why this problem. Glen Sean Coulthard's *Red Skin White Masks* argues that: "Indigenous anticolonialism, including Indigenous anticapitalism, is best understood as a struggle primarily inspired by and oriented around *the question of land*—a struggle not only *for* land in the material sense, but also deeply *informed* by what the land *as a system of reciprocal relations and obligations* can teach us about living our lives in relation to one another and the natural world in nondominating and no exploitative terms" (p. 13). This struggle not only for but *of* land, reconfigures the human "as" land, as a being situated with all other entities, vital and non, on a plane of relations: "we are as much a part of the land as any other element" (p. 61). This ontological axiom allows Coulthard to reject humanist, settler state politics and "the normative status of the state form as an appropriate mode of governance" (p. 36). All of this to say that existing humanist politics pitched at the level of the (settler) state are not compatible with ontologies which position that human as one "element" among many within land. Evans's movement to CST, then, helps us to underscore how difficult it is to think politics and the political outside of the existing humanist frames, even working with alternative, nonhumanist ontologies.

My second text is Bruno Latour's *Politics of Nature*, published in French in 1999 and translated into English in 2004. Latour was, by this point, one of the leading figures in both the field of Science Studies and the theoretical project of Actor Network Theory (ANT). Latour's earlier work had demonstrated, in painstaking detail, that Science (as a quasi-religious entity charged with adjudicating the "facts" to be reckoned with by politics) was a thorough mystification of the sciences—actual practices involving human scientists, the most varied equipment (what physicist Karen Barad [2007] might call "apparatuses"), and myriad human and nonhuman entities and forces that are only very poorly thought about as "objects." Indeed, ANT proposes that all actants—human and non—be treated the same. The actant is, we could say, the deconstruction of the subject/object binary (p. 75). It is any entity that can affect and be affected (which is why ANT and Latour's work can appear in Gregg and Seigworth's rough taxonomy of varieties of affect theory in the *Affect Theory Reader* [2010]). *Politics of Nature* is Latour's attempt to build on all of his earlier work to offer a new political model, one not founded on the distinction between humans and nature (in fact, like Timothy Morton, to whom we'll come in a moment, he wants to eschew the entire concept of "nature").

Latour begins by calling into question the entire nature/culture distinction and the way that it props up a bicameral system where politics (on the side of culture) and "nature" are separated by a "cleavage," where scientists are tasked with turning to the house of nature in order to translate its speech into something that can be heard within the house of culture. The task of his book, then, is:

> To find a successor to the ancient split that separated nature (in the singular) from cultures (in the plural), in order to raise once again that question of collectives and the progressive composition of the common world that the notion of nature, like that of society, had prematurely simplified. (p. 8)

Latour then sets out to imagine a new collective—singular and no longer bicameral—that would be progressive (that is open to change, nontelological) and would include humans and nonhumans jointly in "assembly" or "association" (p. 46). This new collective will have a Constitution (p. 19), and it will require institutions which are "contemporary inventions, unprecedented in history" (p. 43). Perhaps the most straightforward statement of the aims of the project is this:

> Once we have exited from the great political diorama of "nature in general," we are left only with the banality of multiple associations of humans and nonhumans waiting for their unity to be provided by work carried out by the collective, which has to be specified through the use of the resources, concepts, and institutions of all peoples who may be called upon to live in common on an earth that might become, through a long work of collection, the same earth for all. (p. 46)

I like this statement both because it succinctly states the rather ambitious political project, but also because it reveals what seem to me to be the crucial problems. In rejecting the human/nonhuman or subject/object binary axiomatically, it reveals that there is now and always has been a political "banality" in which all kinds of beings and things participate. This imbrication or—to use a word that appears in a lot of new materialist writing following the work of physicist Karen Barad (2007)—entanglement *is* ontological. The entirety of modern political institutions in the West (going back, in many ways, to ancient Greek thought and its re-articulations by the Romans) has disavowed this, and for Latour to foreground it, is an epochal event. The problem appears, to my mind, when Latour categorizes these "banal" associations as pre-political, and in need of the labor of collective collecting: a political autopoiesis that is open—more on that in a moment—but which does not yet exist and must come into being through work. Thus, while Latour comes very close in some ways to the kind of political, ecological praxis that I will affirm shortly under the heading of "bewildering education," he also forestalls its impact by imagining that to be "politics," there must be a single collective (which, whatever else, sounds a bit close to fascism to me) with a Constitution, a singularly modern way of understanding the political as such.

Having directly stated my main disagreements with Latour, let me quickly add two aspects of his project that I also find crucial. The first is that Latour takes communication in some form as equiprimordial with politics, but he does this without proposing a humanist conception of language. He writes, "speech is no longer a specifically human property, or at least humans are no longer its sole masters" (p. 65). This, in its own way, is epochal, and contemporary linguistics and ethology are being forced to reckon with this as well (something I've written about elsewhere [Snaza 2013]). Politics, perhaps, *is* communication, but this doesn't mean it has to take place in *human* language. It is on this point where Latour's proximity to an inhumanist affect theory sounds the loudest. Secondly, Latour also rejects a teleological conception of politics: "This is its great virtue. It *does not know* what does or does not constitute a system. It does not know what is connected to what" (p. 21). This openness, which is I think rather restricted by the insistence on a single collective with a Constitution, is precisely what an ecological politics and education will require. Latour, in a way, demonstrates how difficult it is to stay open.

This general affirmation of openness that sometimes closes due to influence of an enormously pervasive and sticky humanism is also characteristic of my third text, which is not one: Timothy Morton's *The Ecological Thought* (2012) and *Hyperobjects* (2013). In *Hyperobjects*, Morton begins, more or less, where Chakrabarty begins: "we are no longer able to think history as exclusively human, for the very reason that we are in the Anthropocene" (2013, p. 5). That, is the human is both a biological *and* a geological agent, and that this movement causes us to reckon with the fact that history is greater than the human, which means that agency, necessarily, is also something that cannot be restricted to the human (this nonhuman agency is something Chakrabarty doesn't take up explicitly, even as he radically problematizes human agency[4]). As Morton puts it near the end of the book, "what has happened so far during the epoch of the Anthropocene has been the gradual realization by humans that they are not running the show" (p. 164). This metaphor of the show, though, risks something of a teleology, suggesting that someone does or might know where the show is headed, whither it is being "run" (by whomever). This is disappointing given what I think this is one of Morton's most useful insights: "thinking and art and political practice should simply relate directly to nonhumans. We will never 'get it right' completely. But trying to come up with the best world is just inhibiting ecological progress" (p. 109). This ethics, which is profoundly open and agnostic—and which also appears in slightly different form in important books by Matthew Calarco (2008) and Cary Wolfe (2012)—is precisely where an ecological politics and pedagogy will find its praxis, affirming that we do not "know" in advance what will produce joy, or sadness, or collectivities, or negative affects, but those things can not, in fact, be known and so we must *act* (this is, as some of you will recognize, a profoundly Nietzschean proposal).

In *The Ecological Thought* (2012), Morton offers a useful metaphor for thinking about what Latour called the "banality" of associations: the mesh. Rather than a closed system, the earth "is a vast, sprawling mesh of interconnection without a definitive edge or border" (p. 8). In part, perhaps, because this reminds me of one of the uses of the acronym "S.F." so often used by Donna Haraway—it can mean string figure as in a net or mesh, science fiction, speculative feminism, socialist feminism. I like this mesh. (As an aside, let me say that even better might be Andrew Pickering's [1995] "mangle," especially as it gets put to work in the feminist theory of Susan Hekman [2010]. My worry is that mesh might be too "man-made," while the mangle suggests a contingent mess that isn't restrictable to human doing).

All of this leads me to expect that Morton will offer what I've been trying to circle so far: a way out of the Anthropocene, not just out of global warming. He comes very, very close to doing just that, but he doesn't seem to be able to help himself. Referring to a bleak or nihilistic vision of ecological catastrophe, Morton scathingly attacks the idea that we need to "'let Nature/evolution take its course,'" he writes: "This implies that we have no responsibility for, nor should we feel any guilt about, suffering beings and changing ecosystems" (p. 129). I am not so sure that it does imply this, although I concur that it's one possible reading. It could also signify precisely the thing Morton cannot see here: responsibility. There might be some ethical circumstances when, having committed a wrong, or confronted with a terrible double bind, the best course of action might be to demure. The thing is, Morton reads this deferral as profoundly violent: "'Letting be' is just the flip side of laissez-faire ideology. There is something passive-aggressive in the injunction to leave things alone, without drawing human 'interference.' There is something of the hunter in letting be: 'Be vewy vewy quiet,' as Elmer Fudd says, on the hunt for Bugs Bunny" (p. 128). But what if a deferral isn't a lying in wait, a sort of trap that, presumably, *pretends* not to act as a sort of ruse or trap. What if, instead, it might actually be *ethical* to say: I am responsible for suffering and I do not know how to act in the future as a result, so I will direct my energies elsewhere than trying to be in charge. In other words, at the end of the book, Morton's reading strategies—which seem strangely devoid of the attention to pluralities of meaning that one might expect from a professor of literature—all skew hard toward propping up the necessity of humanity taking charge in and of the Anthropocene. Thus, what Morton calls "the ecological thought" is, as both subjective and objective genitive, "of" the Anthropocene. It is not a way out (I'm thinking here both of Franz Kafka's Rotpeter and Hélène Cixous's [1975] "Sorties").

BEWILDERING EDUCATION

So, if we want to think about what a way out from the Anthropocene might look like—and not just an ecological politics and pedagogy that would avert some of the worst outcomes of global warming or climate change—we might begin by re-tracing some of insights from these texts in a way that does not circumscribe them within an anthropocentric, teleological narrative in which the human owns up to the Anthropocene and does a better job running the show (even if, as in Latour, the human only *sets the terms of the show* in which others also participate and steer). I want to offer this sketch under the title of what I have come to call "bewildering education." In an essay that was published in 2013 in *Journal of Curriculum and Pedagogy*, I proposed "that education be reconceived as a process that leads us—teachers, students, researchers, philosophers, etc.—*away from being human*, or at least away from thinking that we have any clear idea about what that means. I propose[d] that it lead us away from the stable, predictable, and cultured world of civilization, of cities, of routine, of politics as we have known it. Wither it should lead us is—and must be—unknown" (Snaza 2013, p. 49). As I did then, I want to pitch this bewildering education now in relation to feminist and queer theory, discourses that have (in certain modes at least) sought to draw forth the considerable potential of unknowing without losing sight of the fact that many "humans" (those excluded, based on their gender, sexuality, race class, and ability, from participation in what Wynter calls Man) have never experienced being part of humanity without having to struggle for recognition. Feminist and queer theory, then, offers me a way of amplifying some of the lessons offered in my analysis of the texts above.

First, "agency" is not something obvious, nor is it something that can be restricted to the human. Let's begin here by noting that any human caught up in the political and pedagogical mesh has a body. Although a great deal of modern thought was driven by the desire to rigidly delineate this body from its "environment" and other bodies, this common sense belies a crucial feature of the facticity of bodies. As Coole and Frost put it in their introduction to *New Materialisms*, "no adequate political theory can ignore the importance of bodies in situating empirical actors within a material environment of nature, other bodies, and socioeconomic structures that dictate where and how they find sustenance, satisfy their desires, or obtain resources necessary for participating in social life" (2010, p. 19). The body, in short, *is not* excempt through entanglements with hydrogen and oxygen, carbon based life forms re-configured as "food," economies, ecosystems, cultures, and trillions of microorganisms. To assert the power of the "individual" to act rationally and independently is to downplay this entanglement. We can think here, of course, of Foucault who would remind us that an agent only ever has power as a result of imbrication in network of relations that authorize certain (limited) forms of acting on behalf of some power situated elsewhere (remembering that there is no discrete agency doing the authorizing either). To the extent that humans "have" agency, it is by virtue of being a site of relay of other, inhuman powers circulating in and around the human body.

Elizabeth Grosz (2004) foregrounds this in *The Nick of Time*, one of her studies of Darwin, Nietzsche, and Bergson. She writes, of linguistic and economic systems (in relation to biological systems) that they "are not human products but are *inhuman*: systems functioning beyond or above the control of their participants, systems that, as much as biological processes, form and produce their subjects" (p. 39). The human is, then, an *effect* of inhuman systems. She too ties this to the body: "we need to understand the body, not as an organism or entity in itself, but as a system, or series of open-ended systems, functioning within other huge systems it cannot control, through which it can access and acquire its abilities and capacities" (p. 3). Between Coole and Frost and Grosz, we can propose that the human as "biological" and as "geological agent" only makes sense in relation a host of relations and entanglements with inhuman systems that produce the human as a particular entity that can *mistakenly understand itself as closed*, bounded, individuated. Stacy Alaimo's notion of "trans-corporeality" is helpful here too, since it refers to "the time-space where human corporeality, in all its material fleshiness, is inseparable from 'nature' or 'environment'" (2008, p. 238). Whatever else human agency is, it is a sort of aftereffect of inhuman agencies, and any education that would have humans reckon with the Anthropocene has to begin with this messy, shifting, ontologically primary openness.

Second, we have to let go of our desire to plan, to act in ways that conform to a priori rules or maxims, to act only in the (false) certainty that our actions are just because they are oriented toward some good which we cherish. We must stop deriving the options for present actions from an anticipated justification of those actions as "right." On the one hand, this is forced upon us by what Chakrabarty diagnoses as the Anthropocene's sundering of our ability to imagine futures from an axiomatic assumption that the future will be continuous with the past. We cannot say what global warming will or will not *actually* do to human cities and forms of inhabitancy, and so our actions cannot, in any way, take for granted that we can know or predict the future. This is likely to be the rub for many educators since we are saturated in a supersolution of ideologies making it seem like teaching cannot possibly proceed except toward an end, a telos, a set of learning outcomes, a vision of a better world. And yet, this is precisely where I think an education attuned to the Anthropocene has to unfold: in a space of radical uncertainty with respect to any future that would make the present "worthwhile."

Grosz finds an "aleatory materialism"[5] in Darwin, Nietzsche, and Bergson, and she uses this aleatory ontology to ground struggles associated with feminist, queer, antiracist, and postcolonial politics. She writes that, "Political and cultural struggles are all, in some sense, directed to bringing into existence futures that dislocate themselves from dominant tendencies and forces.... Political activism has addressed itself primarily to a reconfiguring of the past and a form of justice in the present that redresses or rectifies harms of the past. It needs to be augmented with those dreams of the future that make its projects endless, unattainable, ongoing experiments rather than solutions" (2004, p. 14). Let me repeat that: "*experiments rather than solutions.*" What we need is not, then, a "solution" to the problem of climate change, one that would—at last—redress the violence of "human" civilization that has made the earth, quite literally, behave in ways that may well make human civilization impossible in the future. What we need, instead, is a series of experiments, as many of them as possible, that would test out uncertain, hap-hazard ways of entangled living that attune to global warming or climate change without expecting that they can ever lead to solutions.

This leads me to my third lesson, which is that being responsible for global warming does not necessarily entail striving to figure out how to get the inhabitants of the earth (human and not) organized into some sort of collective that would exert its collective agency in order to avoid some consequences of climate change, especially if these actions end up being guided by the humans Wynter calls Man. I think this involves experimenting with pedagogies that affirm what Jack Halberstam (2011) calls the queer art of failure. Discussing the movie *Chicken Run* and Roderick Ferguson's reading of black lesbian feminism, Halberstam calls for:

> Feminists... [to] think about a shadow archive of resistance, one that does not speak in the language of action and momentum but instead articulates itself in terms of evacuation, refusal, passivity, unbecoming, unbeing... A different, anarchistic type of struggle requires a new grammar, possibly a new voice, potentially the passive voice. (p. 129)

This new grammar and/or new voice is crucial. What is not needed—contra Latour –is a new formalization of political relations with new institutions. This would especially be true of any institutions instituted by those humans—Man—who already have the most say in how the world is run. What is needed is new discourses, new lexicons, new ways of enacting speech—understood, as Latour does, as something that is not only the providence of humans—that would allow us to attune to the *actual*, banal political reality of entanglement that we have hitherto failed to register. We are always already *in* politics with all beings and things. That is, what Latour sees as a pre-political reality that calls for properly political institutions, is, for me, the very scene of politics.

There is no movement that any thing can make that is not political, not about relations among things.

In this context, I don't see much sense in an ecological education that would seek to produce "better" human subjects who act more knowingly in the spheres of electoral politics and consumerist expenditure. I don't rule out that those may well be arenas in which changes must occur, but I think we limit our abilities to imagine ourselves and the world when we take "subjects" and "economies" and (settler colonial) "nation-states" for granted. Instead, I want us to get lost, bewildered, estranged from any and every certainty of our political, economic, and epistemological grammars (including those certainties of "I," subjectivity, gender, race, class, sexuality, species).[6] I want us to figure out as we go how we should act in ways that are driven by as open an attunement as is possible to all those present—human and non—and to their flourishing (to use a word from Donna Haraway's [2008] nontelological, nonhumanist ethics). I want us to learn from Halberstam to embrace our failures: "In certain circumstances failing, losing, forgetting, unmaking, undoing, unbecoming, not knowing may in fact offer more creative, more cooperative, more surprising ways of being in the world" (2011, pp. 2–3). I think what we could use is learning how to *fail* at being geological actors, and let our failures teach us that we never really were "agents" like we thought anyway, that there is a swarming mass of agentic matter all around us and in us, and that letting go isn't a passivity of violent, nihilistic abdication of our duties, but an experimental ethical approach informed by admitting complicity in violences that can't help but make us who we are.

NOTES

1. The phrase "geological agent" is Naomi Oreskes's (2007), an historian of science who reviewed more than a decade of scientific literature on climate.
2. See debates about "differentiated responsibility" and "common but differentiated responsibility" in late 1980s/early 1990s (discussed in Chakrabarty 2009).
3. One can get a sense of how this ontology can be figured into political struggle in Coulthard's (2014) of the politics of land and indigenous sovereignty in Canada.
4. We could note his use of "falling" and "stumbling," which underscore the contingency this and a kind of attentiveness to other possible agencies affecting the human. Chakrbararty's more recent talks (March 2015)

have ended by explicitly posing the problem of human *and* nonhuman agencies), in part through reference to Latour.

5. There is as Coole and Frost note, a history of "aleatory materialism" that was an "underground stream" within Western philosophy (2010, p. 35), one that found expression Epicurus, Spinoza, Marx, Heidegger, and Athusser: "It emphasizes emptiness, contingency, and chance" (p. 35).

6. It is crucial to note that this bewilderment will take enormously different forms given how people are already oriented toward Man and its humanism. For those subjects who have never been admitted into Man (at least not without enormous struggle) (Luciano and Chen 2015), this disorientation will not be the same as it will be for those whose "humanity" has been recognized by the state. Indeed, we have to ask, with Lauren Berlant (2011): "Why do some people have the chops for improvising the state of being unknowing while others run out of breath...?" (p. 37).

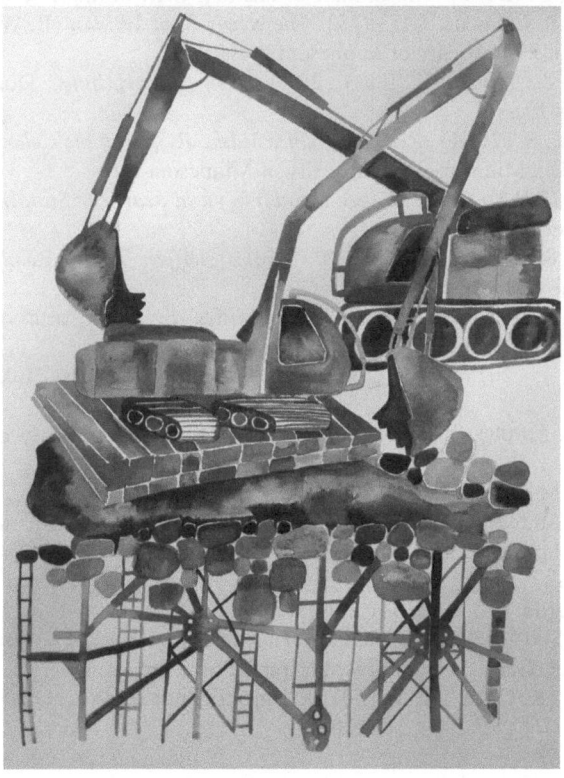

Image 15.1   Cranes (Mia Feuer, watercolor)

REFERENCES

Alaimo, S. (2008). Trans-corporeal Feminisms and the Ethical Space of Nature. In S. Alaimo & S. Hekman (Eds.), *Material Feminisms* (pp. 237–264). Bloomington: Indiana University Press.

Barad, K. (2007). *Meeting the Universe Hallway: Quantum Physics and the Entanglement of Matter and Meaning.* Durham: Duke University Press.

Berlant, L. (2011). *Cruel Optimism.* Durham: Duke University Press.

Bogost, I. (2011). *Alien Phenomenology, or What It's Like to Be a Thing.* Minneapolis: University of Minnesota Press.

Butler, J. (2011). Bodies in Alliance and the Politics of the Street. *Transversal.* Available at http://www.eipcp.net/transversal/1011/butler/en. Accessed 2 January 2017.

Calarco, M. (2008). *Zoographies: The Question of the Animal from Heidegger to Derrida.* Chicago: University of Chicago Press.

Chakrabarty, D. (2009). The Climate of History. *Critical Inquiry, 35,* 197–222.

Cixous, H., & Clément, C. (1975). *The Newly Born Woman* (B. Wing, Trans.). Minneapolis: University of Minnesota Press.

Coole, D., & Frost, S. (Eds.). 2010. *New Materialisms.* Durham: Duke University Press.

Coulthard, G. S. (2014). *Red Skin White Masks: Rejecting the Colonial Politics of Recognition.* Minneapolis: University of Minnesota Press.

Evans, T. L. (2012). *Occupy Education: Living and Learning Sustainability.* New York: Peter Lang.

Gregg, M., & Seigworth, G. (Eds.). (2010). *Affect Theory Reader.* Durham: Duke University Press.

Grosz, E. (2004). *The Nick of Time: Politics, Evolution, and the Untimely.* Durham: Duke University Press.

Halberstam, J. (2011). *The Queer Art of Failure.* Durham: Duke University Press.

Haraway, D. (2008). *When Species Meet.* Minneapolis: University of Minnesota Press.

Hardt, M., & Negri, A. (2012). *Declaration.* Available at https://antonione-griinenglish.files.wordpress.com/2012/05/93152857-hardt-negri-declaration-2012.pdf. Accessed January 2, 2017.

Hekman, S. (2010). *The Material of Knowledge: Feminist Disclosures.* Bloomington: Indiana University Press.

Latour, B. (2004). *Politics of Nature: How to Bring the Sciences into Democracy* (C. Porter, Trans.). Cambridge: Harvard University Press.

Luciano, D., & Chen, M. Y. (2015). Introduction: Has the Queer Ever Been Human? *GLQ, 12*(2–3), 183–207.

Morton, T. (2012). *The Ecological Thought*. Cambridge: Harvard University Press.

Morton, T. (2013). *Hyperobjects*. Minneapolis: University of Minnesota Press.

Oreskes, N. (2007). The Scientific Consensus on Climate Change: How Do We Know We're Not Wrong? In J. DiMento & P. Doughman (Eds.), *Climate Change* (pp. 65–100). Cambridge, MA: MIT Press.

Pickering, A. (1995). *The Mangle of Practice: Time, Agency, and Science*. Chicago: University of Chicago Press.

Snaza, N. (2013). Bewildering Education. *Journal of Curriculum and Pedagogy, 10*(1), 38–54.

Wolfe, C. (2012). *Before the Law: Humans and Other Animals in a Biopolitical Frame*. Chicago: University of Chicago Press.

Wynter, S., & McKittrick, K. (2015). Unparalleled Catastrophe for Our Species? Or, to Give Humanness a Different Future: A Conversation. In K. McKittrick (Ed.), *Sylvia Wynter: On Being Human as Praxis* (pp. 9–89). Durham: Duke University Press.

CHAPTER 16

Dispatch from the Future: Science Fictioning (in) the Anthropocene

Jessie Beier

White Skies Ahead (or, Keep Cool and Carry On)

The sky has been a particular shade of pale for 37 days now. Each morning, after rolling out of bed, smartphone in hand, I open the blinds to see the same alabaster sky, suspended like a dropped ceiling created to hide the meteorological infrastructure that mediates what we have come to call "the weather". Just a few months ago, the low, white sky would have been seen as a harbinger of precipitation, but today, June 27, 2023 marks the forty-second day without rain in the Canadian prairies.[1] Today also marks the first anniversary of the Anthropogenic Albedo Modification Project (AAMP) launch in this area, a multi-stage Solar Radiation Management (SRM) scheme that has been put into place to moderate global warming by reducing the amount of sunlight that reaches the earth.[2] This geoengineering enterprise, hailed as a "Global Clean-up Project!" and the "Final Solution to Global Climate Change!", uses high-tech cannons to shoot sulfur particles into the stratosphere in order to increase the Earth's reflectivity (or albedo), in turn creating a cooling effect on our warming planet. Or, as the campaign has pitched (and sold) it: "Keep Cool and Carry On!".

J. Beier (✉)
University of Alberta, Edmonton, AB, Canada

© The Author(s) 2018   359
j. jagodzinski (ed.), *Interrogating the Anthropocene*, Palgrave Studies in
Educational Futures, https://doi.org/10.1007/978-3-319-78747-3_16

After my reluctant check-in with the weather, I continue with my usual and more full-on checking-in process on my various digital networks. Each morning, I log in with the trustX data management platform, one that I have been using for 13 months now, which provides me with my daily dose of aggregated news and social media updates. trustX houses all of my email and social media accounts, tracks my online interactions, and provides me with updates based on my online history and "preferences". The trustX system is one of the more reputable Knowledge-Based Trust (KBT) networks available today, operating through a dynamic multi-layer probabilistic modeling system that helps to filter the information I receive (based on KBT rankings) so that what I am presented with each morning has been unanimously agreed on as *a reasonable proxy for truth*.[3] Today, for instance, after checking-in with my various social media networks, a process that both aggregates relevant news stories and mines my data, the reliable trustX algorithms spit out upcoming events, product promotions, potential contacts to add to my network, and up-to-date news stories that are ranked according to both popularity and the correctness of factual information provided by the source.

A few of the news stories catch my eye. "Hypercane Brings New Hope" mentions the weather conditions presently sweeping the Eastern shores of both Canada and the USA, suggesting that the most recent wave of hypercanes[4]—the extreme tropical cyclones borne from rising sea levels and warming waters—have provided new opportunities for people to band together and build new communities. A few swipes down are more reports of good cheer and happy affects; the article "Optimism Soars with Food Prices" positions rising food costs as an exciting new opportunity for local businesses, while "Alberta Schools Lead the Pack" discusses how recent educational reforms, characterized by policies that push further privatization and commercialization of once-public schools, have placed Alberta at the top tiers of recent PISA evaluations.[5] Even local weather reports seem optimistic in their predictions: "White Skies Ahead, No Need to Dread" makes brief reference to the extant bleached skies before shifting attention to the celebrations taking place this week in honor of the first anniversary of AAMP. According to this source, the milky skies are nothing to fear, but rather, this is a "normal" side effect of suspending fine aerosol particles in the stratosphere. Despite swelling suspicions that the planet is undergoing uncertain, and potentially catastrophic transformations, indeed, even at the brink of ecocide and calamity where the life support systems of the Earth are being damaged in ways that threaten our very own survival, today's newsfeed assures me that *everything is under control*.

The reality presented to me via this data platform, which has been championed for its ability to reliably compute the true trustworthiness levels of various sources, is one characterized by good cheer, optimism, and ultimately, *hope* for The Future™. This decidedly static tomorrow, which proceeds in a linear and continuous line from today, not only offers a perpetual dream of transcendence—tomorrow *will* be better than today—but also occupies the horizon of the thinkable, what we deem possible in the first place, in turn driving our collective crusade for low-level happiness. That is, this future-oriented sense of hope, this "inconstant pleasure",[6] helps us cover over the threat of even our own species' annihilation, in turn maintaining the current state of affairs. Even in the face of ample "evidence" and scientific "facts" that denote how humans, an exceptionally exceptional species, have ushered in a new geological epoch, it's business as usual on this planet we call Home. It is *as if* this was a science fiction story and we are those unwitting characters being manipulated by concealed alien forces to unquestioningly accept the status quo via subliminal messaging.[7] But this is not science fiction. And, as trustX has guaranteed me for over a year now, the information I am receiving is accurate, reliable, and ultimately, *truthful.*

ON FICTION (OR, POST-POST-TRUTH TIMES CALL FOR POST-POST-TRUTH MEASURES)

KBT networks, such as trustX, were first rolled out in late 2017 as one measure to deal with the growing acknowledgment that we now live in a "post-truth" era. In fact, post-truth was the 2016 Oxford Dictionary word of the year, introduced as an adjective "relating to or denoting circumstances in which objective facts are less influential in shaping public opinion than appeals to emotion and personal belief".[8] In other words, the post-truth era has been described as one in which emotion, affect, and ultimately, *belief,* are more significant factors in shaping public opinion and politics than so-called facts and evidence. Following the 11/9 events of 2016 when a demagogue was elected to the USA Presidency (and the subsequent re-election in 2020[9]), which resulted in the full-on weaponization of stupidity and "alternative facts",[10] the distinctions between truth and lies, and by extension, reality and fiction, seemingly became ruptured beyond repair. To deal with this new post-truth reality, measures such as KBT networks were introduced to help manage, even govern, the general politics and regimes of truth in online spheres in order to enable authoritative distinctions between true and false information. Slaves to the algorithm, we now rely on aggregated and filtered data sources, such as trustX, in order to help manage the "truthiness"[11] of our current lived reality.

What these measures failed to consider, however, is how the world always exceeds our conception of it. No matter how "reliable" or "trustworthy" the KBT scores profess to be, they can only ever provide a partial perspective of reality. After all, things in the world are not cut to our measure; even with the deep and pervasive need for a logical universe that makes sense *for-us*, "the real universe is always one step beyond logic".[12] Reality cannot be wholly represented *in-itself*,[13] be it through language, images, or algorithms, and so what we are presented with is always already *fictional*. Of course, the world itself is not a fiction, but rather our attempt to explain, to conceptualize, and ultimately to *know* the world is characterized by a series of fictions.[14] Where the real is understood as a problematic and substantive multiplicity that exceeds human representations, any reality we construct is necessarily fragmentary, holey, a pervious tissue of inconsistencies. This is not to say that there are no particular facts or regularities between certain certainties, but rather that the "laws" that chase these regularities are always *contingent*.[15] Scientific models, for instance, can only ever be approximate calculations of reality; these deductive fictions merely function to attune reality to particular questions or research aims, necessarily ignoring, obscuring, or simply rejecting other anomalous realities in the process. That is, even "Science", which is often positioned as an authoritative arbiter of truth, cannot observe all things, all the time, from all perspectives, and thus what science is able to tell us about reality functions through a restricted angle of observation and hypotheses that proceed through "as if" postulates.[16] These "as if" presuppositions do not stand in opposition to reality, but instead constitute what is perceived as reality itself by means of fictive designs that are necessarily only ever *partly true*.[17] The aggregate of news I receive daily may, indeed, be true, or at least it is *not not true*, but it is not *wholly* truthful. Rather, it is a *holey* truth, a partial perspective that does not replace or stand opposed to reality, but instead to the *ideal* of a consistent, universal, and totalizing construct of reality. Put briefly, our situation is "a fiction set and invented by us".[18]

In philosophical terms, there is nothing especially timely about this assertion. From Plato's contention that all we ever see is shadows, to the Renaissance sentiment that the world is but a dream, to Baudrillard's assertion that the sign has been absorbed by the signifier, producing a hyperreality that has become more real than reality itself, there have always been thinkers who feel that the world is a misleading fiction.[19] Where speculative philosophers such as Vilém Flusser and Gilles Deleuze depart, however, is with the idea that our fictional understanding of reality has subsumed some sort of original external reality. As Flusser himself writes, "there is no comparative reference for the fiction that surrounds us. Fiction is the only reality".[20] Likewise, Deleuze recognizes the fictional character of those all-too-human regimes of representation that have come to think on our behalf, asserting that "the first presupposition of philosophy is also its first lie".[21] For Deleuze, there are no hard-and-fast facts independent of events, because the significance of any fact depends on how we associate that information with the physical and ideal changes expressing it.[22] In short, and as Deleuze writes, "thought is filled with no more than an image of itself, one in which it recognizes itself more than it recognizes things".[23] Taking these affirmations as a starting point for thinking about what constitutes truth and falsity therefore necessitates a different attitude toward questions of fiction and reality, that is, we must not lament the fictional character of the world, but instead recognize the sense of the fictitious all around us.[24] Rather than bemoaning the loss of some authentic truth, these assertions indicate how fictions both represent reality *and* through such representation actualize *another world*, albeit one that is made invisible.[25] Indeed, "there is another world, but it is this one".[26] That is, through the necessarily fictitious representations that have come to constitute our world, there are always "points of deterritorialization" that are able to "bifurcate worlds and release incompossible and indiscernible elements that enter into new variations around the position of the actual".[27] These bifurcations, inconsistencies, and contradictions must not be denied or "solved", but rather we must recognize both the illusory nature of a universal and stable reality defined by an ideal truth, as well as the illusion of the end of all illusions.

In the post-post-truth context, however, instead of acknowledging the ways in which reality itself is the coincidence of varied fictions, we seem to be searching for a definitive truth. After all, the main impetus behind the implementation of online KBT networks was to provide more legitimacy to the veracity and reliability of information in a time when deception, distortions, disinformation, and "fact-free 'news' stories spread like wildfire".[28] Faced with increasing postures of anti-intellectualism, scientific skepticism, and a cultural milieu where personal opinions had come to outweigh "expert" knowledge, platforms such as trustX were introduced to monitor, control, and ultimately manage truth itself. Not unlike those thinkers that seem eager to preserve a reality that is *not fictitious*, trustX assumes that "the truth is out there". Not only that, but the platform operates as if this truth can also be measured, in a quantitative manner, in order to produce trust rankings, which are then used as the basis for future trust score comparisons and classifications. Although the truth produced by trustX is contingent upon those bits and bytes of information that have been deemed accurate or reliable in the past, it is nevertheless positioned as a dependable and trustworthy mediator of worldly realities and future truths. Further, despite the inherent holiness of the "Knowledge Vaults" on which trustX draws, particular "regimes of truth"[29] are legitimized based on what has been considered "good" or "necessary" in the past–present *as if* this hegemonic reality system is just "the way things are", *as if* this vault of information is natural, neutral, *given*. Not unlike current political imaginaries, where (insidious) propaganda machines and dispersive control systems have come to manage our very perception of the world, the truth indices generated by trustX are legitimized based on the disappearance of those fictions through which the platform operates in the first place.

This morning, for instance, the "truths" I am fed through my newsfeed present a world where ongoing human interventions are capable of providing hope and optimism in spite of the uncertain transformations taking place on the planet. While dire situations produced by anthropogenic climate change continue to unfold across the globe, indeed, while crises gather force and speed, the dominant sentiment advanced by my newsfeed is nevertheless characterized by an unwavering confidence in human ingenuity and a bright future. Extreme weather events, for instance, are seen as the necessary catalyst for people to put petty differences aside and finally come together in order to build new communities. This story proceeds, however, with little to no acknowledgment of the anthropogenic nature of these eco-catastrophes, let alone the motors, including neoliberal petro-capitalism, that have exposed the limits of our arbitrary divisions between so-called "nature" and "culture". What the current onslaught of hypercanes makes apparent is how these seemingly distinct categories are nothing of the sort, but rather "they are seamlessly swept together in counterclockwise rotations".[30] Faced with the unpredictable, yet interrelated, ecological fluctuations over which we actually have little or no control, our Guardians[31] have abandoned all hope of changing current economic and political organizations, opting instead for expensive and disruptive relocation strategies. Likewise, the various food crises, spurred in part by the relocation of climate migrants both nationally and internationally, are understood as a new field of opportunity for Entrepreneurs,[32] that is, opportunity for capitalist expansion and new commodity frontiers.[33] Rather than provoking panic or reflection, the devastation of agricultural yields and subsequent mass migrations and soaring food prices have generated excitement for those who might profit from new market niches and "innovative" product development. This same sense of capitalist opportunism has seeped into all areas of contemporary social and political life, including schools and universities, which now operate under a "business ontology" that subjects something as ephemeral as learning itself to market pressures and government-imposed targets such as PISA scores and Fraser Institute rankings. In turn, schools contribute to the maintenance of a broader fiction wherein competition, individualism, and standardization are not only necessary, but desirable factors for "success" within dominant social, political, and economic organizations. Despite the convergence of crises that now defines contemporary existence on this planet, including declining annual rainfalls and severe droughts, intensified social inequalities, and political polarization and extremism, these bleak realities are *sold* as optimistic sites for community-building, economic development, capital accumulation, and ultimately, "progress".

In the case of my brief morning encounter with reality, then, I am faced with a multi-layered series of fictions, or put otherwise, an *assemblage of partial conceptualizations* that have come to constitute a *common sense* vision of reality.[34] Despite Gaia's intrusion, signifying that "there is no afterwards",[35] despite shouts from the cosmos that "[a] world is in danger, this planet is in great danger!",[36] the information I am presented via my newsfeed assures me that, indeed, danger can be thwarted, and, in fact, everything is going to be A-OK. While trustX produces a series of "as if" postulates that aggregate and compare the trustworthiness of information as if the world can be described and explained based on identifications with online "Knowledge Vaults", mass-media outlets continue to peddle in comforting narratives, in turn creating a climate of unquestioning passivity. Beyond algorithmic aggregates and mass-media homogenization, a series of fictions also underline those common sense assumptions about the world that render certain things thinkable in the first place. These common sense narratives, or, that which "everybody knows", only have to get things *right enough*, for practical purposes, to help us manage the range of situations to which we have become accustomed. That is, these common sense understandings of the world are under no obligation to adequately represent the complexity of the world, opting instead for reductive and familiar representations that operate through a conservatism which limits what we perceive possible, even thinkable, in the first place. In just a few minutes with my newsfeed, for instance, I encounter a world defined by common sense stories of *bounded individualism*, where the human subject is characterized by discrete identities, a stable and rational "I", and a distinct separation from the ecologies it inhabits; stories of *unlimited progress*, where tomorrow proceeds from today in a linear, predictable, and limitless manner; and, stories of *human exceptionalism*, where humans are positioned as a species that is not only categorically different than those non-human bodies with whom we share the planet, but also a species that is capable of controlling and transforming the Earth in concordance with all-too-human will and desires.

While these fictions may help to simplify our orientation to the world, they can only ever enable a partial grasp of reality. These fictions cannot claim to understand reality as a whole, as some sort of "god trick of seeing everything from nowhere",[37] and thus the realities they represent are necessarily holey, fragmentary, incomplete, or, *inadequate*.[38] The idea that we are the all-knowing protagonists in our own stories has fueled an infantile fantasy of omnipotence, one that consoles rather than disturbs us. To think that our understanding of reality "owes its existence to a fictionalized consensus will [therefore] always carry an uncanny charge".[39] It is this uncanniness, the inherent strangeness, and unknowability of the world, that we seem desperate to fight off. But, as Flusser asserts, the known world has always been a counterfeit and spurious fiction because reality itself can never be fully ascertained. Instead of lamenting and/or seeking to overturn this fictional character of the world, we must instead make explicit the speculative conditions of how we have come to construct current realities. Where we must place our focus, then, is not with the question of what is true or false, but instead how "truthiness" emerges in the first place, the worlds it creates, and the worlds it disappears in the process. If "reality is fiction, and fiction is reality",[40] we must not only recognize reality's fictitious character but also adopt an *anamorphic gaze*, a partial perspective that dislodges the true/false binary and recognizes how different things are made legible from different positions.[41] While it may be easier, or at least more comfortable, to simply lament the loss of truth, to become nostalgic for those good ol' pre-post-truth days, this sentimentality is itself founded in illusion. If anything, what the post-post-truth moment makes apparent is a general loss of faith in reality as something discoverable and given, and thus the need to examine and potentially challenge the truths we create. Instead of mourning the loss of truth, which for trustX involves providing measures to manage unanimously agreed upon *proxies for truth*, we must recognize the fictional character of the world and the "as if" postulates that guide our thought and action in order to ask if these are, in fact, the fictions we desire.

ANTHROPOSCHEMES (OR, HOW I LEARNED TO STOP WORRYING AND LOVE SULFUR)

Between stories of opportunity, progress, and new hope, today's news-feed also features several articles highlighting the celebrations and ceremonies taking place in honor of the first anniversary of the AAMP in North America. Commemorating the achievement of the first stage of the AAMP pilot, which was completed in 2022 at Harvard University,[42] as well as the subsequent experiments taking place across North America, including here in central Alberta, today's updates range from press conference briefings and political handshakes to previews of local community gatherings that have been organized to celebrate the AAMP launch and its promises. The AAMP pilot was able to take off, so to speak, based on particular "as if" postulates that positioned human techno-scientific intervention as not only the most effective and desirable way to mitigate climate change, but also the most *plausible*. That is, this particular anthroposcheme, was able to push forward based on the common sense insistence that although the very best option would be to reduce overall greenhouse gas emissions through radical changes to our ways of living on this planet, such an approach would simply not be possible. Or as Nobel Prize-winning atmospheric chemist Paul Crutzen asserted, "if sizeable reductions in greenhouse gas emissions will not happen and temperatures rise rapidly, then climatic engineering, [...], is the only option available to rapidly reduce temperature rises and counteract other climatic effects".[43] What once seemed like science fiction became situated as the only plausible way to deal with impending global temperature rises, and thus projects like AAMP were *actualized* through a particular series of hypotheses and fictive designs, or "as if" presumptions, that propelled this anthroposcheme from fiction to reality.

Geoengineering, also called climate engineering or ecological engineering, was first proposed as early as the 1960s, but it wasn't until the mid-2000s when such strategies were legitimized as *real* options to mitigate global climate change. Described as large-scale and deliberate techno-scientific interventions in climate systems to moderate global warming, geoengineering proposals have taken many forms. While SRM and carbon dioxide removal strategies are the most common schemes discussed,[44] proponents of geoengineering have also proposed ideas such as towing massive icebergs to the coast of desert nations to induce precipitation[45] and wrapping Greenland in a blanket to reflect the sun's rays and prevent the ice caps from melting.[46] Driven by the grim realizations that humanity would have as much as a 50/50 chance of staying below the 2°C threshold *even if* the world hits zero carbon emissions before 2045,[47] geoengineering is now supported by both scientific communities and international organizations, such as the Intergovernmental Panel on Climate Change (IPCC), as the best possible way to mitigate warming temperatures and impending ecological crises.[48] Although geoengineering is now an accepted reality, when schemes such as SRM were first introduced, extensive debate ensued. Critics raised questions about the potential risks and side effects that could result from "hacking" the planet, including threats of further environmental degradation based on the inability to control potential global impacts and the inherent unknowability of the consequences that might arise. In the early literature, even those that championed geoengineering schemes, such as SRM, underlined the potential risks, including decreased precipitation levels, potential ozone loss, increased air pollution (followed by increased pollution-related deaths), and even the onset of acid rain and increased ocean acidification.[49] Others expressed fears of technologies falling prey to mechanical failure, human error, or weaponized appropriation, all of which could not be readily predicted or managed in advance.[50] Beyond potential climactic and environmental risks, detractors also highlighted the ways in which such techno-scientific interventions failed to take into account the potential social and political ramifications of the scheme. In the absence of any sort of international agreements or treaties regulating geoengineering practices, combined with the fact that climate is indifferent to national borders, critics highlighted the risk of new geopolitical tensions that could arise from the negotiation and trade-offs between the amount of geoengineering that would be best for one region as opposed to another.[51] Further, even advocates of geoengineering strategies acknowledged that these interventions could lead to an induced complacency in dealing with the fundamental causes of climate change, namely petro-capitalism and its reliance on unlimited capitalist expansion and accumulation.

Despite the various risks raised in these early conversations, advocates such as Harvard Kennedy School professor David Keith, who literally wrote the book on climate engineering, argued that those risks appeared "relatively small when compared to the risks of rapid climate change".[52] Ten years after Keith's assertions, when carbon dioxide readings have now far surpassed the 410 parts per million milestone,[53] it really does seem that geoengineering may be our only hope for dealing with global climate change and its impending annihilations. Instead of re-imagining our ways of living on this planet, which might look like placing a moratorium on growth alongside massive redistribution of resources and large-scale shifts in social planning, geoengineering schemes are able to advance based on the overwhelming assertion that "there is no alternative". This *realist impulse* is particularly apparent in the articles that populate my newsfeed this morning. Today, on the first anniversary of the project, all conversations of potential risk have been replaced with celebratory and triumphant messaging. "Scientists 'AAMPed' Up By SRM Pilot Project" recaps a press conference and dinner that took place last night, wherein the pilot project was championed for its relatively low financial costs[54] and quick-acting effects. In addition to the apparent success of the project, the article highlights how the first stage has also provided more comprehensive data to model future stages of geoengineering schemes, providing more sufficient information for policymakers to make decisions about prospective climate mitigation interventions. Not unlike the feedback loops through which the trustX network operates, AAMP and its subsequent modeling have provided a new framework for understanding the "reliability" and "trustworthiness" of anthropogenic interventions and the potential for human ingenuity to mitigate impending eco-catastrophe.

This trust in, and even celebration of, human measurements, understandings, and ultimately ingenuity is not unique to the phenomenon of geoengineering, but is indicative of a broader acknowledgment of the role human impact has had on transforming the planet. As early as 2000, when Paul Crutzen first exclaimed "We're no longer in the Holocene but in the Anthropocene!",[55] the role of human impacts on the Earth have been at the forefront of "natural" and "social" scientific discourse alike. The Anthropocene was first introduced to signal the geological era, where, through its activities and its growing population, the human species has emerged as a geological force now altering the planet's climate and environment. In scientific domains, the Anthropocene invoked a new need to not only name, but come to terms with the changing role of human impact on ecological systems. As geologist and chair of the Working Group on the Anthropocene (WGA), Jan Zalasiewicz, contended, "[t]he significance of the Anthropocene is that it sets a different trajectory for the Earth system, of which we of course are part".[56] Beyond scientific discourse, the Anthropocene also drew the attention of humanities and social sciences scholars, catalyzing much-needed interdisciplinary discourse to deal with the growing acknowledgment that the world was undergoing uncertain transformations brought on by human activities. While the Anthropocene provided an "amazing lexical invention",[57] which invigorated ecological, social, and political discourse alike, it was also criticized as yet another master narrative that failed to recognize the scale. and complexity of the situation. Critics suggested that this naming device was just another compensatory charge aimed at reducing the unpleasant and often unfathomable effects of anthropogenic climate change, or worse, just another form of liberal jouissance, a vehicle for "enjoying destruction, punishment, and [ultimately] knowing".[58] Others noted how the label remained faithful to a particular set of fictions defined by human exceptionalism, which in turn reified a new set of grand narratives and naive universalisms. That is, at the same time that Anthropocene discourse provided an opportunity to unravel and potentially challenge thought based on an atmosphere of deep-seated human exceptionalism, it also (re)produced a series of fictions that ultimately re-centered a human species-supremacist planetary politics where "Man" was, once again, positioned as the hero of the story.[59] Furthermore, while these Anthropocene fictions drew attention to the power of human dominance over the planet, they necessarily disappeared other realities, namely those produced and maintained by the dispersive, deterritorialized, and non-human forces of contemporary capitalism. That is, critics noted how the so-called Anthropocene did not begin with "Man" and his ever-increasing needs—#notallhumans—but with capitalism and its necessary enclosure of the commons, destructive mechanisms of extraction and exploitation, and relentless drive for infinite accumulation.[60]

And so, as dissatisfaction with the totalizing narratives of Anthropocene discourse proliferated, so too did a myriad of alternative names, each of which (cl)aimed to propose a more adequate set of fictions to describe the situation: Capitalocene, Eurocene, Anthrobscene, Chthulucene, Plantationocene, Plasticene, Misantanthropocene, Betacene, Planthropocene, Thermocene, Thanatocene, Phagocene, Phronocene, Agnotocene, Polemocene, Anglocene, Metropocene, Gynocene, Manthropocene, Sociocene, Homogenocene, Econocene, Aerocene, Growthocene, and Palaeoanthropocene.[61] Although each naming device drew attention to the particular conditions of power, modes of relation, and historical contexts through which we might understand ecological problematics, what each name made apparent was the difficult task of thinking about, let alone representing, the "hyperobjective",[62] violently slow,[63] and, ultimately, incommensurable character of anthropogenic transformation. Once again, the world always exceeds our conception of it, and thus, "[t]o name is not to say what is *true,* but to confer on what is named the power to make us feel and think in the mode that the name calls for" (my italics).[64] Indeed, *naming is a risk.*[65] In the case of naming the Anthropocene, the risk lies in how particular names work to reconstitute the world by means of fictive design. Naming the Anthropocene frames humanity's impact as incontestable, further enabling human specialness to slip in so as to extend human conquest over the Earth from geographical space to geological time.[66] In short, the Anthropocene delivers us a "Promethean self-portrait" that illustrates "an ingenious if unruly species, distinguishing itself from the background of merely-living life, rising so as to earn itself a separate name (anthropos meaning 'man,' and always implying 'not-animal') and whose unstoppable and in many ways glorious history has yielded an 'I' on a par with Nature's own tremendous forces".[67]

Thinking back on this debate, I can't help but feel a bit nostalgic. At least in the "good ol' days" Anthropocene discourse gave some pause for thought. Now, it seems The Anthropocene™ is not only accepted as both the geological epoch and cultural milieu in which we are currently living, but is also *venerated.* After all, it was the legitimization of the Anthropocene as a new geological boundary that catalyzed our current anthroposcheming in the first place. Or, as one of today's headlines read: "The Anthropocene Offers A Good Path in Turbulent Times".[68] Where Anthropocene discourse once provoked debate and climate change denial was seen as a fringe attitude (despite rather large swaths of non-believers), today's Anthropocene denial, and by extension, climate change denial, is much more ubiquitous, quiet, insipid, *banal.* That is, one of the most troubling things about the Anthropocene is the sheer number of people it fails to trouble.[69] Perhaps, in the end, we are all climate change deniers.[70] What the Anthropocene has made apparent is that there is the world in which we eat, breathe, create waste, and absorb toxins from the air and water, and then there is an invisible world composed of human-made fictions that make malign interventions easier to fathom. Not unlike Doyle's Professor Challenger—the ultimate model of scientific rationality, "the super-scientist: the "arch-pioneer", "the first man of all men whom Mother Earth had been compelled to recognize"—the Anthropocene denotes our unheeding enthusiasm to assert further power over the planet, despite its screams".[71]

The Realist Impulse in/and Education
(or, Is There Really No Alternative?)

As I swipe and swipe and swipe through my newsfeed, I seem to be searching for something specific. What is it that I am looking for? Or, perhaps more fittingly, what is it that I am *waiting* for? In short, I seem to be scanning for some sign that today will unfold differently than yesterday. A glitch. A blip on the radar. A sign that *something else is possible*. Perhaps today will be the day when the clouds finally part and the sun shines through. Or, perhaps today will be the day that our Guardians acknowledge the pain and suffering taking place all over the world and take radical action to transform the situation. Or, perhaps today will be the day that people around the world rise up and say "No!" to the current situation, "No!" to those who tell us to take the world as it is. It is as if my incessant swiping might be the key to change, as if the revolution will come like the stories that populate my newsfeed daily—quickly and without ambiguity. "Could it be that there are no breaks, no 'shocks of the new' to come?"[72] If I keep swiping, scanning, refreshing—if I keep *hoping*—surely change will come? And so, as I oscillate between "the 'weak messianic' hope that there must be something new on the way" and "the morose conviction that nothing new can ever happen",[73] I seem to be left in a perpetual state of "waiting and seeing".

This call to inaction, the call to "wait and see", the imperative to bide our time and wait for events to run their course, not only relies on particular assumptions about the future—tomorrow *will* be better than today—but also the idea that, indeed, *this is it*. In a somewhat paradoxical fashion, we have learned to think and act based on the common sense assumption that the hopeful future is nevertheless determined by "the way things are" in the here and now, in turn quashing the capacity to imagine that something else is possible. Despite the fictitious character of the world, we have learned to ignore, even deny, the fabricated and contingent nature of the conditions through which we have come to understand reality by rendering these fictions as natural, as neutral, as *given*. In turn, the dominant fictions that have come to construct our reality have taken such a firm hold so as to obscure any and all possibility to think otherwise. What makes the AAMP scheme so "realistic", for instance, is its reliance on a particular set of ideological fictions—namely, a deep-seated faith in human-made techno-scientific interventions—which operate successfully based on their naturalization and subsequent disappearance.[74] Through this disappearance, the fictive character of any given situation is no longer thought of as a set of values, or beliefs, but instead positioned as a matter of *fact*. Despite the inherent precarity of our existence in the world and the unknowability of times to come, we seem to have the impression of "having seen it all before"[75] and so while we spend our time "waiting and seeing" we are left stuck in a stupor of *reflexive impotence*.[76] That is, we know things are bad, but more than that, we know that there is nothing we can do about it. This haze of powerlessness, characterized by the conviction that "there *really* is no alternative", has imploded the possibility to think otherwise, in turn paving the way for those hegemonic powers produced and maintained by the status quo to push forward their narrow-minded imaginaries.

In other words, the powerful, if illusory, fiction that "there is no alternative" to "the way things are" has produced what we might call a *realist impulse*. This impulse relies on particular forms of realism that "seamlessly [occupy] the horizons of the thinkable",[77] in turn conflating what we *believe* to be possible with possibility itself. *Capitalist realism*, for instance, posits that capitalism is not only the most viable political and economic system today, but that it is now impossible to imagine any sort of systematic alternative to it.[78] Operating through the fatalistic submission, the unquestioned belief, and the shameless substitution of all cultural, political, and by extension educational, imaginaries by brutal and direct processes of exploitation and accumulation, capitalist realism is "not a particular type of realism; it is more like realism in itself".[79] Likewise, *scientific realism*, or what we might think of as "scientism", constructs the world based on a dogmatic approach wherein capital-S "Science" is always at the ready with a solution to every problem. Contrary to scientific practice, which is acutely and humbly aware of its limits and always contingent on the provisional character of its conclusions, scientism is a dogmatic approach that peddles in certainties and the assumption that the solution to every problem is a scientific one.[80] Both capitalist realism and scientific realism have come to undergird yet another form of realism: *Anthropocene realism*. Evidenced by the fervent support for geoengineering schemes and various forms of "green capitalism" or "clean growth", Anthropocene realism relies on the assertion that "there is no alternative to petro-capitalism" and thus humans must accept our reliance on fossil fuels as an inescapable reality.[81] As AAMP makes apparent, instead of questioning, let alone divesting in, our ways of living (and dying) on this planet, the main goal for mitigating the unexpected and potential damaging impacts of the so-called Anthropocene is to maintain business as usual. Taken together, these dispersive realist attitudes correlate what we deem possible with the limited options provided by those deep-seated fictions that have come to constitute what we think of as "realistic" in the first place. In turn, this implosion of possibility has led to the disparaging conclusion that the end has already come—indeed, *this is it*—and thus the future harbors only reiteration and re-permutation.

Despite the bleak sentiment that "there is no alternative" and the reflexive impotence this foregone conclusion inspires, the imperative to remain *hopeful* nevertheless persists. This optimistic attitude is perhaps most evident within the sphere of educational research and practice. In the name of past habits and future hopes, even so-called progressive pedagogies struggle to hold the line, keeping ordinary (and often unproductive) routines locked in place. The Western educational project, one founded and reliant still on Enlightenment rationality and dreams of unfettered progress, is driven by a deep-seated optimism that works to perpetually repeat and amplify what has been recognized as "good" in the past, in turn (re)producing a future that is always-already determined by what seems possible or acceptable in the past–present. This illusion of a fixed, or at least predictable, future is, in part, made possible by a realist impulse, which conditions what is considered "realistic" in the first place by naturalizing the dominant ideologies that have come to underline curriculum discourse and pedagogical practice as just "the way things are". For instance, where schools now operate under an unquestioned business ontology, processes of teaching and learning are seamlessly absorbed under the auspices of neoliberal capitalism, in turn naturalizing these mechanisms as *given*. Dominant approaches to education most often fail to acknowledge their *complicity* within capitalist reproduction, which relies on systems of oppression, processes of enclosure, and imperialistic, colonial and accumulative logic, and as a result these mechanisms are neutralized, disappeared or *made invisible*. Contemporary curriculum reform and educational policy initiatives, for example, operate on the unquestioned assumption that, above all, schools should be responsive to the demands of "the market". In light of the ever-changing demands of neoliberal capitalism, contemporary curriculum reform initiatives have responded (with zeal) by replacing the once "rigid" structure of the classroom with a new sense of "flexibility".[82] Where schools were once compared to factories that roll out model citizens, they are now shaped by the model of the corporation wherein individuals are trained to engage in various forms of self-motivated modulations, competitive postures, and continuous training.[83] Capitalist realism, indeed.

This business ontology has not only influenced curriculum discourse, but also policy decisions and funding models, which are increasingly defined by both capitalist and scientific realism. Alberta Education, for instance, has followed in the footsteps of American education reform, creating a two-tier education system, one "public" and one "private", under an initiative called "Freedom to Choose". Within this model, schools have been gradually privatized based on the claim that privatization decreases costs for taxpayers and the philosophical assertion that "a private enterprise system is more efficient and produces superior results in all circumstances".[84] As funding for private schools increase, they are able to boast small class sizes, the most up-to-date technologies, and extensive support systems, while public schools are under continuous attack for being "too expensive" or outside of the "affordability zone" for taxpayers. The fiction that public schools are failing is amplified by standardized test results, such as those compiled by PISA, wherein private schools consistently perform better than public schools. Despite the inadequacy of standardized assessments' capacity to represent the actual learning that takes place in classrooms, these scientific measures are taken as truth, thus creating a feedback loop where good results lead to increased funding for privatization, which leads to more private schools, more good results, and yet more privatization, and so on and so on. Within this process, pedagogical practice is understood as that which is always-already submittable to scientific data, which claims to not only describe current conditions, but also accurately predict future phenomena, such as student success. The scientific realism underlining the steady privatization of educational domains derives its power from the circular assumption that the operational success of a particular approach, in this case standardized assessments, lends credence to the continuation of that approach, based on its seemingly effective results.

The recognition of educations' complicity within the dominant economic and social paradigms in which it is situated is, of course, not new. Concerns about the neoliberal hegemony of contemporary North American schooling, for instance, have been raised by numerous educational scholars who critique the ways in which market-driven practices stifle and negate efforts to enact democratic practices, while simultaneously producing systemic inequities, dehumanization, and instrumentalization of teachers and students alike.[85] Likewise, educational privatization has been censured for the lack of evidence that competitive environments improve learning, not to mention the ways in which increasing "educational choice" leads to increased segregation of students by race, social class, and cultural background.[86] What is perhaps new, or at least worth noting given our post-post-truth moment and the advent of the so-called Anthropocene, is the power of dominant fictions to persist *in spite of* their own failings. That is, despite the ongoing stagnation of income levels and the realization that increasingly capable robots and artificial intelligence systems will continue to kill "blue" and "white" collar jobs alike,[87] alongside the growing acknowledgment that the labor force of the future will be defined by increased precarization and part-time work,[88] educational practices remain wedded to the long-held myth that going to school will, indeed, guarantee future economic success. As a response to the transformations occurring within the labor force, as well as the pressing challenges raised by anthropogenic climate change, educational domains have adopted their own form of Anthropocene realism. For example, while curriculum initiatives at the secondary school level now advocate for provision training and competency development in STEM fields, nano-degrees in infrastructure renewal, climate change adaptation, and "green economy" retrofits are now common place in post-secondary spaces. Not unlike those anthroposchemes that have been championed as the only option to mitigate climate change, educational domains have embraced the idea "there is no alternative to (neoliberal) petro-capitalism" by adjusting learning goals and pedagogical structures so as to maintain the fiction that education can and should provide hope for the future. Or, as Alberta's Ministerial Order on Student Learning puts it, education should prepare students to be lifelong learners that are able to "adapt to the many changes in society and the economy with an attitude of optimism and *hope for the future*" (my italics).[89]

No matter how illusory these fictions are, they have become so real, or rather realistic, that they have "colonized the dreaming life"[90] of educational domains, in turn limiting the capacity to imagine pedagogical life unfolding otherwise. That is, the fictions produced and maintained through the dispersive attitudes of capitalist realism, scientific realism, and, by extension, Anthropocene realism are so taken for granted that it seems they are no longer worthy of comment. It is worth recalling, however, that what is currently called realistic was itself once deemed impossible, and conversely, what was once eminently possible is now deemed unrealistic.[91] That is, the realist impulse functions by limiting what can and cannot be thought in the first place, in turn collapsing the potential to *believe* that something else is possible. The powerful assertion that "there is no alternative" purposefully and necessarily obscures, occludes, or otherwise disappears the fictitious character of the world, and thus the potential to *fiction another world*. The realist impulse can therefore only be threatened if it is shown to be in some way inconsistent or untenable; if, that is to say, "the ostensible 'realism' of the situation turns out to be nothing of the sort".[92] Put another way, and returning back to Flusser, we must recognize the sense of the fictitious all around us, and in turn, the collective loss of faith in reality as something given and discoverable. This is not to say that anti-realism is the answer, but rather, this assertion underscores the urgent need to recognize, with humility, that "one cannot 'know the residue' of what would be independent of what is for-us, [and thus] that the world could be *other* than it is for-us".[93] Taking cues from Deleuze, who was neither an anti-realist nor a straightforward realist,[94] we must recognize how reality itself always exceeds human regimes of representation and is thus irreducible to them. Between a reality that is always-already made for-us, in our all-too-human regimes of representation, and a reality that exists in-itself, that is, a world beyond our perceiving it, there is the "and". It is this "and"—this conjunctive figuring—that constitutes "a line of flight" that escapes and eludes common sense assumptions about "the way things are".[95] In short, thinking beyond, or even against, the realist impulse does not just require choosing between anti-realism and realism, between truth and falsity, between fiction and reality, but rather necessitates modes of thought that are capable of *defamiliarizing the given*— what "everybody knows"—in order to expose that something else is, indeed, possible.[96]

Science Fictioning (or, from Critique to Creation)

After my morning scroll, I am often left asking, if the realist impulse is so seamless, if the fictions that have come to define our reality are so deep-seated, and by extension, "if current forms of resistance are so hopeless and impotent, where can an effective challenge come from?"[97] In this post-post-truth era, when a realist impulse has made something as untenable as "hacking" the planet the most plausible response to global climate change, when pedagogical life is seamlessly subsumed under the logics of neoliberal petro-capitalism, indeed, when "it is easier to believe in the end of the world than the end of capitalism",[98] I am left wondering how to "keep the dream of revolution alive in counterrevolutionary times".[99] As my morning check-in highlights, the fictions that have come to dominate current constructions of reality have produced a *stupefying consensus*. The stupidity at play here is not just a concerted "passion for ignorance",[100] although this is part of it, but "stupid" insofar as solutions to any given challenge, be it the challenge of anthropogenic climate change or the challenge of educating a future generation, are detached from their problematic fields and thought in isolation.[101] The inadequacy of present solutions is underscored by a common sense consensus that works to domesticate difference and alterity by dismissing as inconsequential those alternative modes of thinking, sensing, and acting that might lead us to believe that something else is possible. While trustX attempts to manage truth through correlation, in turn ignoring the contingent nature of what constitutes "truthiness" in the first place, educational domains bolster common sense approaches to teaching and learning by narrowing pedagogical life to that which can be recuperated by dominant economic and political demands. These regimes of management maintain a particular set of fictions, namely bounded individualism, unfettered progress, and human exceptionalism, to the point that the singular nature of our own subjectivity has become as endangered as those species disappearing from the planet on a daily basis.[102] Put otherwise, these fictions have produced a confabulated consistency that covers over anomalies and contradictions, to the point that we now accept the incommensurable and the senseless without question.[103] The task facing us as we spin into the future is thus not one of producing further agreement, unity, or consensus, but rather one of cultivating *dissensus*.

Despite the seemingly totalizing nature of those perpetual and habitual fictions within which we are today caught, the potential for dissent lingers on. After all, systems are never fully totalizing, and are thus open to de/re/territorializations and "lines of flight" that might offer new horizons for both thinking and action. For instance, in response to the hollowing out of educational domains via privatization, alongside surging student debt crises and growing economic inequality, students have literally and metaphorically *turned their backs* on those Guardians that tell them to blindly accept "the way things are".[104] As recent polls indicate, swells of young people in North America and Europe no longer support dominant economic and political systems, with many youth expressing that they are ready to take part in large-scale uprisings against the generation in power.[105] Young people around the world have been organizing a range of protests under the banner of what has been nicknamed "Neither Fish Nor Fowl" actions, which aim to refuse the limited options provided by current political and social organizations. Operating under the slogan "Another End of the World is Possible!", these youth have come together to reject the current state of affairs, some going so far as to propose systematic plans that advocate processes for "learning to die" in the Anthropocene. Ranging from anti-natalist groups who petition under the catchphrase "Make Kin, Not Babies!",[106] to radical organizations who suggest it is time to "Torch the Dusties!",[107] what these student groups have in common is the underlying assertion that perhaps the Earth is not ours to save.

While these dissent formations demonstrate a swelling desire for change, a desire to refuse the options given and actualize something else, they also indicate the power of dominant fictions to recuperate and/or disappear resistance itself under the guise of "possibility". That is, the power of these micropolitical resistances is that they *do not make sense* to the dominant fictions that have come to think on our behalf; the idea of rejecting current modes of schooling, and further, the radical proposals that advocate learning to die simply do not compute given the current system codes under which life operates. While these discordant lines of flight demonstrate how even within dominant social organizations "everything flees",[108] these modes of dissent are thrown out as frivolous, foolish, or simply unrealistic. Despite protests, collective outcries, and even radical critique, the reality of the situation is simply not questioned, let alone challenged. Indeed, even in my own work as an undergraduate instructor and educational researcher, I have been warned numerous times that speculating on the potential to think otherwise is nothing more than frivolous time-wasting. But, to those that tell me "we don't have the luxury to fabulate, there are more important issues at play", I say, "we don't have the luxury not to". As extreme weather events, planetary temperature ascensions, and the depletion of global water sources surge in tandem with continued financial crises, the dissolution and privatization of social services, and a growing sense of precarity in relation to the future of work, we must come to terms with the idea that, indeed, *the world as we know it* is ending. Further, given the complexity and contradictions of the post-post-truth anthropocenic moment, we must confront the distressing realization that perhaps past and current modes of resistance will no longer suffice.

As distinctions between so-called facts and fabrications become increasingly muddied, alongside the realization that our all-too-human logos cannot adequately represent the world in which we exist, resistance and dissent based in a logic of critique have become exposed as insufficient. Concerned with discursive categorical sets and discrete organizations, as well as the systematic and analytical practice of "intelligible", "rational", and/or "enlightened" modes of debate, critique itself must be critiqued in the name of more affirmative, generative, and/or creative modes of thinking. Or, as Deleuze writes, what we lack most in the contemporary moment is *belief in the world*, or rather, the means by which new ways of believing in the world might be created: "we need both creativity and a people".[109] If late capitalism has taught us anything, it is that any intelligent or useful insight, even those that offer radical critique, can be instantly coded, co-opted, and commodified according to the seemingly inexhaustible rhythms of capitalist exchange. Likewise, in political spheres, criticizing the credibility and integrity of those Guardians whose currency is founded in fabrication and falsehood is to mistake the situation entirely. That is, while critique aims to analyze and potentially expose the (in)validity and limits of particular political ideologies, it is an ineffective strategy for dealing with the virtual nature of power. While critical approaches are busy seeking out and defending something like "truth", the Guardians have already departed from the here and now "as if" the world is the way they believe it to be. In the case of current anthroposcheming, for instance, critique is embedded within the various projects being proposed. Even the most ardent geoengineering supporters are critical of the potential side effects and unpredictable nature of anthropogenic intervention, yet this critique is obscured, and ultimately ignored, based on the powerful belief that there *really* is no alternative.

In the context of new ecological problematics, where stupidity abounds and isolated solutions push forward through seemingly unanimous consensus, where truth is established based on the limited purview of what has been considered "good" or "possible" in the past–present, where we live in a "veritable cemetery for destroyed practices and collective knowledge",[110] the only way to resist is to *introduce a crisis of a different kind*. That is, following Deleuze and Guattari, it is the invention of "a people to come"—the attempt to articulate the voice of a collectivity that does not yet exist—that holds the capacity to induce processes of becoming-other, and by extension, a crisis of a different kind.[111] Beyond merely critiquing the situation, such a crisis must fabulate a fundamentally different register of possibility wherein the experimental invention of new repertoires, new collectivities, and new practices of living, are able to actualize a sense of *becoming possible of the impossible* that might work to overturn the powerful rhetoric that "there is no alternative" to "the way things are". Indeed, "the possible has been tried and failed. Now its time to try the impossible".[112] Put otherwise, thought itself must be pushed to the limits of possibility so as to mobilize a break with orthodox referents and habits of repetition, toward the *fabulation* of new images of life. This affirmation of the *unthought* does not entail a rejection of the actual world, nor an attempt to escape the present for some Utopic vision of the future, but instead, the practice of fabulation seeks to intensify those hitherto obscured or dejected forces that flee dominant organizations, in turn dilating the possibility to think otherwise. In short, what needs to be reclaimed is not just a revolutionary attitude, but the *art of paying attention*.[113] (Are *you* paying attention?) This call is not just another appeal to recognize diverse voices and struggles, although this too has value, but to develop strange, even alien, modes of attunement that might dilate thinking, sensing, and acting beyond common sense consensus and accord.

For Flusser, the art of paying attention might take the form of acknowledging the fictional character of the world, without lament, in order to make explicit the speculative conditions of thinking itself. As exemplified in his own poetic-philosophizing, by explicating the conditions of fictionality in, say, philosophical discourse, we might actualize new ideas, new fictions, and, by extension, new worlds.[114] Likewise, for Deleuze and Guattari, in order to disrupt those fictions that have come to organize the very field of possibility, determining what we are capable of thinking in the first place, we must *probe* for stranger and more fluid modes of organization. "Probe-heads" are Deleuze and Guattari's name for these "nonsignifying, nonsubjective, essentially collective, polyvocal and corporeal"[115] modes of organization that are able to rupture dominant fictions, not through modes of critique and negation, but through the formation of "strange new becomings, new polyvocalities".[116] This mode of probing, which necessarily involves non-human becomings and different connections to the world, is itself a form of fictioning insofar as it involves a different way of thinking the world beyond common sense organizations and the realist impulse. In this way, and following Deleuze and Guattari, philosophical thought itself might be read as "a kind of science fiction" that does not seek to imagine the future of philosophy, but rather aims to invent a philosophy of the future.[117]

Positioned as a science fiction, or what we might think of as the active process of science fictioning, such a philosophical approach requires a certain kind of openness to strange probes and unknown futures. Put otherwise, science fictioning might be understood as those modes of thought that are able to imagine what we are unable to know, in turn dilating what counts as knowledge in the first place. Devoted to the "not yet", the "otherwise", and/or the "Outside", science fictioning aims to launch thought into those domains and dimensions wherein we can no longer quantify the world in terms of measurable, repeatable, and/or readily recognizable knowledge, but instead it is through the *creation* of concepts that the present might bifurcate into other worlds and unprecedented—yet unthought—possibilites.[118] That is, science fictioning does not function to create visions of Utopia, but instead, displaces the promise of a "better" future in favor of the production of alternative fictions within the present. This mode of thinking takes seriously the idea that "every creation must incorporate the knowledge that it is not venturing into a friendly world but into an unhealthy milieu",[119] and thus the time of struggle cannot postpone the time of creation. Where prototypical philosophical questions ask *"what is?"*, in turn tempting us to turn to common sense responses, science fictioning asks *"what if?"*, in order to pursue an experiment in experience that deterritorializes concepts of representation and visions of the future that are determined in advance. The signs and events produced through science fictioning therefore hold the capacity to draw attention to, and ultimately outstrip all-too-human measures, rendering epistemology itself mute and in need of new means of expression. In short, science fictioning rebuffs the pretense of reality by affirming the "as if" postulates on which the reality of pretense is itself founded, in turn proposing counterintuitive scenarios that seek to unsettle and singularize otherwise diminished conditions of possibility through concomitant movements of extrapolation and speculation.[120]

To summarize, science fictioning does not seek out evidence or proof to explain and defend "the way things are", but instead extrapolates from the current state of affairs in order to suggest new lines of inquiry that science, analytic reasoning, and/or inductive generalizations would not otherwise stumble upon. Where science, and particularly scientific realism or scientism, seeks to settle upon predictable and repeatable results, science fictioning seeks to fold the problematic of an unknown future into the interval of the present, so as to create transversal lines and alternative becomings.[121] That is, science fictioning does not rely on a logic of critique or scientific expertise where the goal is to *prove* facts and positions, but instead pushes us to challenge realist assumptions in order to explore other modes of existence through active forms of speculative creation. Taking flight from those trajectories that are always-already defined by "the way things are", science fictioning dilates our sense of possibility by asking how things *might be otherwise*. This speculative approach seeks to "articulate and enable the contingencies of the given, armed only with the certainty that what is, is always incomplete".[122] By experimenting with the innately porous, nontotalizable set of givens that have come to define those fictions that structure our sense of reality, science fictioning affirms that, indeed, "there's nothing new under the sun, but there are new suns".[123] In short, by drawing on those metaphysical anomalies, baffling mutations, and cracks in time that are inherent to our holey understandings of reality, science fictioning provides a site wherein our precious, self-important, and all-too-human sense of truth is rendered all but irrelevant.

Coda (or, Beyond All Hope)

If we can learn anything from our current situation, it is that we cannot just hope the revolution into existence. While everybody is sure that change is right around the corner, that our human power and ingenuity will steer us the right way, while everybody is "waiting and seeing", assured that time is on our side, while we refresh our newsfeeds, again and again and again, hope is exposed for what it really is: Not a motor for change, but a brake. Beyond hoping for some day when the skies will finally part, when realism is exposed as yet another ambit of human philosophical privilege, when we realize that perhaps the Earth is not ours to save, we must produce modes of thinking and acting that are capable of "short-circuiting the 'here and now' in order to play out the scene differently".[124] In light of the post-post-truth moment, where the Anthropocene is welcomed as yet another sign of human triumph over the Earth, where a realist impulse has occupied the horizon of the thinkable, in turn disappearing possibilites for something else to emerge in political, economic, and educational spheres alike, science fictioning might provide a more adequate mode of mapping the various hyperobjects that increasingly determine our lives but that are too vast to "see", let alone resist.[125] That is, science fictioning approaches the question of The Future™ differently via its confrontation with the inherent incalculability and incommensurability between our all-too-human regimes of representation and that which we have come to perceive as reality itself. Positioned as an active process—a verb—science fictioning does not seek to represent, or even critique, the fictional character of the world, but instead *fiction another world*. Science fictioning therefore shifts the question of hope from visions of a transcendent, Utopic, or "better" tomorrow, toward the capacity to imagine that something else is possible in the present.

What the so-called Anthropocene and its coming annihilations make apparent is that now, more than ever, is the time to learn to "say 'no' to those who tell us to take the world as it is".[126] This negativity is not characterized by melancholy and lamentation, the point is "not to replace angelic message with arcane ones",[127] but rather this mode of dissent renders reality, as it is currently perceived, as a retrospective, yet consensual, confabulation that is therefore vulnerable to alternative fabulations. If "reality is a fiction, and fiction is reality", if both the past and the future are not actual, but virtual entities that are open to alternative potentials, then science fictioning is not only a response to what is already established, but also holds the capacity for that which is already established to reconfigure itself in response to divergent and unexpected fictions. Indeed, "the only thing that makes life possible is permanent, intolerable uncertainty; not knowing what comes next".[128] In this way, science fictioning is not only future-oriented, but also concerns itself with a re-engineering of the "the way things are" by offering a re-writing of past and present fictions, so as to project forward into strange and yet unthought futures.

Notes

1. The average "Length of Dry Period" in central Alberta has risen steadily since 1950. In the area surrounding Edmonton, Alberta, the precipitation trend summaries indicate that the "Length of Dry Period" has increased by approximately 5 days between 1950 and 2010, from approximately 27 out of 80 days in 1950 to 32 out of 80 days in 2010 (Clark & Kienzle).

2. Solar Radiation Management (SRM) or albedo modification is a geo-engineering strategy that proposes using balloons, aircraft, or cannons to shoot reflective particles or engineered nanoparticles into the stratosphere in order to intentionally manipulate climate forcings with the goal of counteracting undesired climate change (Crutzen 2006).
3. Proposed by Google Inc. in 2015, Knowledge-Based Trust (KBT) is a method that computes trustworthiness scores in order to rate the quality of web sources. Whereas web sources have traditionally been evaluated using exogenous signals (such as a source's popularity or relation to other hyperlinks), the KBT method proposes a new approach that relies on endogenous signals, namely, the correctness of factual information provided by the source (Dong et al. 2015).
4. A hypercane is a hypothetical class of extreme tropical cyclone that could form if ocean temperatures reached 50°C. In his short story "Hypercane," Eric Holthaus (2015) shows how the promise of geoengineering could bring about such extreme weather events if we don't aggressively investigate the technologies behind it.
5. PISA, or The Programme for International Student Assessment, is a triennial international survey led by the Organisation for Economic Co-operation and Development (OECD), which aims to evaluate education systems worldwide by testing the skills and knowledge of 15-year-old students. PISA assessments are standardized assessments and include both multiple choice and long answer questions. Following the academic testing, participating students also answer a questionnaire on their background including learning habits, motivation, and family background. School directors also provide information describing school demographics and funding ("Frequently Asked Questions").
6. As Spinoza writes in *The Ethics* "[h]ope is nothing else but an inconstant pleasure, arising from the image of something future or past, whereof we do not yet know the issue" (Part III, Prop. XVIII, Note II, p. 144).
7. In John Carpenter's (1988) satirical science fiction horror film *They Live*, an unnamed drifter discovers the ruling class are, in fact, aliens concealing their appearance in order to manipulate people to spend money, breed, and accept the status quo via subliminal messages in mass media.
8. "Word of the Year 2016 Is... | Oxford Dictionaries". Oxford Dictionaries | English. n.p., 2017. Web. 18 May 2017.
9. According to Musa al-Gharbi (Paul F. Lazarsfeld Fellow in Sociology, Columbia University), although polls indicate that most Americans do not like President Donald Trump, he will most likely be elected again in 2020. Citing his own research in what he calls the "default effect", al-Gharbi asserts that even when people are unhappy with a state of affairs, they are usually disinclined to change it (al-Gharbi 2017).

10. The phrase "alternative facts" was first used by US Counselor to President Donald Trump Kellyanne Conway during a Meet the Press interview on January 22, 2017. In the interview, Conway defended White House Press Secretary Sean Spicer's false statement about the attendance at Donald Trump's Presidential inauguration by stating that Spicer was just providing "alternative facts".

11. In 2005, American comedian Stephen Colbert popularized the term "truthiness" to describe "the quality of seeming or being felt to be true, even if not necessarily true" (*Oxford Dictionary*).

12. Frank Herbert, *Dune* (Philadelphia and New York: Chilton Books, 1965), 363.

13. In his 2008 book *After Finitude: An Essay on the Necessity of Contingency*, Quentin Meillassoux offers a critique of correlationism that holds that we cannot know reality as it is in-itself but only as it is for-us, as a correlate of human consciousness, language, culture, and conceptual schemes. Meillassoux's critique of correlationism is not wagered as a form of an anti-realism but, rather, as a form of an anti-absolutism, which is invoked to "to curb every hypostatization, every substantialization of an object of knowledge which would turn the latter into a being existing in and of itself" (Meillassoux 2008, p. 11).

14. Vilém Flusser, "On Fiction," trans. Derek Hales and Erick Felinto, *&&& Journal* (12 May 2016), http://tripleampersand.org/flusser-on-fiction/.

15. Quentin Meillassoux, *After Finitude: An Essay on the Necessity of Contingency*, trans. Ray Brassier (London: Continuum, 2008).

16. In referencing the authority of "Science", I am drawing on Deleuze and Guattari's (1987) distinction between Major/Royal Science and minor/nomad science, and thus the distinction between "problematics" and "axiomatics". While axiomatics, or what Deleuze and Guattari call Major or Royal science, necessitates invention and innovation in response to the problematic nature of reality, problematics, as minor or nomad science, calls upon axiomatics to actualize solutions to the problems it lays out. As Deleuze and Guattari (1987) write: "Major science has a perpetual need for the inspiration of the minor; but the minor would be nothing if it did not confront and conform to the highest scientific requirements" (486). Put briefly, and as Deleuze argues throughout his oeuvre, the substantive multiplicity that constitutes what we consider reality is real but not actual, or what he also calls "the virtual" and thus Deleuze's logic of expression, which is linked to the power of fiction here, does not entail a rejection of the actual world, nor a rejection of (good) scientific practice, but is rather an attempt to intensify and problematize the actual world while remaining fully in it.

17. Anke Finger, Rainer Guldin, and Gustavo Bernardo, *Vilém Flusser: An Introduction* (Minneapolis: University of Minnesota Press, 2011), 110.
18. Flusser, "On Fiction," para. 1.
19. Ibid., para. 3.
20. Ibid., para. 3.
21. Gregory Flaxman, *Gilles Deleuze and the Fabulation of Philosophy* (Minneapolis: University of Minnesota Press, 2012), 187.
22. James Williams, "Event," in *Gilles Deleuze: Key Concepts*, Charles J. Stivale, ed. (Montreal and Kingston: McGill-Queen's University Press, 2011), 88.
23. Gilles Deleuze, *Difference and Repetition*, trans. Paul Patton (New York: Columbia University Press, 1994), 138.
24. Finger, Guldin, and Bernardo, *Vilém Flusser: An Introduction*, 110.
25. Ibid., 116.
26. Translated from "Il y a un autre monde mais il est dans celui-ci", this dictum is most often attributed to surrealist poet Paul Éluard.
27. Gregg Lambert, "The Deleuzian Critique of Pure Fiction," *SubStance* 26.3, Issue 84 (1997): 141.
28. Hal Hodson, "Google Wants to Rank Websites Based on Facts Not Links," *New Scientist* (Technology Blog), February 25, 2015, https://www.newscientist.com/article/mg22530102.600-google-wants-to-rank-websites-based-on-facts-not-links/.
29. Michel Foucault, *Discipline and Punish: The Birth of the Prison* (New York: Vintage Books, 1979).
30. Nancy Tuana, "Viscous Porosity: Witnessing Katrina," in *Material Feminisms*, Susan Hekman and Stacy Alaimo, eds. (Bloomington: Indiana University Press, 2007), 192.
31. I borrow Isabelle Stengers' (2015) use of "guardians" to denote the elected and appointed "officials" who have the task of "making the population understand that the world has changed [...] and thus that "reform" today is a pressing obligation" (27). Originally written in French as "nos responsables", Stengers' asserts that faced with a situation wherein politics and democracy has shrunk, "there is nothing much to expect on the part of our guardians, those whose concern and responsibility is that we behave in conformity with the virtues of (good) governance" and that, for the most part, our guardians seem to "dread the moment when the rudder will be lost, when people will obstinately pose them questions that they cannot answer, when they will feel that the old refrains no longer work, that people judge them on their answers, that what they thought was stable is slipping away" (29–30).
32. As Isabelle Stengers (2015) writes, "capitalism knows how to profit from every opportunity" (11). Moving beyond the figure of the Exploiter,

the "bloodsucker who parasitizes the living power of human labor" (65), Stengers asserts that it is the figure of the Entrepreneur that consistently "demands the freedom to be able to transform everything into an opportunity—for new profits, including what calls the common future into question" (66).

33. In his 2015 book *Capitalism in the Web of Life: Ecology and the Accumulation of Capital*, Jason Moore describes commodity frontierism as the development of new frontiers of appropriation through the creation of "Cheap Nature". As Moore writes, "Cheap Nature, as an accumulation strategy, works by reducing the value composition—but increasing the technical composition—of capital as a whole; by opening new opportunities for the investment; and, in its qualitative dimension, by allowing technologies and new kinds of nature to transform extant structures of capital accumulation and world power" (45).

34. The idea of "common sense" I am referring to is that developed by Deleuze and Guattari, which, combined with "good sense" constitute the fundamental *doxa*, or common belief or popular opinion, of representation (Poxon and Stivale 2011, p. 67). Expressed through the formulation of what "everybody knows", common sense assumes that all human faculties can be brought together under the banner of transcendental and/or pre-determined identifications.

35. In an effort to provide a name for a conception of nature that is "neither vulnerable nor threatening nor exploitable" (47), Isabelle Stengers (2015) draws on James Lovelock and Lynn Margulis' term "Gaia" to name "the one who intrudes" as that which is blind or indifferent to the damage she causes (43). As Stengers writes, because Gaia is indifferent to her impacts on the human species, and is herself not threatened despite impending eco-catastrophes, this naming is able to draw attention to that which "our thinking must succeed in bringing itself to do: it is a matter of thinking successfully, the event of a unilateral intrusion, which imposes a question without being interested in the response" (46).

36. In a 1989 interview with The Stony Brook Press, musician and "cosmic" philosopher Sun Ra discusses his approach to shutting out life's distractions in order to play music: "[i]t's more demanding than being a priest; there is no freedom. A world is in danger, this planet is in great danger! You gotta have somebody who can do things. Ministers and priests have done the best they can, but they are blocked by their standard of righteousness and their standard of "truth". The folks I'm dealing with don't care about the "truth" they don't care anything about righteousness; they are only interested in results They want people to rise out of the immature states they are in, and to leave the self-destruct part of

themselves behind. They feel it must be done or else humanity can be considered an experiment that failed" (Franza 1989).

37. Donna Haraway, "Situated Knowledges: The Science Question in Feminism and the Privilege of Partial Perspectives," *Feminist Studies* 14.3 (1988): 581.

38. Borrowing from Baruch Spinoza (1955), I use the term "inadequate" to describe how modes of knowledge production are inherently limited in terms of how they understand relationships between bodies, including their effects and causes. As Spinoza posits, in order to understand what constitutes knowledge, we must distinguish between *adequate* and *inadequate* ideas. As Spinoza writes: "[i]t follows, [...], that the ideas which we have of external bodies indicate the condition of our own body more than the nature of the external bodies" (Part II, Prop. XVI, Corollary 2). That is, even through sense experience itself knowledge is obscured by the correlation of experience to the limits of our own bodies. Further, and as Spinoza writes, "the knowledge of an effect depends upon and involves the knowledge of its causes" (Part I, Axiom 4), and thus because the mind cannot process or contain all of the possible causes of a given body at once, the mind's ideas of external bodies, whether physical or conceptual, are always inadequate.

39. Mark Fisher, *Capitalist Realism: Is There No Alternative?* (London: Zero Books, 2009), 56.

40. Flusser, "On Fiction," para. 9.

41. Jodi Dean, "The Anamorphic Politics of Climate Change," *e-flux Journal* 69 (January, 2016), http://www.e-flux.com/journal/69/60586/the-anamorphic-politics-of-climate-change.

42. In 2017, Harvard University began testing the world's biggest solar geoengineering program to date in order to establish whether climate engineering technology can safely induce atmospheric cooling effects to counter the impacts of climate change. According to project plans as of May 2017, scientists hope to complete a pilot project, which will include two small-scale dispersals of water and then calcium carbonate particles into the atmosphere by 2022 (Nelsen 2017).

43. Paul Crutzen, "Albedo Enhancement by Stratospheric Sulfur Injections: A Contribution to Resolve A Policy Dilemma," *Climatic Change* 77 (2006): 216.

44. Renee Cho, "The Double-Edged Sword of Geoengineering," *State of the Planet: Earth Institute, Columbia University* (blog), May 1, 2012, http://blogs.ei.columbia.edu/2012/05/01/the-double-edged-sword-of-geoengineering/.

45. In 2017, Abu Dhabi-based National Advisor Bureau Limited (NABL) put forward plans to tow icebergs approximately 7800 miles from the

southern pole to the United Arab Emirates in order to provide access to clean water and create "microclimates" that could lower temperatures and induce rain (Gramer 2017).

46. In the episode "Wrapping Greenland" (part of the Discovery series "Ways to save the planet"), Dr. Jason Box proposes that using reflective blankets to cover glaciers in Greenland could mitigate glacial melting and subsequent flooding (Next Nature Network).

47. Citing Kevin Anderson (an expert on climate mitigation scenarios), Andreas Malm (2015) asserts that "humanity might retain a 50 per cent change of staying below 2 degrees if emissions hit zero before 2045, but then 'flying, driving, heating our homes, using our appliances, basically everything we do, would need to be zero carbon—and note, zero carbon means zero carbon'" (187).

48. In its 2013 summary of the "Effects of Stratospheric Aerosol Injection Geoengineering", the Intergovernmental Panel on Climate Change (IPCC) concluded that there is medium confidence that radiative forcing could be achieved through the injection of sulfur (S) annually into the stratosphere.

49. David Keith, *A Case for Climate Engineering* (Cambridge: MIT Press, 2013).

50. Cho, "The Double-Edged Sword of Geoengineering."

51. Keith, *A Case for Climate Engineering*, 55.

52. Ibid., 72.

53. In April 2017, the Mauna Loa Observatory in Hawaii recorded its first-ever carbon dioxide reading in excess of 410 parts per million (Kahn 2017).

54. Ryo Moriyama, Masahiro Sugiyama, Atsushi Kurosawa, Kooiti Masuda, Kazuhiro Tsuzuki, and Yuki Ishimoto, "The Cost of Stratospheric Climate Engineering Revisited," *Mitigation and Adaptation Strategies for Global Change* (2017): 1207–1228.

55. Christophe Bonneuil and Jean-Baptiste Fressoz, *The Shock of the Anthropocene: The Earth, History, and Us* (London: Verso, 2016), 3.

56. Damian Carrington, "The Anthropocene Epoch: Scientists Declare Dawn of Human-Influenced Age," *The Guardian*, August 29, 2016, https://www.theguardian.com/environment/2016/aug/29/declare-anthropocene-epoch-experts-urge-geological-congress-human-impact-earth.

57. Bruno Latour, "Waiting for Gaia: Composing the Common World Through Art and Politics" (Lecture, French Institute, London, November 2011), 3. http://www.bruno-latour.fr/fr/node/446.

58. Dean, "The Anamorphic Politics of Climate Change," para. 3.

59. Donna Haraway, *Staying with the Trouble: Making Kin in the Chthulucene* (Durham, NC: Duke University Press, 2016).
60. Eileen Crist, "On the Poverty of Our Nomenclature," in *Anthropocene or capitalocene? Nature, History and the Crisis of Capitalism*, Jason Moore, ed. (Oakland, CA: PM Press, 2016): 14–33.
61. The following terms can be attributed as follows: Capitalocene (Moore 2013); Eurocene (Grove 2016); Anthrobscene (Parikka 2015); Chthulucene (Haraway 2015); Plantationocene (Haraway 2015); Plasticene (*The New York Times* 2014); Misantanthropocene (Clover and Spahr 2014); Betacene (Howe, n.d.); Planthropocene (Myers 2016); Thermocene, Thanatocene, Phagocene, Phronocene, Agnotocene, and Polemocene (Bonneuil and Fressoz 2016). The other terms listed do not have a specific referent/origin, but have been used on blogs, social media, and other informal communications.
62. Timothy Morton (2013) defines "hyperobjects" as those entities of such vast temporal and spatial dimensions that they challenge traditional ideas about how we understand objects in the first place. Morton suggests that global warming is a prime example of a hyperobject.
63. Rob Nixon (2013) asserts that the violence brought on by large-scale environmental transformations (such as anthropogenic climate change) takes place gradually and often in an invisible manner. Nixon uses the concept "slow violence" to describe such environmental threats and to draw attention to the ways in which this form of violence exacerbates the vulnerability of ecosystems and of people who are poor, disempowered, and often involuntarily displaced.
64. Isabelle Stengers, *In catastrophic times: Resisting the Coming Barbarism*, trans. Andrew Goffey (London: Open Humanities Press, 2015), 43.
65. Ibid., 119.
66. Crist, "On the Poverty of Our Nomenclature."
67. Ibid., 16.
68. In a 2014 interview with ethicist Clive Hamilton and journalist Andrew Revkin, Nathanael Johnson highlights Hamilton's critique of how some thinkers involved in climate change discourse seem optimistic, even excited, about the advent of the Anthropocene. Responding to Revkin's optimistic video entitled "Seeking a Good Anthropocene," Hamilton argues that he "cannot see how, in a world warmed by four degrees, anything can be described as good".
69. Heather Anne Swanson, "The Banality of the Anthropocene," *Cultural Anthropology*, February 22, 2017, https://culanth.org/fieldsights/1074-the-banality-of-the-anthropocene.
70. Citing Clive Hamilton's argument in "Requiem for a Species", Bruno Latour (2011) suggests that in a sense "we are all climate deniers, since

we have no grasp of the collective character—the anthropos of the anthropocene, the "human" of the "human made" catastrophe. It is through our own built-in indifference that we come to deny the knowledge of our science" (4).

71. Ian Pindar and Paul Sutton, "Translator's Introduction," in *The Three Ecologies*, trans. Ian Pindar and Paul Sutton (London: Athlone Press, 2000), 2.
72. Fisher, *Capitalist Realism: Is There No Alternative?* 3.
73. Ibid., 3.
74. Ibid., 16.
75. Flaxman, *Gilles Deleuze and the Fabulation of Philosophy*, 392.
76. Fisher, *Capitalist Realism: Is There No Alternative?* 21.
77. Ibid., 8.
78. Ibid., 2.
79. Ibid., 4.
80. Leon Wieseltier, "Perhaps Culture Is Now the Counterculture," *New Republic*, May 27, 2013, https://newrepublic.com/article/113299/leon-wieseltier-commencement-speech-brandeis-university-201.
81. Morgan Adamson, "Anthropocene Realism," *The New Inquiry*, November 30, 2015, http://thenewinquiry.com/essays/anthropocene-realism/.
82. The current curriculum reforms within Alberta Education draw on research outlined in a document entitled "From Knowledge to Action: Shaping the Future of Curriculum Development in Alberta" (Parsons and Beauchamp 2012), which highlights the word "flexibility" (and its various grammatical formulations) 177 times. Throughout the document, flexibility is positioned in various ways including: Flexible timing and pacing in a variety of learning environments (22); flexibility in the classroom delivery of differentiated learning (22); flexibility in curriculum organization (166); flexibility of learning goals, methods, and assessment (176); flexibility of multimedia technology (204); and perhaps most importantly, flexibility as a key competency and "enduring characteristic to help Albertans become engaged thinkers and ethical citizens with an entrepreneurial spirit" (280).
83. Gilles Deleuze, "Postscript on the Societies of Control," *October* 50.2 (1992): 3–7.
84. "Charter Schools, Private School and Vouchers", The Alberta Teachers' Association, November 20, 2015, https://www.teachers.ab.ca/News%20Room/IssuesandCampaigns/Ongoing%20Issues/Pages/Charter%20Schools%20Private%20Schools%20and%20Vouchers.aspx.

85. John Portelli and Christina Patricia Konecny, "Neoliberalism, Subversion, and Democracy in Education," *Encounters/Encuentros/Rencontres on Education* 14 (2015): 89.
86. "Charter Schools, Private School and Vouchers."
87. Jason Koebler, "If Schools Don't Change, Robots Will Bring on a 'Permanent Underclass': Report," *Motherboard*, August 6, 2014, https://motherboard.vice.com/en_us/article/if-schools-dont-change-robots-will-bring-on-a-permanent-underclass-report?utm_source=mbfb.
88. As Isabell Lorey (2011) asserts, *precarization*, or what might be defined as the ongoing process of social and economic destabilizing, is no longer an atypical or marginal phenomenon of society, but instead precarious living and working conditions are normalized as an inevitable characteristic of contemporary labor conditions.
89. Alberta Education, "Ministerial Order on Student Learning (Order #001/2013)," Edmonton: Alberta Education, 2015, https://education.alberta.ca/policies-and-standards/student-learning/everyone/ministerial-order-on-student-learning/.
90. Fisher, *Capitalist Realism: Is There No Alternative?* 8.
91. Ibid., 17.
92. Ibid., 16.
93. Jeffrey Bell, "Between Realism and Anti-realism: Deleuze and the Spinozist Tradition in Philosophy," *Deleuze Studies* 5.1 (2011): 5–6.
94. Ibid., 13.
95. Ibid.
96. My use of the word "possible" throughout this investigation draws from Deleuze's notion (following Gilbert Simondon) that what constitutes the possible is not determined in advance, but is instead an empty form, a field of virtual potential energies, wherein something new may emerge in an unforeseeable way. For Deleuze "the possible is an empty form, defined only by the principle of non-contradiction" and thus "[t]o say that something is possible is to say nothing more than that its concept cannot be excluded a priori, on logical grounds alone" (Shaviro 2007, p. 5). In this way, possibility itself is neither productive, nor does it have a claim to a pre-defined existence, and thus the problem is how to move from conditions of possibility to conditions of actualization, "from the possible to the virtual" (Shaviro 2007, p. 5). It is this conception of possibility that is taken up below in relation to what I develop as "science fictioning".
97. Fisher, *Capitalist Realism: Is There No Alternative?* 16.
98. Ibid., 2.
99. Andrew Culp, *Dark Deleuze* (Minneapolis: University of Minnesota Press, 2015), 19.

100. Education scholar Kent den Heyer (2014) uses the phrase "passion for ignorance" to denote how people learn to "avoid that which challenges [their] cherished visions or ideals and or implicates [them] as benefiting from, albeit in unequal ways, the many horrors of the present" (para. 5).

101. I use the term stupidity in relation to Deleuze's (1994) formulation wherein he asserts that stupidity is not the result of cognitive deficiencies or poor development, but is rather an effect of certain political orientations that perpetuate a "spectre that perpetually threatens thought" through defensive postures that render certain ways of life tolerable (or not). In this way, Deleuze links stupidity to a particular way of approaching problems: "[t]he problem or sense is at once both the site of an orginary truth and the genesis of a derived truth. The notions of nonsense, false sense and misconstrual must be related to problems themselves (there are problems which are false through indetermination, others through overdetermination, while stupidity, finally is the faculty for false problems; it is evidence of an inability to constitute, comprehend or determine a problem as such)" (159).

102. Felix Guattari, *The Three Ecologies*, trans. Ian Pindar and Paul Sutton (London: Athlone Press, 2000), 6.

103. Fisher, *Capitalist Realism: Is There No Alternative?* 56.

104. In the spring of 2017, Education Secretary Betsy DeVos faced an auditorium of jeering graduates at Bethune-Cookman University who proceeded to turn their backs to her throughout the commencement address.

105. A 2016 Harvard poll found that "[f]ifty-one percent of people between 18 and 29 no longer support the system of capitalism" while a survey conducted by the European Broadcasting Union's "Generation What?" in Ireland found that fifty-four per cent of 18–34-year-olds said they would take part in a "large scale uprising against the generation in power if it happened in the next days or months".

106. Donna Haraway (2016) asserts that the current ecological situation, what she terms the Chthulucene, needs at least one slogan. As she writes: "still shouting "Cyborgs for Earthly Survival," "Run Fast, Bite Hard," and "Shut Up and Train," I propose "Make Kin Not Babies!" Making—and recognizing—kin is perhaps the hardest and most urgent part" (102).

107. Margaret Atwood, "Torch the Dusties" in *Stone Mattress: Nine Tales* (Toronto: McClelland & Stewart, 2014).

108. Gilles Deleuze, "Desire and Pleasure," in *Two Regimes of Madness*, trans. Ames Hodges and Mike Taormina (New York: Semiotext(e), 2006).

109. Gilles Deleuze, *Negotiations 1972–1990*, trans. Martin Joughin (New York: Columbia University Press, 1995), 34.

110. Stengers, *In Catastrophic Times: Resisting the Coming Barbarism*, 98.
111. Gilles Deleuze, *Essays Critical and Clinical*, trans. Daniel W. Smith and M.A. Greco (Minneapolis: University of Minnesota Press, 1997), 90.
112. On the 1965 album *Secrets of the Sun* by Sun Ra and his Solar Arkestra, Art Jenkins was introduced as a new "Space Vocalist". After auditioning with some rhythm-and-blues tunes, Sun Ra told him that he had a nice voice, but he was looking for a singer who could do the impossible: "the possible has been tried and failed; now I want to try the impossible".
113. As Isabelle Stengers (2015) notes, when matters of what one calls "development" or "growth" define political and scientific responses, "the injunction is above all to not pay attention" (61). In light of current ecological problematics and the lack of political will from our Guardians, Stengers asserts that what needs to be collectively reclaimed is the art of paying attention, with the understanding that paying attention "requires knowing how to resist the temptation to separate what must be taken into account and what may be neglected" (62).
114. Finger, Guldin, and Bernardo, *Vilém Flusser: An Introduction*.
115. Gilles Deleuze and Felix Guattari, *A Thousand Plateaus: Capitalism and Schizophrenia*, trans. Brian Massumi (London: Athlone Press, 1987), 17.
116. Ibid., 190–191.
117. Flaxman, *Gilles Deleuze and the Fabulation of Philosophy*, 295.
118. Ibid., 295.
119. Stengers, *In Catastrophic Times: Resisting the Coming Barbarism*, 104.
120. Steven Shaviro, *Discognition* (London: Repeater Books, 2015).
121. Flaxman, *Gilles Deleuze and the Fabulation of Philosophy*, 20.
122. Patricia Reed, "Reorientate, Eccentricate, Speculate, Fictionalize, Geometricize, Commonize, Abstractify: Seven Prescriptions for Accelerationism," in *#ACCELERATE: The Accelerationist Reader*, Robin Mackay and A. Avanessian, eds. (Berlin and London: Urbanomic and Merve Verlag, 2014), 527.
123. Octavia E. Butler, *Parable of the Talents* (New York: Seven Stories Press, 1998).
124. Culp, *Dark Deleuze*, 24.
125. Steven Shaviro, "Hyperbolic Futures: Speculative Finance and Speculative Fiction," *The Cascadia Subduction Zone*, 1.2 (2011).
126. Culp, *Dark Deleuze*, 17.
127. Ibid., 18.
128. Ursula K. Le Guin, *The Left Hand of Darkness* (New York: Ace Books, 1969).

Image 16.1   Tower topple (Mia Feuer, watercolor)

BIBLIOGRAPHY

Adamson, A. (2015, November 30). Anthropocene Realism. *The New Inquiry.* http://thenewinquiry.com/essays/anthropocene-realism/.

Alberta Education. (2015). Ministerial Order on Student Learning (Order #001/2013). Edmonton: Alberta Education. https://education.alberta.ca/policies-and-standards/student-learning/everyone/ministerial-order-on-student-learning/.

al-Gharbi, M. (2017, May 10). Trump Will Likely Win Reelection in 2020. *The Conversation.* http://theconversation.com/trump-will-likely-win-reelection-in-2020-77362.

Atwood, M. (2014). Torch the Dusties. In *Stone Mattress: Nine Tales.* Toronto: McClelland & Stewart.

Bell, J. (2011). Between Realism and Anti-realism: Deleuze and the Spinozist Tradition in Philosophy. *Deleuze Studies, 5*(1), 1–17.

Bonneuil, C., & Fressoz, J.-B. (2016). *The Shock of the Anthropocene: The Earth, History, and Us.* London: Verso.

Butler, O. E. (1998). *Parable of the Talents.* New York: Seven Stories Press.

Carrington, D. (2016, August 29). The Anthropocene Epoch: Scientists Declare Dawn of Human-Influenced Age. *The Guardian.* https://www.theguardian.com/environment/2016/aug/29/declare-anthropocene-epoch-experts-urge-geological-congress-human-impact-earth.

"Charter Schools, Private School and Vouchers", The Alberta Teachers' Association. November 20, 2015. https://www.teachers.ab.ca/News%20 Room/IssuesandCampaigns/Ongoing%20Issues/Pages/Charter%20 Schools%20Private%20Schools%20and%20Vouchers.aspx.

Cho, R. (2012, May 1). The Double-Edged Sword of Geoengineering. *State of the Planet: Earth Institute, Columbia University* (blog). http://blogs.ei.columbia.edu/2012/05/01/the-double-edged-sword-of-geoengineering/.

Clark, C., & Kienzle, S. W. (n.d.). *Alberta Climate Records: Visualizing Temperature Change from 1950–2010.* http://albertaclimaterecords.com.

Clover, J., & Spahr, J. (2014). *#Misanthropocene: Twenty-Four Theses.* Oakland, CA: Commune Editions.

Crist, E. (2016). On the Poverty of Our Nomenclature. In J. Moore (Ed.), *Anthropocene or Capitalocene? Nature, History and the Crisis of Capitalism* (pp. 14–33). Oakland, CA: PM Press.

Crutzen, P. (2006). Albedo Enhancement by Stratospheric Sulfur Injections: A Contribution to Resolve a Policy Dilemma? *Climatic Change, 77,* 211–219.

Culp, A. (2015). *Dark Deleuze.* Minneapolis: University of Minnesota Press.

Dean, J. (2016, January). The Anamorphic Politics of Climate Change. *e-flux Journal, 69.* http://www.e-flux.com/journal/69/60586/the-anamorphic-politics-of-climate-change.

Den Heyer, K. (2014). Badiou and the Educational Situation. *Politics and Culture* [Special Edition: The Politics of Alain Badiou]. https://politicsandculture.org/2014/09/01/badiou-and-the-educational-situation-by-kent-den-heyer/.

Dong, X. L., Gabrilovich, E., Murphy, K., Dang, V., Horn, W., Lugaresi, C., et al. (2015, February 12). Knowledge-Based Trust: Estimating the Trustworthiness of Web Sources. *Proceedings of the VLDB Endowment.* https://arxiv.org/pdf/1502.03519v1.pdf.

Deleuze, G. (1992). Postscript on the Societies of Control. *October, 50*(2), 3–7.

Deleuze, G. (1994). *Difference and Repetition* (P. Patton, Trans.). New York: Columbia University Press.

Deleuze, G. (1995). *Negotiations 1972–1990* (M. Joughin, Trans.). New York: Columbia University Press.

Deleuze, G. (1997). *Essays Critical and Clinical* (D. W. Smith & M. A. Greco, Trans.). Minneapolis: University of Minnesota Press

Deleuze, G. (2006). Desire and Pleasure. In A. Hodges & M. Taormina (Trans.), *Two Regimes of Madness* (pp. 122–134). New York: Semiotext(e).

Deleuze, G., & Guattari, F. (1987). *A Thousand Plateaus: Capitalism and Schizophrenia* (B. Massumi, Trans.). London: Athlone Press.

Finger, A., Guldin, R., & Bernardo, G. (2011). *Vilém Flusser: An Introduction.* Minneapolis: University of Minnesota Press.

Fisher, M. (2009). *Capitalist Realism: Is There No Alternative?.* London: Zero Books.

Flaxman, G. (2012). *Gilles Deleuze and the Fabulation of Philosophy*. Minneapolis: University of Minnesota Press.

Flusser, V. (2016, May 12). On Fiction (D. Hales & E. Felinto, Trans.). *&&& Journal*. http://tripleampersand.org/flusser-on-fiction/.

Foucault, M. (1979). *Discipline and Punish: The Birth of the Prison*. New York: Vintage Books.

Franza, R. (1989, February 16). Exploring the Omniverse. *The Stony Brook Press, 10*(9). https://ir.stonybrook.edu/jspui/bitstream/11401/60570/1/Stony%20Brook%20Press%20V.%2010%2C%20N.%2009.PDF.

"Frequently Asked Questions." PISA: Programme for International Student Assessment. Accessed June 05, 2017. http://www.oecd.org/pisa/pisafaq/.

Gramer, R. (2017, May 5). This Country Wants to Tow Icebergs from Antarctica to the Middle East. *The Cable* (Foreign Policy). http://foreignpolicy.com/2017/05/05/this-country-wants-to-tow-icebergs-from-antarctica-to-the-middle-east/.

Grove, J. (2016, January 11). Response to Jedediah Purdy. *Forum: The New Nature, Boston Review*. http://bostonreview.net/forum/new-nature/jairus-grove-jairus-grove-response-jedediah-purdy.

Guattari, F. (2000). *The Three Ecologies* (I. Pindar & P. Sutton, Trans.). London: Athlone Press.

Haraway, D. (1988). Situated Knowledges: The Science Question in Feminism and the Privilege of Partial Perspectives. *Feminist Studies, 14*(3), 575–599.

Haraway, D. (2015). Anthropocene, Capitalocene, Plantationocene, Chthulucene: Making Kin. *Environmental Humanities, 6*, 159–165.

Haraway, D. (2016). *Staying with the Trouble: Making Kin in the Chthulucene*. Durham, NC: Duke University Press.

Hebert, F. (1965). *Dune*. Philadelphia and New York: Chilton Books.

Hodson, H. (2015, February 25). Google Wants to Rank Websites Based on Facts not Links. *New Scientist* (Technology Blog). https://www.newscientist.com/article/mg22530102.600-google-wants-to-rank-websites-based-on-facts-not-links/.

Holthaus, E. (2015, October 29). Hypercane. *Motherboard*. https://motherboard.vice.com/en_us/article/hypercane.

Johnson, N. (2014, July 7). Is the Anthropocene a World of Hope or a World of Hurt? *Grist*. http://grist.org/climate-energy/is-the-anthropocene-a-world-of-hope-or-a-world-of-hurt/.

Kahn, B. (2017, April 20). We Just Breached the 410 Parts Per Million Threshold. *Climate Central*. http://www.climatecentral.org/news/we-just-breached-the-410-parts-per-million-threshold-21372.

Keith, D. (2013). *A Case for Climate Engineering*. Cambridge: MIT Press.

Koebler, J. (2014, August 6). If Schools Don't Change, Robots Will Bring on a 'Permanent Underclass': Report. *Motherboard*. https://motherboard.vice.

com/en_us/article/if-schools-dont-change-robots-will-bring-on-a-perma-
nent-underclass-report?utm_source=mbfb.
Lambert, G. (1997). The Deleuzian Critique of Pure Fiction. *SubStance, 26*(3),
128–152.
Latour, B. (2011, November). Waiting for Gaia: Composing the Common
World Through Art and Politics. Lecture, French Institute, London. http://
www.bruno-latour.fr/sites/default/files/124-GAIA-LONDON-SPEAP_0.
pdf.
Le Guin, U. K. (1969). *The Left Hand of Darkness.* New York: Ace Books.
Lorey, I. (2011). Governmental Precarization (A. Derieg, Trans.). *Transversal:
EIPCP Multilingual Webjournal.* http://eipcp.net/transversal/0811/lorey/
en.
Malm, A. (2015). Socialism or Barbecue, War Communism or Geoengineering.
Some Thoughts on Choices in a Time of Emergency. In K. Borgnäs,
T. Eskelinen, J. Perkiö, & R. Warlenius (Eds.), *The Politics of Ecosocialism:
Transforming Welfare* (pp. 180–194). Basingstoke: Routledge.
Meillassoux, Q. (2008). *After Finitude: An Essay on the Necessity of Contingency*
(R. Brassier, Trans.). London: Continuum.
Moore, J. (2013, May 13). Anthropocene, Capitalocene, and the Myth of
Industrialization, Part I. World-Ecological Imaginations. https://jasonw-
moore.wordpress.com/2013/05/13/anthropocene-or-capitalocene/.
Moore, J. (2015). *Capitalism in the Web of Life: Ecology and the Accumulation of
Capital.* London: Verso.
Moriyama, R., Sugiyama, M., Kurosawa, A., Masuda, K., Tsuzuki, K., &
Ishimoto, Y. (2017). The Cost of Stratospheric Climate Engineering
Revisited. *Mitigation and Adaptation Strategies for Global Change, 8*(22),
1207–1228. https://doi.org/10.1007/s11027-016-9723-y.
Morton, T. (2013). *Hyperobjects: Philosophy and Ecology after the End of the
World.* Minneapolis: University of Minnesota Press.
Myers, N. (2016, January 21). Photosynthesis. *Theorizing the Contemporary,
Cultural Anthropology.* Website https://culanth.org/fieldsights/790-
photosynthesis.
Nelsen, A. (2017, March 24). US Scientists Launch World's Biggest Solar
Geoengineering Study. *The Guardian.* https://www.theguardian.com/
environment/2017/mar/24/us-scientists-launch-worlds-biggest-solar-
geoengineering-study.
The New York Times. (2014, June 14). Notes From the Plasticene Epoch: From
Ocean to Beach, Tons of Plastic Pollution. *Editorial.*
Next Nature Network. Wrapping Greenland. Last modified October 26, 2009.
https://www.nextnature.net/2009/10/wrapping-greenland/.
Nixon, R. (2013). *Slow Violence and the Environmentalism of the Poor.*
Cambridge, MA: Harvard University Press.

Oxford Dictionary, s.v. Truthiness. (n.d.). https://en.oxforddictionaries.com/definition/truthiness.

Parikka, J. (2015). *The Anthrobscene*. Minneapolis: University of Minnesota Press.

Parsons, J., & Beauchamp, L. (2012). *From Knowledge to Action: Shaping the Future of Curriculum Development in Alberta*. Edmonton: Alberta Education. https://open.alberta.ca/dataset/bc0bd7df-2bfe-4b8b-8af0-db19b17a7721/resource/5f11d83e-3074-408b-bff7-bcc27987864a/download/5976960-2012-From-Knowledge-Action-Curriculum-Development-Alberta.pdf.

Portelli, J., & Konecny, C. P. (2015). Neoliberalism, Subversion, and Democracy in Education. *Encounters/Encuentros/Rencontres on Education, 14*, 87–97.

Poxon, J. L., & Stivale, C. J. (2011). Sense, Series. In C. J. Stivale (Ed.), *Gilles Deleuze: Key Concepts* (pp. 67–79). Montreal and Kingston: McGill-Queen's University Press.

Reed, P. (2014). Reorientate, Eccentricate, Speculate, Fictionalize, Geometricize, Commonize, Abstractify: Seven Prescriptions for Accelerationism. In R. Mackay & A. Avanessian (Eds.), *#ACCELERATE: The Accelerationist Reader* (pp. 521–536). Berlin and London: Urbanomic and Merve Verlag.

Shaviro, S. (2007). The 'Wrenching Duality' of Aesthetics: Kant, Deleuze and the 'Theory of the Sensible'. http://www.shaviro.com/Othertexts/SPEP.pdf.

Shaviro, S. (2011). Hyperbolic Futures: Speculative Finance and Speculative Fiction. *The Cascadia Subduction Zone, 1*(2), 3–5 and 12–15.

Shaviro, S. (2015). *Discognition*. London: Repeater Books.

Spinoza, B. (1955). *Ethics* (R. H. M. Elwes, Trans.). New York: Dover Publications.

Stengers, I. (2015). *In Catastrophic Times: Resisting the Coming Barbarism* (A. Goffey, Trans.). London: Open Humanities Press.

Swanson, H. A. (2017, February 22). The Banality of the Anthropocene. *Cultural Anthropology*. https://culanth.org/fieldsights/1074-the-banality-of-the-anthropocene.

They Live. Directed by John Carpenter. USA: Universal Pictures, 1988.

Tuana, N. (2007). Viscous Porosity: Witnessing Katrina. In S. Hekman & S. Alaimo (Eds.), *Material Feminisms* (pp. 118–213). Bloomington: Indiana University Press.

Wieseltier, Leon. (2013, May 27). Perhaps Culture is Now the Counterculture. *New Republic*. https://newrepublic.com/article/113299/leon-wieseltier-commencement-speech-brandeis-university-201.

Williams, J. (2011). Event. In C. J. Stivale (Ed.), *Gilles Deleuze: Key Concepts* (pp. 80–90). Montreal and Kingston: McGill-Queen's University Press.

Word of the Year 2016 Is… | Oxford Dictionaries. Oxford Dictionaries | English. n.p., 2017. Web. 18 May 2017.

INDEX

© The Editor(s) (if applicable) and The Author(s) 2018   401
j. jagodzinski (ed.), *Interrogating the Anthropocene*, Palgrave Studies in
Educational Futures, https://doi.org/10.1007/978-3-319-78747-3